SCIENCE AND TECHNOLOGY IN CANADA

Science and Technology in Canada

Editors:
Paul Dufour
and
John de la Mothe

Science and Technology in Canada

Published by Longman Group UK Limited,
Longman Industry and Public Service Management Publishing Division,
Westgate House, The High, Harlow, Essex CM20 1YR.
Telephone: (0279) 442601
Facsimile: (0279) 444501

First published 1993

A catalogue record for this book is available from the British Library

ISBN 0-582-10106-9

PREFACE

Janet E. Halliwell

Canada is at a crossroads. Its economic strength, still heavily based on the export of raw commodities from the natural resource sector, is being challenged. The social fabric is being weakened by political uncertainty and inability to cope with rising costs in its medical, educational, and social support systems. Many of these pressures result from Canada's slow transformation to a technology based economy; increasingly, however, Canadians are coming to appreciate that in the judicious use of science and technology lie some, at least, of the answers to these challenges.

Canada has a rich tradition of research and educational strength in fields as diverse as geotechnical engineering, neurosciences, cosmology, and geophysics. The relatively mature system of higher education attracts students from virtually all countries of the world, and produces first class practitioners and researchers in a broad array of fields of science and engineering. Yet, the private sector, overall, makes little use of science and technology. The immediate domestic employment opportunities for graduates in most disciplines are correspondingly sparse, or at the very least, cyclical. This situation is in sharp contrast to long-term projections of a critical shortfall in highly qualified personnel in Canada.

It is this dissonance between the science and engineering base and the functioning of the economy that sets the stage for this rich and timely review of science and technology in Canada. This review holds more than national interest in its characterization of the evolving structures and policies of an S&T system for a country that is undergoing a dramatic transition. Canada is a country undergoing economic restructuring, a country in transition, a country that is struggling to realize the benefits of its investment in the people and institutions of science and technology.

The forces affecting Canada are the fruits of the knowledge revolution, a revolution that challenges the traditional basis of economic and political strength. Power has shifted towards those nations that have successfully and constructively deployed science and technology. The traditional relationship between economic prosperity and natural resources has been sundered, a message that Canada has been slow to appreciate. It is clear, however, that the key to Canada's success in this new world order is the quality and richness of its intellectual resources, and the manner in which they are combined with plentiful natural and physical resources to produce a vibrant society and robust economy that is at ease in an increasingly interdependent world.

But attitudes, policies, and institutions–public and private – that were structured for the old order are not those required of the knowledge-intensive order.

The classical structure and strategy of firms that resulted in the separation of marketing and research is no longer successful. There is a merging and integration of technology, marketing, sales and the shop floor within a firm and a new awareness of a firm's responsibility for stewardship of fragile natural resources. Universities and community colleges are increasingly being drawn into a larger and denser network of knowledge institutions that extends deeply into industry, government and society, well beyond national boundaries. Governmental structures, programs and policies are shifting their focus towards activities that are more enabling, diffusion and innovation oriented. Global vision and interaction is becoming commonplace.

This review is a snapshot of the Canadian science and technology system during this vital global transformation. It is a case history for those interested in the interplay of science, technology, innovation, economic restructuring and performance, social and cultural aspirations and rapidly changing trade regimes. It provides an opportunity for the practitioners of science and technology policy to observe the process of re-orientation and renewal.

This review is of more than national interest. The issues herein are being fiercely debated by many other nations of the world. Canada, in its push to foster innovation and more effective deployment of science and technology, has launched new mechanisms, new programs, and new policies that will be of interest to many readers. This book provides insight to the evolving structures of a science and technology system in a country that is undergoing a dramatic transformation. Patterns of public and private investment in science and technology are changing, there are experiments with new institutions, such as networks of centres of excellence that bring together university, industry and government researchers, there are new partnerships between business and education. At a different level is the increasing presence of sub-national entities in science and technology in the restructuring of educational and research structures.

At the time of writing this preface, Canada was drawing to a close an exercise that may well lead to further change. Launched in 1991, a year-long 'Prosperity Initiative' spearheaded by an independent steering committee has functioned in a broadly consultative manner in order to examine means of enhancing Canada's prosperity. A key actor in this examination of competitiveness and learning was a national Task Force on Science and Technology and Related Skills. As Co-Chairman of this Task Force, I was deeply embroiled in debates on how we might better nurture and deploy science and technology for national, and, indeed, global well-being. The recommendations of the Task Force reveal the views of a diverse group – including educators, entrepreneurs, research managers, students – that effective use of the new determinant of growth – knowledge – requires change in three essential areas:

- creating advantage with people
- fostering a climate for innovation
- initiating institutional renovation

Of these, the most profound and far reaching are the proposals for renovation and renewal of institutions that enhance public participation in and support of the policy-making process. There is recognition of a need for a comprehensive and accessible policy research capability grounded in professional interpretation of information and comprehensive data and meaningful international comparisons. All are essential for effective decision-making and well-founded, public dialogue – public dialogue that is a vital force in those nations that profit from their investments in science and technology.

Contents

Acknowledgements

In preparing this book, we received valuable support from a number of sources. In addition to the supportive environment from the Science Council of Canada (which was unfortunately eliminated following a federal budget decision of February 25, 1992), we received assistance from faculty of Administration of the University of Ottawa, The Kennedy School of Government of Harvard University and from Statistics Canada. In preparing tables in the appendices, our thanks in particular to Fred Gault and Bert Plaus at Statistics Canada and to Hollis Whitehead and Adam Holbrook in Industry, Science and Technology Canada. For their kind permission to reprint sections of the Technology Networking Guide: Canada in the Appendix, we wish to thank James Kelly and Jan Caroleo at Industry, Science and Technology Canada. Secretarial support is gratefully acknowledged from Nicole Menard, Cheryl Fortier, and Ann Langlois. Nutritional and jocular support is gratefully acknowledged from all those at the Elgin Street Ritz.

A special thanks to Paul Cunningham of the University of Manchester for his active encouragement of this project. We are grateful to the authors of the Chapters in this book for agreeing to submit to this tortuous exercise and for putting up with constant harassment from the editors. Their good-naturedness is much appreciated in addition to their obvious intellectual contribution. All material written for this book has the usual caveats: *i.e.*, the views expressed are those solely of the authors; they do not represent the views of any of the governments of Canada nor do they represent necessarily the policies of the institutions with which they are affiliated.

Finally, a special and heartfelt appreciation is extended to our families: to whom this book is dedicated. To Karen and Donna, to Philippe, Laura and Julien, our thanks and love.

PD/JdlM
Stowe, Vermont

Contributors to Science and Technology in Canada

FRANCES ANDERSON is a Program Officer with the Biotechnology Program Office of the National Research Council of Canada (NRC). She has a PhD. in the history of science from the Université de Montréal. Involved in research evaluation, policy, and planning for the NRC's four biotechnology institutes, Dr Anderson's personal research interests include the uses of bibliometrics for evaluating strategic research and policy.

DOUG ANGUS is the Director of the Queens University of Ottawa (formerly the Economic Council of Canada's economic projects on the Cost Effectiveness of the Canadian Health Care System). Mr. Angus was educated in business and economics at St. Patrick's College and the University of Ottawa. With more than 20 years experience in research, teaching and consulting on health care and related social issues, he has consulted for numerous community, provincial, national and international organizations – most recently for the Nova Scotia Royal Commission on Health Care.

RICHARD BOTHAM is a policy analyst in the Industrial and Technology Policy Branch of the Ontario Ministry of Industry, Trade and Technology. He received political science degrees from Carleton University and Queen's University.

JEFFREY CRELINSTEN is a science writer and consultant specializing in s&t. Educated in physics, astronomy and history of science at McGill University, the University of Toronto, and the Université de Montréal, Dr Crelinsten was a professor of science and human affairs at Concordia University before becoming a co-founder of The Impact Group, a Toronto-based firm specializing in science and technology policy, education, and communications. He works regularly as a writer, editor and publisher of popular and educational materials for radio, television and the print media. Most recently, he has been involved in communications planning for the National Research Council, and curriculum design for the Algonquin Space Campus, a summer program for teenagers. He is the past-President of the Canadian Science Writers' Association.

ROBERT DALPÉ is professor of political science at the Université de Montréal and researcher at the Centre for Research on the Development of Industry and Technology (CREDIT). He works on the evaluation of research and science policy and the role of public demand in technological innovation. He is the author of several articles on research policy and co- edited (with Réjean Landry) *La Politique technologique au Québec.*

CHARLES DAVIS is a Principal Program Officer at the International Development Research Centre in Ottawa. He was educated at the University of Notre Dame and the Université de Montréal. Dr Davis has worked as Science Adviser for the Science Council of Canada and the Quebec provincial government, and has published numerous articles on S&T policy and international development.

PAUL DUFOUR is a Senior Fellow of the International Science Policy Foundation and a former Science Adviser to the Science Council of Canada. Educated in political science and science policy at Concordia University and the Université de Montréal, he has worked in science policy for over a decade in such federal government agencies as the Science Council of Canada, the Ministry of State for Science and Technology, and the Department of External Affairs. He lectures regularly on science policy and has authored numerous articles on Canadian science and technology.

JOHN DE LA MOTHE is a professor of science and government in the Public Policy and Management Group of the Faculty of Administration at the University of Ottawa and Research Fellow in the Centre for Science and International Affairs, the John F. Kennedy School of Government, Harvard University. Educated at Concordia University, Oxford University and the University of Sussex (SPRU), Dr de la Mothe has worked for several government agencies in science policy including the Science Council of Canada, Investment Canada, the Ministry of State for Science and Technology, and the Natural Sciences and Engineering Research Council of Canada. He has written numerous articles and edited both books and journals on aspects of science and technology policy including *Science, Technology and Free Trade* (edited with Louis Marc Ducharme, 1990) and *C.P. Snow and the Struggle of Modernity, 1992*. He is a co-editor of the journal *Science and Public Policy*.

RON FREEDMAN has worked in the field of science and technology policy for over a dozen years in both the public and private sectors. Educated at York University and the University of Sussex, he is a co-founder of The Impact Group, a Toronto- based firm specializing in s&t policy, education, and communication. Most recently, he has been involved with the Ontario Centres of Excellence program to design a centres-wide communication plan, the Canadian Research Management Association on a study of industrial research contracted to university and federal government labs and the province of Ontario on the future role of ORTECH International.

MICHEL GIGUÈRE is Senior Research Officer for the Advisory Committee of Francophone Affairs to the Ontario Minister of Colleges and Universities. Educated at the Université de Montréal in computer science, mathematics, and the history/sociology of science, he has worked as a policy analyst in s&t policy at the Ontario Ministry of Industry, Trade and Technology.

YVES GINGRAS is professor of history at the Université du Québec at Montréal and researcher at the Centre for Research on the Development of Industry and Technology (CREDIT). He has published several papers on energy R&D, science policy and the history of science in Canada. He is author of *Physics and the Rise of Scientific Research in Canada* and co-editor (with Richard Jarrell) of *Building Canadian Science: The Role of the National Research Council of Canada*.

JANET E. HALLIWELL is Chair of the Nova Scotia Council on Higher Education and former Chairman of the Science Council of Canada. Educated in chemistry at Queen's University and the University of British Columbia, she has held positions in university research administration and research policy at the national level, serving on many national advisory boards and committees. She has received several honourary degrees and distinctions.

ROBERT KAVANAGH is Director General of Scholarships and International Programs at the Natural Sciences and Engineering Research Council of Canada. Educated in Electrical Engineering at the University of New Brunswick, the University of Toronto, and Imperial College (London), Dr Kavanagh has held positions at all three universities including Dean of Graduate Studies and Research at UNB. Dr Kavanagh has published extensively in areas relating to both science policy and electrical engineering and he is a Fellow of the New York Academy of Science.

MICHEL LECLERC who was educated in political science and the history/ sociology of Science at the Université du Québec at Montréal and at the Université de Montréal, was a researcher at the Centre d'études politiques administratives du Québec (ENAP). He is now responsible for s&t indicators within the Ministère de l'Enseignement supérieur et de la science (Québec). With more than 10 years of experience in the field of science and technology policy, his most recent book is *Les Enjeux économiques et politiques de l'innovation* (1990).

PATRICIA LEMAY has served as a Science Advisor on pharmaceuticals and health to the Science Council of Canada. as an analyst in pharmaceutical market research to the federal government, and as a pharmacist in a number of teaching hospitals. Educated at the University of Toronto and the University of Ottawa, she is now with National Health and Welfare Canada where she is working on a national pharmaceutical strategy. Previously she was with Peat Marwick Stevenson and Kellogg in the s&t practice group. She has participated on a number of national task forces studying pharmaceuticals R&D, risk assessment and the drug approval process in Canada.

ANDRÉ LEMELIN has worked for many years both as a freelance writer specializing in science and education and as an administrator in the Quebec provincial government in research policy. Educated in German studies at Université Laval he is currently in the Ministère de l'enseignment supérieur et de la science. He is the co-author of the Québec government report entitled *Science et technologie – Conjoncture* (1988) and is collaborating in writing a book on the periodic table.

CAMILLE LIMOGES is currently Director of the Research Centre for the Social Assessment of Technology (CREST) and Director of the Science Technology and Society Program at the Université du Québec at Montréal (UQAM). He has taught at the Université de Montréal, Johns Hopkins University and Harvard University. He has also served as Deputy Minister for Higher Education, Science and Technology in the Québec provincial government during the 1980s. He has published widely in the history of biological sciences as well as in science and technology policy. His current research is on public technological controversies, particularly as they relate to risks arising from biotechnology, electromagnetic fields, the use of pesticides in forestry management, and water fluoridation.

RICHARD LIPSEY is currently a professor of economics at Simon Fraser University and Alcan Fellow of the Canadian Institute for Advanced Research for whom he is directing a large- scale, international research project on Economic Growth and Policy. Dr Lipsey received his B.A. from the University of British Columbia in 1951, M.A. from Toronto in 1953 and PhD from London School of Economics in 1957. He has held a chair in Economics at the London School of Economics and was chairman of the Department of Economics and dean of the

Faculty of Social Science at the new University of Essex, England from 1964 to 1970. From 1970 to 1986, he was Sir Edward Peacock professor of Economics, Queen's University. From 1983 to 1989, he was senior economic advisor for the C.D. Howe Institute. Dr Lipsey has authored several textbooks in economics and published over 100 articles on various aspects of theorectical and applied economics.

PETER MACKINNON is the Special Advisor for Investment at the Canadian High Commission in London, England. Educated in earth sciences, Mr. MacKinnon has more than 15 years experience in R&D, business planning, policy design and high- technology corporate development. He has represented Canada at the OECD, was a co-founder and CEO of Canada's largest speciality artificial intelligence company (CAIP Corporation), was special advisor on advanced information technology to the Ministry of State for Technology, and was – until recently – in charge of corporate development for Cognos Inc., a leading international software firm.

ROGER VOYER has been involved in the development of Canadian science, technology and industrial policies for more than 25 years. Educated in chemical engineering and solid state physics at Queen's University and the University of Grenoble, France, Dr Voyer has been Director General in the Ministry of State for Science and Technology, Canadian representative at the permanent Canadian delegation to the OECD in Paris, Director of Research at the Science Council of Canada, and Executive Director of the Canadian Institute for Economic Policy. He is currently a senior partner in NGL Consulting Ltd.

Chapter 1

Geography, economy and culture: an introduction to Canada

John de la Mothe and Paul Dufour

Canada's presence on the North American continent is perhaps best expressed in terms of its eloquent, and highly original, discourse on science and technology.[1] This Canadian discourse is neither American nor European, but it is an oppositional culture conditioned by its geography, its economy and its history. Indeed, at work in the Canadian mind is a perennial polarity between technology and landscape.[2]

Central to the Canadian identity are such space-bridging techniques as the transcontinental Canadian Pacific Railway, the telegraph, microwave and radio. George Grant and Marshall McLuhan are emblematic figures in Canadian thought. The competing perspectives on technology represent both the limits and the possibilities of the Canadian discourse on technology. Indeed, Grant and McLuhan stand to one another as bi-polar opposites on the question of technology. McLuhan stands out as a leading exemplar of technological humanism; Grant is the most important Canadian representative of thought on technological dependency and continentalism. In between these perspectives lies a mediating avenue of technological realism as best expressed by the political economist Harold Innis. While McLuhan wrote with utopian insight of the 'global village' in which Canada is intimately integrated, and Grant wrote of a 'technological lament', Innis's ideal was a balance between technology, empire and culture. To McLuhan's celebration of 'space', and Grant's concentration on the recovery of 'time' against the spatialising qualities of electronic technologies, Innis appealed for a reintegration of time and space in the Canadian experience.

Together, these thinkers typify and structure the Canadian discourse on science and technology. The elements of the discourse – the nexus of geography, economy and culture – can be seen easily from the fact that Canada is the world's second largest country. It has a land area of 10,000,000 square kilometres. It stretches 4,600 kilometres from north to south and nearly 5,500 kilometres from east to west. Travelling across Canada would thus be the rough equivalent of travelling (West to East) from London to Pakistan or (North to South) Copenhagen to Namilra. Canada borders on the United States to the south, and the Atlantic, Pacific, and Arctic Oceans. It has the world's longest undefended border and coastlines. As a circumpolar nation, Canada's climate is dominated by its long, harsh winters. And yet, it is still ranked first by the United Nations as the most desirable place to live.

In 1992, Canada's population topped 27 million. More than 80 per cent of the population lives within 100 kilometres of the USA/Canada border (the 49th parallel) in urban areas, and nearly 34 per cent of the population lives in one of Canada's three largest cities – Montreal, Toronto, and Vancouver. (The nation's capital is Ottawa.) On 89 per cent of its territory, there are no permanent inhabitants. More than 80 per cent of Canadians were born in Canada, but Canadians represent a diversity of national and cultural groups. The Federal Government has an official policy of bilingualism – the Official Languages Act (French and English). People of British and French descent compose the majority (34 per cent and 24 per cent respectively), while those of Ukrainian, German, and Italian origins make up significant minorities. The native population constitutes less than three per cent of the total. In the 1980s, the principal regions of origin for immigrants were south-east Asia and northern Europe.

Life expectancy for Canadians is 78 years and infant mortality is 7.2 per 1000 live births. Canada ranks first in the United Nations Human Development Index, which is a composite of population longevity, knowledge, and decent living standards. Its overall educational attainment rate is 70 per cent – this measure being a combination of the adult literacy rate and mean years of schooling. The education sector in Canada accounts for roughly five per cent of GDP and employs close to one million people (roughly seven per cent of the total work force). Over five million Canadians are enrolled in elementary and secondary schools, half a million students are to be found in the college system (200,000 are in Quebec's CEGEP system), and well over one million are in various post-secondary educational institutions – twelve per cent of whom are in graduate programs. In 1989, Canada had 69 universities leading to the award of 100,000 bachelor's degrees, 17,000 master's degrees and 2,500 doctorates.

Canada's political structure is in evolution. The current situation is typified by a sharing of federal and provincial powers. (There are ten provinces and two northern territories.) As this book goes to press, discussions are underway to create a third northern territory. The British North America Act of 1867, which established the terms of Canada's Constitution and which was repatriated in 1982, has provisions which recognised the nation's multi-cultural heritage, affirms the existing rights of native peoples, confirms the principle of equalisation of benefits among its provinces, and strengthens provincial ownership of natural resources. Among other traditional responsibilities, the provinces have control over such matters as education, property laws, and natural resources (such as forestry and mining), while responsibility for agriculture, immigration, social welfare and health care is divided between the federal and the provincial governments. The federal government has formal authority over such areas as foreign policy, defence, navigation, postal services, banking and macro-economic policy. Federal programs provide unemployment insurance, war veterans' and old age pensions, and family allowances; other types of welfare are covered by provincial programs. The vast majority of the population adheres to the federal medical care insurance program, which provides coverage for medically-required services rendered by a physician or surgeon. Provincial hospital insurance programs, funded in part by the federal government, cover almost all of the population. Recent constitutional negotiations between the federal and provincial governments have put a number of these traditional jurisdictions into question. Under this new 'co-operative federalism', several changes to the division of power among the country's thirteen authorities were contemplated (with a national referendum to vote on the new constitutional proposals slated for October 26, 1992). The referendum motion was defeated. The political impacts of this are, as this book goes to press, unclear. If legitimised via the referendum, the federal government will have the authority to transfer to provincial governments it's competence in urban affairs, tourism, housing, culture and training. The federal government will maintain its existing cultural institutions such as the Canada Council, the National Art Gallery and the Canadian Broadcasting Corporation.

The economic and social self-determination of Canada's aboriginal peoples is also an outcome of these negotiations. In addition, various agreements between the federal and provincial governments were to be concluded in the areas of immigration, communications and regional economic development.[3]

The Canadian Parliament consists of the Senate and the House of Commons. The House of Commons has 295 elected members and the Senate has 104 members, appointed by the Governor General of Canada. As a result of the new Constitutional arrangements, it was proposed that the House of Commons will be increased to 337 members, and the Senate reduced to 62 elected representatives (six per province and one each for the two territories). The federal political party structure is dominated by the ruling Progressive Conservative Party, and – in opposition – the Liberal Party and the New Democratic Party. Potentially significant and emerging parties include the ultra-conservative Reform Party, the *independentiste* Bloc Québécois, and the Maritime province-based Confederation of Regions. It is certain that Canada will be a changed country in the coming decade.

Canada is a member of the G-7, the Commonwealth, *la Francophonie*, the OECD, the United Nations, the Asia Pacific Economic Conference, NATO, the Organization of American States and numerous other bilateral and multilateral agencies. Because of Canada's unique geopolitical status, it is also concerned about such transactional and transgenerational issues as the environment. It has taken – wherever possible – a strong and active role in such issues as aerospace law and regulation, and the environment. Evidence of the latter can be seen in the 1992 UNCED "Earth Summit", held in Rio de Janeiro, which was chaired by Canadian businessman/philanthropist Maurice Strong, and in the fact that Canada was the first country to agree to sign the Rio Biodiversity Convention. As a circumpolar nation, Canada has several alliances with its Arctic neighbours, including the USA, the former Soviet Union, Sweden, Norway, Iceland, Finland, and Denmark. In 1990, the Canada Polar Commission was created in order to provide a strong leadership, information and monitoring role on polar research issues. The commission represents Canada on the International Arctic Science Committee.

In 1989, Canada entered into a Free Trade Agreement with the USA. The principal Canadian motives were (i) to avoid American protectionism (which during the mid-1980s was a looming threat to the high portion of Canadian exports that were destined for the U.S.), and (ii) to improve productivity and reduce production costs at home. As this book is being prepared, an expanded economic space is being negotiated between Canada, the USA, and Mexico (with Mexico acting as *demandeur*). Indeed, in August 1992, a proposal for the North American Free Trade Agreement (NAFTA) was initialed and submitted for ratification by the partner countries. If accepted by all three parties and ratified, the agreement is scheduled to come into effect in January 1994. Canada's broad objectives in the NAFTA are;

(a) barrier-free access to Mexico for Canadian goods and services
(b) improved access to the U.S. market in such areas as financial services and government procurement
(c) improved conditions under which Canadian businesses can make strategic alliances within North America to better compete internationally
(d) ensuring that Canada remains an attractive site for foreign and domestic investment
(e) the establishment of a fair and expeditious dispute settlement mechanism

In addition, such a North America Free Trade Agreement **could** provide a continued impetus for the U.S. "Enterprise for the Americas" initiative, and thus help in the development of hemispheric trade arrangements. The overall aim is to create a trade-liberalising agreement which will cover trade in goods and services, as well as investment and intellectual property. These objectives and aims were in support of goals which the Canadian Conservative government set out at the outset of its administration in 1984. In addition to the platform of national economic renewal and national reconciliation, these were to: remove barriers to growth and to encourage entrepreneurship and risk-taking.

Within Canada, the federal and provincial governments have been active in liberalising inter-provincial trade barriers. Such barriers have been found to represent about one per cent of the country's Gross National Product (GNP). These barriers, examples of which can be found in transportation (trucking and drayage), construction and the distribution of wines and spirits, are considered by business and governments alike to be detrimental, as they impede domestic trade in goods and services. It is **not** therefore surprising that regional development policies and issues have played a significant formative role in Canada's industrial, scientific and technological policies. The 1170km corridor from Windsor, Ontario to Quebec City has often been referred to as the economic heartland of Canada representing, during the 1980s, over 55 per cent of Canada's population, 60 per cent of its income and production and 70 per cent of its manufacturing employment. In contrast, the Western provinces of British Columbia, Alberta, Saskatchewan and Manitoba accounted for about 32 per cent of Canada's Gross Domestic Product (GDP), with the Atlantic provinces of Newfoundland, Nova Scotia, New Brunswick and Prince Edward Island making up about six per cent of the national total.

In 1992, Canada's GDP was more than $700 billion. Its real GDP per capita ranks second only to the USA, and as a result is ranked with Switzerland, Sweden, and the USA as one of the most affluent countries in the world. Canada has a developed market economy that is export-directed and highly-integrated with the U.S. economy. The U.S. accounts for 72 per cent of trade with Canada, followed by Japan, the UK, Germany, South Korea, the Netherlands, and France. Exports account for close to 30 per cent of total GDP. Canada's 50 leading industries, in terms of world market share, account for 22 per cent of total exports. Automobiles accounted for 24.5 per cent of total export values. Most of the remainder is in the resource-based sectors. Resource-based exports account for 46 per cent of total exports.

Agriculture accounts for approximately five per cent of the GDP and employs a comparable percentage of the work force. (Despite its size, only 12 per cent of Canada's available land mass is arable.) Canada is one of the world's major grain producers. Wheat is its chief export crop. The timber industry in Canada is extremely well developed, and forest resources are vast. Canada is the leading exporter of wood and wood products. Other very important resource industries include fishing, mineral extraction, and energy. In the mineral sector, Canada is in the top five world producers of uranium, zinc, gypsum, potassium, nickel, cobalt, titanium concentrate, asbestos, molybdenum, platinum, sulphur, aluminium, copper, lead, cadmium, silver and gold.

Manufacturing and other secondary industries account for one third of Canada's GDP, and employ approximately one quarter of its work force. The province of Ontario accounts for over 50 per cent of Canadian manufacturing. Petroleum refining and petrochemicals, and the production of motor vehicles are among the dominant industries, along with aerospace.

An important aspect of the Canadian resource and manufacturing sectors is the high degree of foreign ownership or influence. (Foreign owned firms account for 61 per cent of Canadian manufactured exports, and 64 per cent of imports.) Although exploitation of the country's mineral resources has been a major factor in Canada's economic development, it has also been a major cause of the massive inflow that has resulted in the domination of Canadian industry by foreign corporations. The stock of foreign direct investment (FDI) in 1990 was $126 billion – 80 per cent of this is due to reinvested earnings by foreign corporations operating in Canada.

The service sector accounts for more than 60 per cent of the GDP and employs more than two thirds of the work force. Like other G-7 countries, Canada is – at the time of writing – still trying to climb out of recession. Nevertheless, during the 1990–1992 recessionary period the decline in the service sector was significantly less than with goods. This has been largely led by the finance, insurance and real estate sectors, even though the dampening effects of the Olympia and York difficulties over the Canary Wharf development plan in London, and the insolvency of the Robert Campeau group of companies have been felt.

Overall, Canada's industry consists of more than 932,300 registered companies, of which 97 per cent have fewer than 50 employees. These firms have a strong impact on job creation and on regional development. In the period between 1979 and 1989, small businesses with fewer than 50 employees created 81 per cent of all the new jobs in Canada. In addition to traditional macroeconomic policies, overall government policies and programs for business development, which are important in Canada, include very generous capital gains tax treatments (including important r&d tax credits), corporate tax at parity with the USA, strengthening of competition laws, privatisation of Crown Corporations, and improvements to Canada's skills development and training sectors. These, together with Canada's program of social security, system of education, health care, and other welfare programs, have made Canada an attractive nation in which to live and invest.

Along with the constitutional debate, the nature of competitiveness, the environment and the economic union are all being debated. (See Chapter Thirteen for a more detailed discussion of the new directions for s&t). These factors are being triggered by Canadians' concerns with such issues as eroding quality of life, declining productivity growth rates, declining world market share, increases in unemployment (which at the time of writing stands at 11.8 per cent nationally), and the long term sustainability of an economic structure which is skewed towards primary resource-producing industries. Throughout these deliberations – which involve not only governments in formal talks but also educators, students, employees, and business owners – the role of science, technology, and innovation is increasingly becoming the focus of policy, labour and corporate attention.

Notes

1 See Kroker A 1984 *Technology and the Canadian Mind* Montreal New World Perspectives
2 The trajectory of thought represented by such authors as Margaret Atwood, Northrop Frye and Pierre Trudeau has been labelled 'technological humanism' and 'technological nationalism'. The flip side 'technological dependency' has been discussed by M Patricia Marchuk 1979 *In Whose Interests: an Essay on Multinational Corporations in a Canadian Context* Toronto McClelland and Stewart

Chapter 2

The historical conditioning of s&t[1]

Paul Dufour and John de la Mothe

Introduction

Just as Canada's s&t have been shaped by its geography – its sheer space and resources – so too has it been shaped by its history. It is important to realise that the
science system in Canada is a young one relative to the other G-7 nations. Nevertheless, Canada has had a strong history of inventive activity. The telephone, the
variable pitch propeller, kerosene, the electron microscope, the Canadarm (an
extremely succesful robotic attachment to the US space shuttles), the self-
propelled combine harvester, the CANDU nuclear power system, the ski-doo
and even pablum are all on the long list of successful Canadian inventions and
innovations. At the level of fundamental research, numerous Canadians have
distinguished themselves. Gerhard Herzberg, John Polanyi, Richard Taylor,
Roland Hoffman, and Henry Taube have all been awarded – or shared in – the
Nobel Prize.

These signs of strength notwithstanding, the youth and fledgling development of Canada's formal science system is important to underscore. Some of this
can, of course, be suggested by the fact that by 1917 only two Canadian universities – McGill University and the University of Toronto – offered work leading to
the PhD. degree in science and engineering. It is important to note, as socio-
historian of Canadian science Yves Gingras has done[2], that both these doctoral
programs were modelled strongly on the US system – noteably the program at
Johns Hopkins University. Between 1898 and 1910, only eleven doctorates had
actually been conferred by both Canadian universities. Total industrial expenditures in 1917 were less than $150,000, a survey of 8,000 firms showed that only
37 had any kind of research lab whatsoever, and the federal government expenditure on R and D in 1937 (just prior to World War II) amounted to less than
one million dollars.

Deeper assessments of Canada's scientific and technological structure can be
found in the fact that the development of science and research was greatly conditioned by Canada's cultural and colonial ties both to Europe – especially to Great
Britain and France – and its reaction against the anti-loyalist/'frontier' ideology
that was – in the mid-19th century – shaping the still emergent USA. Both sets of
cultural conditions served to promote and strengthen the development of Canadian ties with researchers in Europe and the USA.

This occured at both an individual and institutional level. In the biological sciences, for example, collaboration between University of Toronto researchers

Frederick Banting and Charles Best working with John James McCleod resulted in the discovery of insulin. Ernest Rutherford, who is most remembered for being at the Cambridge Cavendish laboratories and the Manchester University laboratories, was at McGill University in Montreal from 1902 to 1905 – just before he was awarded his Nobel Prize in 1908. Otto Hahn, who won his Nobel Prize in 1944, was also a student at McGill University, and Frederick Soddy, who was an associate of Rutherford's also at McGill – won his Nobel Prize in 1921. These, as well as many other prominent American and European researchers in physics, chemistry, medicine, biology, and geology, maintained longstanding contacts with Canadians, and indeed attracted numerous Canadian post-graduates to Cambridge, London, Oxford, Heidelberg, Paris, New York, Berkeley, and Boston. For example, between 1891 and 1917, nearly 50 Canadian students benefited from the '1851 Scholar's Program', which was set up by the UK in part from the profits (the value of which was only determined by a Royal Commission in 1890) of the 1851 Universal World Exposition in London. Additionally, between 1875 and 1900, more than 300 Canadian students received graduate education in the sciences and engineering from Cornell University, Johns Hopkins University, the University of Chicago, and Harvard University.

Even as early as 1857, 1882, 1884, and 1889, institutional ties were being forged and strengthened when, at the invitation of Canadian members, the annual meetings of both the American and British Associations for the Advancement of Science were held in Montreal and Toronto.[3]

While Canadians thought of themselves in terms of a British heritage and allegiance to the Crown, their lives were continually affected by the presence of a burgeoning, turbulent and energetic American republic. Their technology mirrored the difficulty of being caught between two powerful, and often opposing, influences. Evidence of these forces can be easily found. For example, the Canadian Society of Civil Engineers (later re-named the Engineering Institute of Canada), which was established in 1887, used ingredients that were similar to those active at the origins of US engineering societies. The engineering profession in Canada was strongly influenced by the competition with British and American engineers for jobs.[4] The establishment of Canada's National Research Council in 1916 very much paralleled the design, function, and organisation of the Department of Scientific and Industrial Research in Britain, which was also founded in 1916.[5] (Indeed, the national scientific structures in Australia, New Zealand, South Africa, and India all possess structural similarities and can, it is often argued, be traced directly to the activities of the British government in 1917.) Other influences on the development of Canada's educational and scientific infrastructure came from France, Ireland, and Germany.

At the national level, the Royal Society of Canada, created in 1882, helped to foster the professionalisation of the early Canadian research community, as did several other local scientific societies such as the Mechanics Institutes in St. John, Halifax and Montreal, the Literary and Historical Society of Quebec (1824) and the Natural History Society of Montreal (1827). Canada's first truly international scientific journal, *The Canadian Journal of Research*, emerged from the NRC in 1929.

At the provincial level, experimentation was under way to establish economic development and industrial research institutes such as the Research Council of Alberta (1921) and the Ontario Research Foundation (1928). Within the educational system, the research, training and technical programs were modelled after the US land grants college system, as well as the French polytechnic systems, and the Irish, Welsh, and Scottish liberal arts traditions. Together, these resulted in such institutions as McGill University, Laval University, École Polytechnique,

Acadia University, and the University of Toronto. This system evolved into a distinctly Canadian network of institutions offering high level research training. Paralleling this institution building, the Dominion of Canada, as it was then referred to, slowly began to design it's own s&t policies and structures. Amongst these were the creation of the Honourary Advisory Council for Scientific and Industrial Research (the early National Research Council) and a number of parliamentary commissions such as the Cronyn Commission of 1919 on scientific research, and the Royal Commission on Industrial Training and Technical Education (1910).

Prior to World War I, most of Canada's activities in s&t were focused on it's natural resources. Agriculture, fisheries, forestry, and mining were of primary concern.[6] The Geological Survey of Canada, which was created in 1842, laid the foundations of the mining industry. In 1885, following an investigation of American agricultural stations, the Canadian government took steps to establish a number of experimental farms to conduct scientific work in agriculture. (Canada's contributions to the field of agricultural innovation – for example, in the form of Marquis Wheat – have since become well known). Concern for Canadian fisheries and fish stocks resulted, in 1898, in the formation of the Board of Management of the Biological Stations (later the Fisheries Research Board). The Department of Agriculture and, later, the Forestry Branch of the Department of the Interior undertook some experimental nursery research that led to the establishment of the first formal forest experimental station in 1917. These early initiatives resulted in a major involvement by government in scientific activities to support the development of resource industries that has continued to this day. However, as was intimated above, virtually no industrial r&d was carried out in Canada before World War I. World War I stimulated both industrial activity in Canada and industrial r&d.

In 1916, the National Research Council was established -ostensibly as a result of Britain's concern that the dominions should each have an organisation akin to its own, newly created, Department of Scientific and Industrial Research. It had come as a sobering shock in 1914 to the nation that had given birth to the industrial revolution that, when the war called for highly qualified personnel, none to speak of could be produced. In fact, 'there were more trained scientists in a few of the great German industries than could be found in the entire British Empire.'[7] On the eve of World War I, Britain found herself still importing vast quantities of minerals, metals and dyestuffs from Germany. Worse still, as events transpired, dependence on German technology included vital supplies for industrial needs, such as zinc, smelted from ore originating from within the empire. Even the army artillery came equipped with gunsights by Goetz of Berlin.[8]

What Canada actually created in response to the British request was a committee of cabinet ministers – the Privy Council Committee on Scientific and Industrial Research (PCCSIR) – which in turn appointed an Honourary Advisory Council or National Research Council for the purposes of being 'caretaker' of the nation's efforts in these endeavours. Some confusion would later arise from the fact that the research laboratories with which the name NRC is now synonymous did not materialise during the first sixteen years of the Council's existence. Their establishment was only to come during the depression years, and only then partly as a competitive response to the creation by the Ontario government of the Ontario Research Foundation.

The creation of the advisory body was warmly welcomed by universities across the country and the Canadian Manufacturers' Association, both of which provided men – eleven in all – to serve on the Council as honorary members. The Council was charged with 'the promotion of researches which would develop industrial production and the utilisation of natural resources in Canada'. How to implement this lofty order was another matter. All members concurred, however, that Canada's immediate and future development in both resources and industry would require a more solid scientific foundation than then existed. The Council's proposed mandate was thus as broad as it was ambitious.

One of the Council's first tasks was to survey the state of scientific affairs in the country, starting with industry. Out of the Dominion's ten thousand industrial firms which were approached, and of the twenty four hundred firms which responded, the figure then spent per year on industrial research appears to have amounted to just $135,000 (salaries not included). The Council Chairman, Dr A.B. Macallum, who was then a biochemist at the University of Toronto, estimated that the whole of Canada at that time possessed less than fifty persons engaged in scientific research. Things did not look promising. To the men assembled for the purpose of coordinating science and research in Canadian industry, and who saw their role as very much one of nation building, there appeared to be precious little to coordinate.

Moreover, the resources that were initially made available by the government to the Council contrasted miserably with the immensity of the problem at hand. No central office existed, no clear budget had been assigned, and the entire staff consisted of one (very busy) civil servant with priority obligations elsewhere. In fact, all members of Council were simultaneously occupied with other full-time posts and no arrangements had been made to even reimburse their travelling expenses.

Nevertheless, these bottlenecks were slowly removed and in its first twenty years of existence, the NRC succesfully focused on a series of domestic issues and problems relating to natural resources. Witness, in the year 1934–35: studies on the temperature, relative humidity and air movement in celery and poultry storage rooms; weather and wheat yield in Western Canada; the determination of the freezing point of canned fruit; the cleaning and handling of barley; studies on Alberta bitumen; sugar-beet table syrup; the properties of honey; industrial utilisation of potatos; and studies on laundering and dry-cleaning. Other studies focused on the adhesive for postage stamps and floor detergents, aircraft skis and farm windmills. Other work that year, which would have surprisingly far reaching results, involved X-rays and radium.

Between 1916 and 1939, the NRC had grown from an initial official budget of $91,600 and one full time staff member to a budget of just under $1 million and a staff of three hundred. From its first cramped and temporary laboratory on the top two floors of a commercial building on Queen Street in downtown Ottawa, it now occupied a building of classical grandeur (known colourfully as the Golden Temple of Science) on Sussex Drive, not far from the Prime Minister's residence.

In an assessment on the organisation of research just prior to World War II, the NRC noted that of the 600 firms reporting scientific research, just under $9 million was spent on research. Of that number, the scientific research expenditures by firms involved in development of new products was only $355,838.[9] In another survey by the Canadian Manufacturers' Association in August 1945, 218 firms reported that they had spent $4.5 million on all forms of scientific research in 1938. 280 firms reported that they had spent $10.75 million on such activity in 1944 (admittedly, some of the increase can be attributed to reimbursements to the firms from the federal government).

This is in some ways not surprising, since during World War II, much of Canadian and American research was conducted at government expense. Unlike the American situation, however, Canada had never gone through a period where there was a large body of research being undertaken in industry, and a small amount in government institutions. This situation was seen as being largely due to the 'branch plant' situation, which meant that most Canadians firms did no research at the beginning of the war and thus were not even capable of undertaking research during the war. Those firms that did have a research capability were foreign-based subsidiaries, and conducted their activity outside Canada. This clearly prevented the placement of research contracts with industry by government, as was done increasingly in the United States, since it was only possible to place a research contract with organisations which already had experience in research, along with a competent research staff.

With a general fear that the scientific capability which did emerge after the war might not be well-organised or supported, there was also concern that the experience of the war had not created any widespread interest amongst Canadian men of business. Indeed, as early as December 1943, prominent men of science were growing afraid that the end of the war would see a great slump in research activity in the country. Some of them said that despite the work done by the NRC on behalf of the war effort, the government had no real interest in the subject.[10]

As late as April 1944, when it issued its White Paper on Employment and Income, the federal government did not appear to be contemplating any large-scale incursions into peacetime scientific research. However, the White Paper did say that the government would foster scientific research as a means of creating employment. This was to be done by establishing a technical and scientific information service (which became known as the Technical Information Service) making available to industry, and particularly small industry, the results of recent research. The Paper also said that the government would establish and coordinate long-term research programmes. To assist in the achievement of these rather ill-defined goals, control of the NRC was transferred, in April 1945, from the Department of Trade and Commerce to the Department of Reconstruction.

Conscious of its critical role in undertaking peacetime research for the country, the NRC attempted to consolidate its activities. To this end, the NRC asked Parliament in 1946–47 for $6.4 million. In making this request, the then Minister C.D. Howe made the plea that 'scientific research in Canada required that the activities of the NRC be maintained on the same order of magnitude as in the war'. Mr. Howe was careful to point out that, despite the impressive demonstration of applications from atomic fission during the war (and Canada had played a significant role in the early r&d on the atomic project), the budget for which he was asking did not cover 'very sizeable expenditures on atomic energy activities'. (These activities were then directed by the President and administrative council of the NRC.)

The 1944 White Paper also noted the government's intention to expand post-war research in the universities. But this statement was made a full twelve months before it was announced that peacetime activities of the NRC would continue on it's war-time scale. It is significant that in announcing the post-war plans for research of the government, Mr. Howe made no reference to the training of scientific personnel.

In the fiscal year 1944–45, the amount allocated by the NRC to those fifteen Canadian universities whose research facilities were considered suitable was only $648,180. This apparent failure of the federal government to take full cognizance of the need for adequate federal assistance to universities in carrrying out fundamental research is remarkable, especially when one remembers that the papers of the day were filled with stories concerning the potential brain drain of qualified personnel to the USA., and had there been no supply of trained research workers in Canada, the country would not have been well-placed to play a worthwhile part in war-time research.

When compared with American and British figures, the amount expended for research by Canadian universities appeared to be inadequate. Figures show that in the fiscal year 1944–45, a total of $1.25 million was spent on scientific research by the universities of Alberta, British Columbia, Manitoba, McMaster, Queen's, Saskatchewan, Toronto, and Western Ontario. Moreover, of this sum, $441,900 was provided by the NRC for war-related research. Thus, eight out of Canada's ten leading universities (McGill University was not included in this survey) appeared to have spent among them the total of only $809,100 of their own funds on scientific research in that year. For it's part, the NRC funded a total of $284,000 in university grants and scholarships. By 1946–47, this amount had virtually doubled to $537,000, and within a decade (just prior to Sputnik), the total value of NRC grants and scholarships had risen to $3 million. This paralleled total federal support for university research which, in the same period, rose from $1.06 million to $7 million.(For a further discussion on the development of the university research infrastructure see Chapter Six.)

Canada's involvement in the two World Wars and the post-war era served to stimulate a broad band of activity in science, engineering, and research policy. The NRC and several Canadian universities were very active in various Allied scientific projects, including those involving atomic fission research, sonar development, nylon and synthetic textiles, chemical warfare, and medical research. The first experiments in bringing together various research competences and expertise together under one specific coordinated effort was founded in the mid-1920s by the NRC in the form of 'Associate Committees'. These 'Committees' brought together, for the purposes of planning a programme of research, representatives of government departments, universities and industries who were in a position to make personal contributions to the particular problem being studied. These Committees achieved a great deal during the war years and beyond, and covered a broad spectum of issues ranging from aeronautical research to magnesium products to radio research.

It was during the ferment of the immediate post-war period that a number of key institutions for the future health of Canadian research were also inaugurated. The Division of Medical Research was established in the NRC, a body that would eventually lead to the creation in 1960 of the Medical Research Council. The Atomic Energy Control Board was set up in 1946 to guide the uses of atomic energy; the NRC's Division of Building Research came into being; and the Defence Research Board took over the NRC's defence responsibilities. The coordination of policy for scientific research followed a new trajectory as science became a major contributor to the health of the nation. A Scientific Advisory Panel was appointed in 1949 to aid the Privy Council Committee on Scientific and Industrial Research. Two Royal Commissions examined the role of science and it's future through the 1951 Massey Commission on the Arts, Letters and Sciences and the 1957 report of the Royal Commission on Canada's Economic Prospects; the latter set the stage for a more intense national debate on the role of s&t in the country's economic development.

The mid to late 1950s and early 1960s saw the emergence of specific measures by the government to encourage r&d in industry. Initially, these took the form of tax incentives. However, following the cancellation of the Avro Arrow aircraft project in 1959 (which had been responsible for a significant proportion of the country's r&d expenditure), and the signing of the Canada-US Defence Production Sharing Agreement, several programmes were introduced, together with policies for contracting-out r&d. The NRC's Industrial Research Assistance Program was established, and remains today one of the most successful industrial r&d support programs.

The economic, political, and international forces which emerged from that formative period led, as it did in other industrialised countries, to a formalisation process which culminated in the Glassco Commission and the Lamontagne Committee Report.

The origins of Canada's s&t apparatus

The Royal Commission on Government Organization, known more generally as the Glassco Commission, reported its findings to the government in January 1963. The Commission noted that, 'one of the original purposes in devoting money to research was to encourage and stimulate Canadian industry. From being a primary goal, this has, over the years, been relegated to being little more than a minor distraction....'

Highlighted for blame in the diminution of the role of research in public affairs, the armed services were singled out by the Commission. In 1961–62, the armed services, along with the Defence Production Board and the Defence Research Board, had at their disposal just under one-third of federal r&d funds ($77 million). The Commission complained that Canadian industry's participation in defence research activity had been reduced to nominal proportions compared with its counterparts in the UK and the USA. and added that;

> 'senior officers of the Armed Services display a marked lack of confidence in Canadian s&t as a whole... that this lack of confidence is unwarranted is amply demonstrated by a record of first rate achievement by Canadian science over the past few years.'

In addition, the report highlighted the long-term advantages of using government contracts to stimulate industrial research. The absence of a strong industrial research tradition, and the relative lack of interest on the part of management in research, though a consequence of Canada's economic history, was seen to threaten Canada's further economic development. Attention was drawn to the changing balance in the Canadian economy,

> 'Our renewable resource industries – forestry, agriculture, fishing – are declining industries. By 1980... probably only ten to twelve per cent of the workforce in Canada will be employed in these industries as against thirty-five per cent or more in the late 1920s... We live among nations engaged in an unprecedented drive towards diversification and a larger degree of self-sufficiency... Canada needs a broadly-based industrial structure—the resource industries need vigorous and successful industries behind them.'

The Commission considered that the system had failed to function as intended. It observed that the Privy Council Committee and the Advisory Panel for Scientific Policy (established in 1949) met infrequently, and that the NRC concentrated on its own, more specific, responsibilities. The Commission commented that 'in view of this evident breakdown of the system as designed, it is remarkable that science in the government has from some points of view flourished as never before.' And after noting with approval a number of scientific and technological developments (as well as some of the factors contributing to them) the Commission added wryly that 'the failure to build on the basis of a cohesive program has not inhibited the spending of public money.'[11]

To provide for more cohesive governmental organisation, the Commission made a series of recommendations. In respect to R&D policy, four specific recommendations were made;

- the President of Treasury Board who, without any departmental operating responsibility, be designated as minister responsible for scientific policy in the country and the coordination of existing activities in the field of R&D. (It was also recommended that this position have general responsibility for the quality of administration in the public service)
- a Scientific Secretariat be established for the Cabinet under an officer to be known as the Scientific Secretary reporting to the proposed President of the Treasury Board
- a National Scientific Advisory Council be established, with membership drawn from the scientific disciplines, the universities and industry and the community at large to review and submit independent advice with respect to national scientific policy
- the Scientific Secretary act as secretary and the Scientific Secretariat serve as the secretariat for the National Scientific Advisory Council.

Before adopting these recommendations, the federal government asked a former President of the National Research Council, C. J. Mackenzie, for his comments and advice. Mackenzie had served as chief science adviser to the government during the war, and had found then the value of having the right people conduct proper decisions. His overriding philosophy was therefore that getting the right people was more important than drafting an ideal organisation chart. Essentially, he agreed with the Commissioners with regard to the proposed Secretariat. He supported, with further qualifications, the setting up of an Executive Council, but had no recommendation for a role in scientific policy for the President of the Treasury Board.[12] Nevertheless, Mackenzie offered a number of suggestions for study by the proposed National Scientific Council, including the issue of total support for r&d over the coming five years and the balance between sectors in its performance.

The government took action. A Science Secretariat was established within the Privy Council Office (PCO) in the spring of 1964, and the Science Council of Canada, after some delay, was established in 1966 to provide independent advice. The Minister who chaired the Privy Council Committee on Scientific and Industrial Research became the *de facto* Science Minister, but without designation. There was no thought at this time of establishing a separate science policy portfolio within Cabinet.

The Science Secretariat provided services to the Science Council during the first two years of the Council's operation, in addition to its dominant function of serving the confidential needs of the PCO and the Cabinet. This situation was ended in late 1968 when the duties of the Secretariat were confined to the PCO and the Cabinet, and the Science Council became a separate entity (and subsequently a Crown corporation) with its own staff.

In 1967, a Special Committee on Science Policy was established, which issued 4 major volumes on the organisation of science policy, spanning a ten year period (1968–1977). This Committee was known as the Lamontagne Committee (so-called after its chairman, the Harvard-educated economist, Maurice Lamontagne). Although it can in no way be said that the Senate Committee reports offered a complete answer to the problems of Canadian science, its authors had the courage to question many of the dogmas with regard to what public affairs had to do with s&t.[13] Moreover, the Senate Committee treated s&t as a social activity whose pattern of development is just as open to analysis as any other human activity. In so doing it debunked science, its special claim to status, and especially its mode of organisation in Canada. More than anything else, its recommendations attempted to make scientific research accountable for its activities. As a result, it came under considerable attack by the scientific community which recognised the need for directing their creative energies along lines that would benefit the nation, but also emphasised the need for balance that would not destroy creativity by over-direction. As Gerhard Herzberg, Canada's Nobel Prize winner in chemistry, was to summarise the tone of this and the previous Glassco Commission;

> 'The Glassco Commission was really not interested in good science. It was interested in good accounting. There is, of course, nothing wrong with good accounting, except that it does not necessarily lead to good science.'

Indeed, the Senate Committee received considerable testimony from national organisations and individuals. For its sheer scope and scale, it is unmatched in the annals of Canadian science policy, even today. The reports' findings also served to trigger the emergence of a science policy journal known as *Science Forum* (which published articles on current science policy issues in both English and French. French Canada already had a well-established society for the advancement of science, which had been meeting regularly since the early 1920s).

Briefly summarised, the Lamontagne Committee made the following points;

- investment in scientific research must follow the pattern of national priorities and must give special importance to industrial research
- there must be a more precise evaluation of scientific research so as to terminate what is no longer valid and shift investments to new priorities when these present themselves
- a coherent organizational system must replace the present conflicting pattern of responsibilities so that a dynamic, on-going process can be achieved that will take into account both the needs of science and the needs arising from the changing pattern of Canadian priorities

To achieve such a coherence, the Senate Committee made a series of recommendations. Although it is not possible to list all of these recommendations, some of these are summarised in Figure 2.1.

What this diagrammatic representation illustrates (based on the recommendations from Volume II) most clearly was the 'logical' sequence in the system recommended by the Senate Committee. The system began with the idea that the planning of science must be determined by goals that anticipate Canada's future needs. The Lamontagne report attached great importance to the forecasting of these needs and stressed the necessity of creating the 'forecasting' instruments required to do this. Furthermore, recognising the fact that without some degree of economic prosperity the social and cultural needs of Canada could not be met, the report stated that;

> 'Canada's growth strategy must rely mainly on a high and sustained flow of technological innovations introduced by the secondary manufacturing sector of the economy. This will not succeed unless it becomes a major national objective.'

The priority given to innovation through market-oriented technologies came from a belief, on the part of the Committee Chairman, in the importance of industrial innovation as a cause of economic, cultural and social growth. This meant, therefore, that first priority was to be given to industry in the organisation of the new system of r&d. (Today, the debate on this question is virtually identical; see Chapter 13).

To the Senate Committee, the reason for placing industrial research as the top

Figure 2.1: *Recommendations on Science Policy by the Senate Committee (as reported in Volume II of its Reports)*

Level of Activities	General Recommendations			Specific Recommendations		
	Long-Range Recommendations	*Middle-Range Recommendations*	*Short-Range Recommendations*	*Minister for Science and Technology*	*Department of Industry, Trade and Commerce*	*Department of Energy, Mines and Resources*
General Orientation	Creation of an office of industrial reorganization	Creation of an information system on technological and science forecasting	Economic Council creates a Committee on Forecasting, Senate sponsors a conference on forecasting	Responsible for the integration of all scientific and technical research activities	Organization of a task force in each industry	Organization of a task force in each industry
Industrial Research		Increase proportion of industrial research in R&D expenditure to 60 percent by 1980		Preparation of administrators for R&D industrial projects	The Canadian Innovation Bank	
Government Research	Creation of a National Research Academy			Review of all government research	Canadian Industrial Laboratories Corporation	
University Research	Creation of a Canadian Research Board	Increase proportion of basic research to 10 percent of R&D expenditure by 1980		Responsible for the preparation of QSE required for R&D. Control of scholarships		
R&D	Canadian government adopt a long-term plan for R&D	R&D expenditure should reach 2.5 percent of GNP by 1980	National conference on science, technology, and industrial innovation	R&D inventory and control		

Source: Phillippe Garigue, *Science Policy in Canada*, 1972.

priority in the new system was that science policy could not be considered in isolation from an industrial policy, and this, in turn, was dependent on an over-all national economic policy.[14]

The report went on to stress that s&t could not be limited to one government department but that all concerned (even provincial governments) had to be in-volved. What was needed was not a general goal definition for science but the development of broad strategies for national r&d.

Among the major reports consulted, and influencing the view of the Senate Committee, was the OECD review of Canada's national science policy (the tenth in a series of reviews that the OECD had conducted on member countries) and the 1968 report of the Science Council of Canada *Towards a National Science Policy for Canada*. The OECD review was an extensive survey of Canada's infrastruc-ture for science and technolgy and commented on the organisation of science policy in the country. Among its suggestions, the report noted the need for a Minister for Science and a high level Science Policy Council reporting directly to the Minister.[15] The report also pointed to the major challenges of defining na-tional objectives for science in a confederation where federal and provincial jur-isdictions in s&t were not clearly defined.

The Science Council report of 1968 remains the only statement that attempts to outline explicit science policy goals and frame them within a broader context of social and economic strategy. The Science Council national goals were: national prosperity, physical and mental health and high life expectancy, a high and ris-ing standard of education readily available to all, personal freedom, justice and security for all in a united Canada, increasing availability of leisure and en-hancement of the opportunities for personal development, and world peace based on a fair distribution of the world's existing and potential wealth.[16] (This report was also instrumental in recommending that the government use the cap-ability of its laboratories and their missions to work more closely with industry and university, and that government procurement contracts be used to upgrade the technological level of Canadian industry; for more on this subject, see Chap-ters Four and Five).

The Senate Committee also heard from international experts. In May 1969, for example, the Committee met with the US House Committee on s&t, and in Feb-ruary 1970 the invitation was reciprocated. A number of the Canadian senators were quite impressed with the American system of control of science, with Sen-ator Allister Grosart opining on one occasion, 'we ascribe it to that well-known American genius in developing checks and balances and coming up with a sys-tem which in theory makes absolutely no sense, but usually works much better than some of the theoretical structures that on paper make more sense.'[17]

Among the key recommendations from the Lamontagne Committee was the need for a Ministry for Science (contrary to the advice from the OECD review of 1969). Created in 1971, the Ministry of State for Science and Technology (MOSST) was responsible for formulating policies and providing advice on s&t to the federal government. It replaced the Science Secretariat. MOSST was the lead department within the federal ambit on s&t policy and coordination. It played a central role in relation to the government's s&t decision-making and provided an overview of s&t activities to facilitate the efforts of Cabinet and Treasury Board.

In 1980, the Prime Minister designated MOSST as the lead department for space research policy and the coordination of space activities among federal government departments and agencies (eventually, this was devolved to the Ca-nadian Space Agency when it was created in 1989; the recommendation of which had been made in the very first Science Council report of 1967). In 1983, the

Prime Minister assigned the additional role of the Chief Science Adviser to the federal government to the Secretary of MOSST, with responsibility for providing expert advice on the priorities for, and the planning of, Canada's overall s&t effort. This function still remains within Industry, Science and Technology Canada, though it has not been filled in quite some time. The Prime Minister's National Advisory Board on Science and Technology is the *de facto* chief science adviser of the government; and has become more so with the abolition, in 1992, of the Science Council of Canada.

Among its other duties, MOSST was responsible for preparing an annual overview and framework paper that analysed the entire range of the government's s&t activities and reviewed existing and proposed s&t policies and programmes. It was charged with providing advice on methods to strengthen the effectiveness of federal s&t expenditures and improve their linkage to economic development. MOSST was given the task of strengthening and maintaining international linkages for the acquisition, transfer and exploitation of commercial technology (a task that ran into some jusrisdictional difficulties with the Department of External Affairs (see Chapter Seven). It was also responsible for developing and implementing a rationalised and focused strategy for all industry support programmes for R and D. Finally, MOSST was given the job of working with the provinces in developing subsidiary agreements in s&t for regional economic development purposes and to help frame a National Science and Technology Policy.

Such a broad mandate with little clout in Cabinet and with powerful line departments protecting their vested interests would eventually cripple such a structure. And it did. MOSST's power came almost exclusively from its influence and arguement based on knowledge. It lacked the powers that usually accompany ministers of operating departments: tradition, statutory responsibility, financial authority, political prestige, large staff, research facilities. The federal structure for science, technology, and industry has undergone several reorganisations since, with the most recent having the ministries responsible for regional industrial expansion and s&t merged into a new flagship Department of Industry, Science and Technology (1989).

Paralleling these developments, provincial governments began to institutionalise their respective science policy machineries, often in response to federal initiatives, and to assist in regaining control over their own local economies. For example, the province of Quebec was the first to establish a science policy structure in the early 1970s; a functional equivalent of the Science Council of Canada. This development eventually led to the formation of Quebec's science/education policy, and later, to it's technological development strategy. No other province has a richer legacy of the examination of the role of s&t for economic and social objectives.

The rise of interest in s&t policy by provincial governments was as much an attempt to strengthen and diversify the local economic base as it was a deliberate, focused response to the central, federal government's thrusts in the s&t sphere (see Chapter Three for more description of these provincial developments).

Achieving national consensus

With all of the activity during the past three decades in new s&t initiatives, extensive studies of the making of s&t policies, and widespread experimentation with new institutions, it would be reasonable to expect that the governments in Canada were well equipped with the tools necessary to use s&t in a concerted manner for the future development of the country. This was not the case, and it took nearly twenty years after the seminal Science Council report *Towards a Science Policy for Canada* for the country to begin arriving at a national consensus on the general priorities of a nation-wide policy on s&t. Even then, the results have been mixed.

As in the European situation, a combination of global industrial re-structuring through science, technology, and innovation and domestic demand provided much of the stimulus that Canada required to pool it's national skills and natural resources in a manner that could meet the international technology challenge.

There are many reasons why Canada's s&t policy renaissance did not occur until the early 1980s. Part has to do with the natural evolution required to obtain a minimum threshold and infrastructure in r&d. Another has to do with the poor valuation of science in the socio-economic fabric of the nation.

While Canada's situation is still somewhat deficient regarding resources devoted to r&d, the increased funding of r&d has been particularly marked over the past five years, especially from the corporate sector, which has benefited from a series of fiscal (r&d tax credits) incentives designed to encourage investment in knowledge-based activities (see Chapter Five for more on this).

Besides the apparent lack of an "external threat" to catalyse increased government attention on matters scientific and technological, the domestic situation in Canada had been characterised by an inability to conceive what constituted the 'national interest' (with the possible exception of Quebec, which has pursued a strong tradition of s&t policy). Much of the debate in s&t policy revolved around tensions between the two levels of government in identifying their respective spheres of interest and influence.

Therefore, a national perspective was long in coming. With a federal government that, in the 1970s, had given a very low priority to s&t, and with the rise of such provincial movements as the separatist Parti Québécois that espoused a *'maître chez nous'* (master at home) ideology, nation-building through s&t was a difficult challenge.

The vision of Canada as a unified country, with little attention to the provinces and their respective interests and constitutional rights, had gradually become anachronistic. With a new federal Conservative government elected in 1984, a major re-ordering of the national perspective was undertaken. Its roots were found in policies initiated by the previous Liberal adminstrations. In 1978, the Liberal government of the day had cobbled together a statement on r&d in preparation for a federal-provincial ministerial meeting on industrial R and D. This report took a broad view of the domain of science policy, looking at policies for the support of science, policies for the application of s&t, and the role of science in public policy.[18] The paper attempted to capture some of the national priorities of the day (eg., employment, unity, energy, sovereignty, northern development, international aid) and link these with research objectives. Among its issues, the paper noted new policies of the government, including a target of 1.5 per cent of Gross Domestic Product for r&d expenditures by 1983; the use of federal government procurement to stimulate industrial r&d; the establishment of centres of excellence responsive to national needs; and the need to develop r&d priorities.

Echoes of this paper were to be found in 1983 when the Liberal government issued a Technology Policy for Canada.[19] Buoyed by its success in displaying its technological prowess in the Canadarm on the space shuttle Columbia, and fresh from its strong participation in the 1982 Summit Meeting in Versailles (that established the first working group on s&t among G-7 partners), the federal government was anxious to build on the momentum. The new policy defined technology objectives for Canada and seggregated these by sector (business, university, labour, government, provinces). The objectives laid out were;

- to strengthen the Canadian economy through the development of new technologies for producing goods and services and the widespread adoption of new and existing technologies
- to manage the process of technological development so as to ensure that Canadians are aware of both the opportunities and the problems that might arise
- to ensure that the benefits to technology development are shared equitably among all Canadians in every region
- to create a social climate that places a premium on scientific and technological excellence, curiosity and innovation

To help implement these, the Prime Minister established a sub-committee of Cabinet on technology development. Announced were new programmes in support of research facilities (fifteen new research facilities to be built), human resource development (to help workers acquire new skills), and university research assistance (through increased funding to the granting councils). Also instituted was assistance for new strategic technology fields in microelectronics, communications, and biotechnology. To elicit feedback at the national level, the government sponsored a large conference called 'Canada Tomorrow' to debate emerging s&t issues for Canada. At the Conference, the then Prime Minister Pierre Elliott Trudeau outlined the government's role in s&t;

> 'The Government must be more than a patron of technological enterprise, more than a source of funding, for even more fundamental is the government's responsibility to help manage the impact of technological change, and to act as an honest broker between competing forces in the movement towards a technologically-sophisticated society. Government's preoccupation must be to ensure that the benefits of this revolution outweigh the costs.'

On the heels of this 1983 Conference were two reports that examined the role of the federal government in support of technological development. The Task Force on Federal Policies and Programs for Technology Developpment (appointed by the Minister of State, Science and Technology) reported in July 1984 and the Standing Senate Committee on National Finace published its report, *Federal Government Support for Technological Advancement: An Overview* in August of the same year. In both reports, the trend of having public research facilities and public support for s&t follow market responses was established. Relevance (of university and government research to private sector needs) was clearly identified as the guidepost for future policy measures. Canada has not looked back since.

In 1984, with the newly-elected Progressive Conservative government, a concerted effort began, through the Ministry of State for Science and Technology, and along with other governments, to develop a national policy for s&t.

Intensive negotiations followed between both levels of government and, by December 1986, ministers had agreed to the broad outlines of a national policy and to the establishment of a Council of Science and Technology Ministers – a mechanism that would serve to formalise a consultative and collaborative forum between the two orders of government in the area of science policy.

At the centre of the national policy process was the recognition that there was a necessity to take account of federal, provincial and territorial priorities in s&t. The governments also recognized that Canada's provinces and territories had differing scientific and technological capacities in different sectors.[20]

The Policy defined six major objectives that could be attained through cooperation between governments, business, universities, and labour. These were to;

- improve industrial innovation and technology diffusion through public and private mechanisms
- develop strategic technologies for manufacturing, service, and resource-based sectors
- assure the necessary pool of highly qualified people
- support basic and applied r&d
- control the impact of technological change on society
- promote a more science-oriented culture

These objectives are remarkably similar (though with different emphases) to another set of objectives developed for the Chief Science Adviser in May 1984 (before the Conservative election). In that assessment, the following priority issues were identified;

- accelerating the adoption and development of key technologies
- ensuring the availability of highly qualified human resources
- increasing the access to world markets for Canadian high technology
- ensuring the relevance, quality and appropriate management of government intramural r&d
- reducing the impact of technological change on society

There has been some concern that the National Science and Technology Policy signed in 1987 implies a drift in responsibility at the federal level in the face of increased provincial activity in s&t, and that because of the ongoing debate on the constitution and the federal/provincial division of powers, new arrangements are in the offing concerning national cost-shared s&t programmes.

The process of negotiating with provincial and territorial governments, and consultation with business, labour, and higher education, has had a remarkable learning effect within the Canadian s&t infrastructure. Provincial governments have become much more adept at understanding, and making a case for, the introduction of s&t policy within their respective cabinets, as is demonstrated by Premiers and Ministers in a number of provinces through their respective appointment of consultative mechanisms in s&t (every province now has a science policy advisory council of some form, described in the next Chapter).

The federal government has learned to appreciate the vital regional development dimension associated with s&t, and government departments are increasingly involved in extensive bilateral and multilateral consultations with provincial counterparts. Business, labour, and educational interest groups have substantially increased their voice in the s&t arena, though considerably more still needs to be done in this area.

This said, it should be noted that the National Science and Technology Policy remains a desireable objective. It has very few 'teeth' to it, and, to date, has had few successes in driving the national s&t agenda. For this to happen, a major shift in vision is required (see Chaper Thirteen). This brief historical overview provides a backdrop for the the organisation of s&t in Canada that is discussed in Chapter Three.

Notes

1 The reader should note that this is not a chapter on the history of s&t per se; rather, it is about the conditions that shaped the development of the people, programmes, policies, apparatus and institutions that conduct s&t in the country. For more of the history of science, see any of the issues of *Scientia Canadensis*, the journal of the

Canadian Science and Technology Historical Association. A good review of these papers is found in Jarrell R A Hull J P (eds) 1992 *Science,Technology and Medicine in Canada's Past* Ontario Scientia Press

2 Gingras Y 1991 *Physics and the Rise of Scientific Research in Canada* Montreal McGill-Queen's University Press

3 In 1884, fourteen Fellows of the AAAS were Canadian, as were twenty-four Fellows of the BAAS. The AAAS also visited Ottawa in 1938 and revisited Toronto with it's annual meeting in 1981

4 Among the most notable achievements in Canadian engineering (as defined by the 1987 Centennial Board of the Engineering Institute of Canada) are: development of railway networks across Canada, building of the St. Lawrence Seaway, the DCH-2 Beaver bush plane, creation of the world's largest transmission network, the Alouette satellite, the Bombardier snowmobile, the James Bay very-high-voltage transmission system, the Athabasca oil sands development, the Canadian nuclear power system and the industrial installations at Polysar Ltd.

5 It is important to realise that factors of causality, *viz* the creation of such institutes as the NRC, are still undergoing significant debate and revisionist history. After all, it is clear that Canadians were participants in the Washington and London discussions to set up their structures as Americans and Europeans were in Canada's.

6 For a useful summary of science in Canada circa 1886, see Bovey H T *The Progress of Science in Canada* lecture delivered at the St. Paul's YMCA, Montreal 22 Feb. 1886

7 National Research Council of Canada Handbook, c1948 p1

8 Melville Sir H 1962 *the Department of Scientific and Industrial Research* London Unwin and Allen p24

9 For an informative assessment of the health of several disciplines in 1939 (including geology, chemistry, botany, zoology, medicine, astronomy, mathematics and physics) see Tory H M (ed) 1939 *a History of Science in Canada*Toronto Ryerson Press

10 An Interim Memorandum on Scientific Research in Canada, 1 October 1946

11 The reader should note that there are a number of myths associated with the level of industrial research conducted in Canada that have considerable longevity as a result of the Glassco and Lamontagne reports. Some of these are under examination by Canadian historians who are intrigued by how science policy could develop as a result of certain unchallenged assumptions. See, for example, Phillipson D J C The Steacie myth and the institutions of industrial research *HSTC Bulletin* 25:117–134. Also, see the papers in Gingras, Y Jarrell R (eds) 1992 *Building Canadian Science: the Role of the National Research Council of Canada* Ottawa CSTHA

12 Mackenzie C J 1964 *Report ot the Prime Minister on Government Science* Ottawa. Mackenzie's report is notable for its analysis of the structural differences in r&d expenditures between Canada and the USA. For example, he argued that Canada's comparative advantage over the USA and other countries devoting considerable funding to military research was its fairly well developed civilian base. As he noted 'If, as appears likely, Canadian research expenditure in the space, military and atomic fields will be far less in proportion to civil fields than in the United States, it follows that there will be only limited overflow of military research benefits to civil industries and, in the main, Canadian prosperity will depend primarily on industrial productivity in the civil field'

13 Unfortunately, Senator Lamontagne, despite all his efforts and commitment to change, received little attention or support for his reports from the then Prime Minister Pierre Elliot Trudeau. As a result, the full impact of the Lamontagne reports was not felt on the Canadian s&t policy scene. Today, Canada is still debating many of the same policy questions raised by Maurice Lamontagne in his pioneering study.

14 see Garigue P 1972 *Science Policy in Canada* Montreal Private Planning Association of Canada

15 The concept of a Minister for Science had received considerable play in Canada during this period; the New Democratic Party had suggested such a position as early as 1967. The position, like much of Canada's s&t policy apparatus, was modelled on the British system, where a Minister for Science had existed since the early 1960s. Indeed, the first British Minister for Science, Lord Hailsham, was prompted to admit

modestly in 1963 that 'I am, so far as I know, the first and possibly the only Minister for Science (or of Science, for that matter) in the Universe'.

16 When asked by the Senate Committee the reason why the Science Council had undertaken to specify national goals, the Science Council Chairman Omand Solandt said, quite simply, 'somebody has to start'.

17 Hechler K 1980 *Toward the Endless Frontier: History of the Committee on Science and Technology, 1959–79* Washington D.C. US Government Printing Office p393

18 Ministry of State for Science and Technology 1978 *Research and Development in Canada; a Discussion paper* Ottawa

19 Ministry of State for Science and Technology 1983 *Towards 1990: Technology Development in Canada* Ottawa.

20 For a more detailed description of the origins of the National Science and Technology Policy, see Dufour P Gingras Yv 1988 Development of Canadian Science and Technology Policy *Science & Public Policy* 15

Chapter 3

The organisation, structure and decision processes of science and technology

Paul Dufour and John de la Mothe

Introduction

Canada's s&t policy infrastructure has been in constant flux since the post-World War II era. Science and technology grow not in a vacuum but within a complex of interrelated support institutions and decision-making systems. This complex, if well integrated with the broad canvas of other public policy instruments such as education and monetary, fiscal and regulatory mechanisms, and if well stocked with creative and entrepreneurial individuals, provides the necessary support conditions for both technological innovation and growth in science.

Ultimately, the link between science and cultural and economic activity is a critical one. The central question is : science policy for what? The debate in Canada has not changed substantively (though the context has) over the past five decades. What is the objective of Canadian s&t as seen by citizens and translated by politicians? Is it national prosperity (the new currency)? If so, for whom? Is it to overcome regional disparity, to bootstrap the communities throughout the country with the more judicious implementation of the fruits of scientific advance and technological innovation? Is it for international prestige? Is it for industrial competitiveness and increased market acccess?

The inability of policy-makers to answer such questions has been a perennial frustration of public policy and the bane of academic, business and labour leaders. Even more frustrating for the scientific community and the technical and business groups has been the constant re-thinking and re-structuring of government support programmes and policies. Over the past five decades the Canadian approach to science policy-making has swung from a strong interventionist stance to mobilise resources and people in a united cause (such as the war effort) to a more *laisser-faire* position, requiring little strong central control and limited long-term vision. The institutions designed for these periods and the central policy machinery to accompany them therefore have tended to reflect these concepts. We illustrate these shifts diagrammatically with the following series of Figures that trace the evolution of the national system for adminstering s&t. Figure 3.5A outlines the present organizational structure for s&t in Canada.

The Canadian government's chief science adviser over the years 1969–71, Robert Uffen, commented in 1972 on how science policy was made in government. As he noted;

'*Viewed from Cabinet level, the problems of s&t are only some of many that make demands*

Figure 3.1: *A schematic representation of changes in the Essential Structures for Science and Technology Policy in the Government of Canada, 1945–1950**

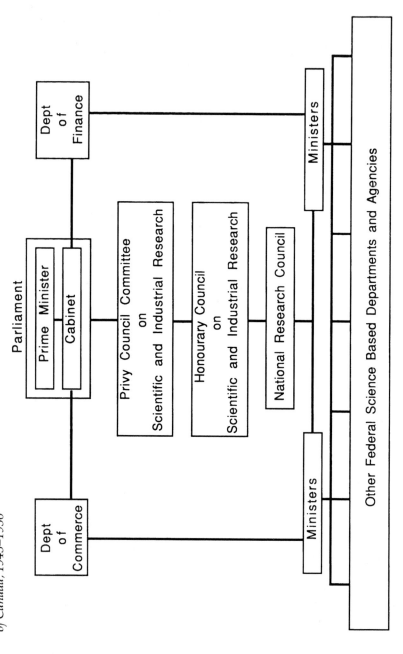

* Typified by the centrality of the National Research Council : a 'strong intervensionist phase'.

Figure 3.2: *A schematic representation of changes in the Essential Structures for Science and Technology Policy in the Government of Canada, 1950–1968**

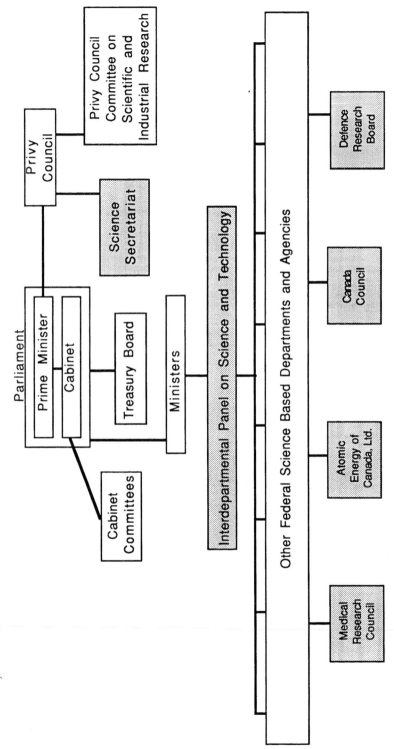

* Typified by a modified National Research Council role, a 'permissive' phase with central political machinery, pre-Lamontagne

Figure 3.3: *A schematic representation of changes in the Essential Structures for Science and Technology Policy in the Government of Canada, 1968–1983**

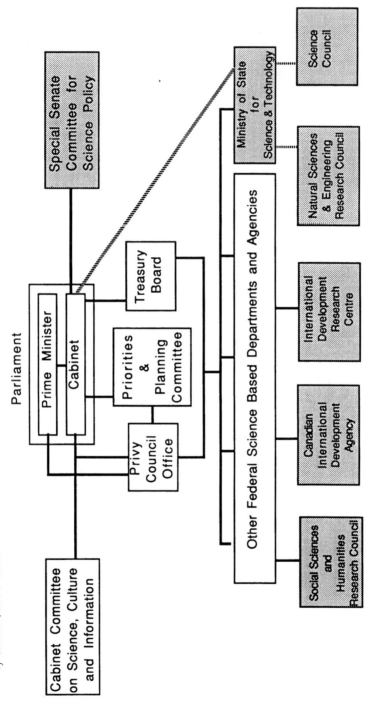

* Typified by the 'restructuring' under Prime Minister Trudeau, experimentation with 'Ministries of State'
  * Science Secretariate transformed into MOSST
  * NSERC created out of NRC
  * SSHRC created out of Canada Council

Figure 3.4: *A schematic representation of changes in the Essential Structures for Science and Technology Policy in the Government of Canada, 1983–1991**

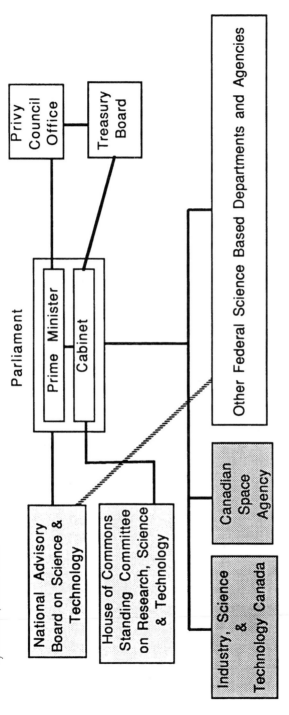

* Typified by the revised structure, with emphasis on federal-provincial sharing of responsibilities in S&T, Phase I
• ISTC created out of MOSST and Department of Regional and Industrial Expansion

Figure 3.5: *A schematic representation of changes in the Essential Structures for Science and Technology Policy in the Government of Canada, 1992–* *

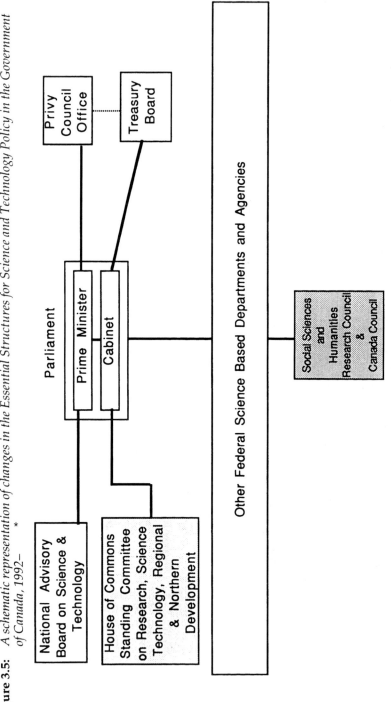

* Typified by the revised structure, with emphasis on federal-provincial sharing of responsibilities in S&T, Phase II
• Science Council eliminated

Figure 3.5A: *Key Elements in Canada's National (Federal, Provincial and Territorial) Science and Technology Policy System, 1992*

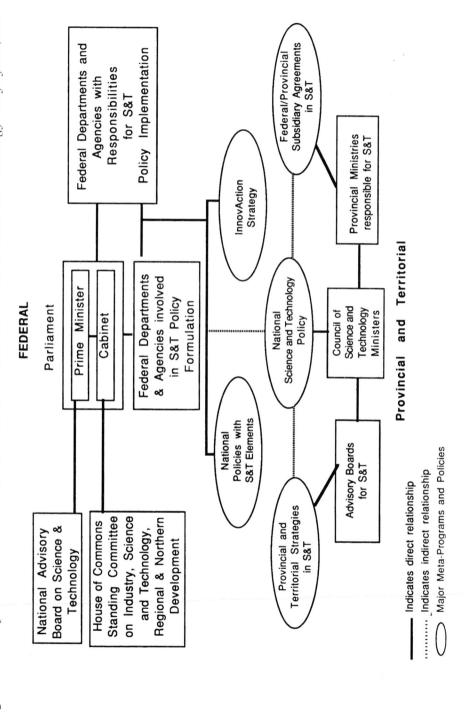

upon resources of money, manpower and time.... Whether scientists like it or not, science does not rate a very high priority, and technology just barely makes most lists of pressing problems.[1]

Uffen traced the process of how science decisions were made with attention to the various consultative phases required within government and outside in the community. While government has grown much larger over the twenty years, and the raw information inputs have ballooned exponentially, the process of getting the attention of Cabinet for new expenditures in s&t has not changed much since. The landscape and institutions for mediating s&t have, nevertheless, led to a new architecture.

This chapter highlights some of key elements of this now rapidly changing landscape.

The Federal Structure

At the centre of Canada's federal science organisation is the Department of Industry, Science and Technology (ISTC), created in 1988. As noted in Chapter Two, it was the result of a merger of the Ministry of State for Science and Technology (MOSST) and the Department of Regional Industrial Expansion (DRIE). Its objectives are to promote international competitiveness and industrial excellence in Canada, and to review and expand Canada's scientific, technological, managerial, and production base. Its 1991 budget was approximately $1 billion and it employed roughly 3,000 people. Expenditure on scientific activities amounted to $239 million in 1992–93, four-fifths of which are targetted to Canadian industry.

ISTC has three Ministers: a senior minister responsible for industry, science, technology, and international trade; a minister responsible for science (to whom a 'chief science adviser' reports, though no such person has yet been appointed); and a minister responsible for small business and tourism. Other areas of responsibility of the Department are regional development, small business and entrepreneurship and aboriginal economic development.

Among its larger s&t support programs, ISTC has a Strategic Technologies Program which stimulates industrial r&d investment by encouraging Canadian firms to form alliances in developing technological capabilities in biotechnology, informatics, and advanced materials. Consultative committees have been established to advise the Minister on policy issues affecting the development of strategic technologies. Thus, the National Biotechnology Advisory Committee and the National Advisory Panel on Advanced Industrial Materials both provide expertise to guide the direction for these technologies.

ISTC also has Sector Campaigns led by its industry branches to enhance the competitiveness of Canadian industries that have high growth prospects (*e.g.*, medical devices, information technology, environmental industries, ocean technology, automotive parts). Support is also maintained for the fifteen Networks of Centres of Excellence which were announced in 1988 (see Chapter Six). ISTC also supports technology diffusion through such mechanisms as the Technology Outreach Program (which funds twenty-three private sector organisations that advise and counsel firms on the sourcing and integration of technology), the Advanced Manufacturing Technology Application Program, and the Manufacturing Visits Program. Extensive support is provided to the aerospace and defence sector through the $215 million Defence Industry Productivity Program.

To bolster support for education and human resources, ISTC has introduced a

Canada Scholarships Program for senior secondary school students to entice them to enter university in the natural sciences and engineering streams, as well as for technicians and technologists studying at community colleges and technical institutes. A Public Awareness Campaign promotes the development of a science culture in Canada through a variety of initiatives, including a National Science and Technology Week, and funding for various organisations involved in science awareness activities.

ISTC is also responsbile for ensuring the federal-provincial coordination in science and technology. Through a federal-provincial relations secretariat, ISTC has been active in elaborating the action plan of the Council of Science and Technology Ministers, a group of thirteen Ministers responsible for implementing the specific initiatives emanating from the 1987 National Science and Technology Policy.

ISTC also houses the Secretariat for the Prime Minister's National Advisory Board on Science and Technology (NABST), a body responsible for providing advice to the Prime Minister on s&t issues of national concern. It consists of twenty-four senior industrialists, labour, government, academic, and financial representatives, as well as the Minister for Science and the Deputy Minister of Industry, Science and Technology. It meets three to four times a year, with the Prime Minister chairing most sessions. Created in 1987, its functions are to;

- Advise on the appropriate use of government instruments for encouraging the development of s&t
- Propose means to sensitize Canadians to the profound changes resulting from the technological revolution, and to help them make the necessary adjustments
- Identify changes that may be required in Canada's educational and training institutions
- Develop methods by which government can assist industry in responding to the challenges of international competition
- Advise on how best to coordinate the efforts of industry, labour, universities, and government in pursuing national goals
- Recommend priorities for the support of scientific disciplines, strategic technologies, and national programmes
- Respond to specific questions or tasks assigned by the Prime Minister

In its short lifespan, it has already been reorganised and has produced close to twenty reports, on subjects ranging from women's participation in science to Canada's involvement in big science projects. While its advice is confidential, much of what it advises is eventually released publicly once the Prime Minister has had the opportunity to review the reports. Current work is under way on technology diffusion, human resources, government procurement, s&t priorities, and competitiveness. One of the more interesting changes to NABST when it was restructured in the fall of 1991 is that members now have access to confidential government documents in order to gain a better appreciation of policies under development by the government. This has similar overtures to the early relationship between the Science Secretariat and the Science Council of Canada described in Chapter Two.

ISTC is the prime client for s&t indicators produced by Statistics Canada and other international statistical agencies. It has extensive data analysis capability and generates a number of analytical pieces describing the Canadian s&t indicators scene[2]. The Department is also charged with helping to coordinate the government's overall investments in s&t, and as such, has developed a Decision Framework policy designed to provide a strategic overview of the impact of government expenditure patterns in r&d. It is also charged with responsibility for the federal policy on extramural performance (intended to increase the amount of federally-funded s&t performed by the private sector).

ISTC, until very recently, was the home of the Canadian Space Agency, a $417 million organisation that now has departmental status and is responsible for the government's space operations. (See Chapter Nine for more on the Canadian Space Agency).

Among its' other remits, ISTC provided financial assistance to the Royal Society of Canada ($5 million over five years) to help the society implement its corporate plan and research evaluation. (The Society has undertaken large scale studies on the future of advanced industrial materials research and molecular biology). ISTC has also given matching grants (to complement private sector and provincial government support) for a unique institution called the Canadian Institute for Advanced Research (CIAR). CIAR sponsors world-class research projects on social, economic and scientific issues. Among its research areas are artificial intelligence, cosmology, evolutionary biology, population health and the impact of technology on economic growth.

At the time of writing, ISTC is coordinating a response to the Prosperity Initiative, a private-sector steering group appointed by the federal government to provide it with an action plan on improving the quality of life of Canadians as well as the country's overall competitiveness (see Chapter Thirteen for more on this Initiative). One of the central elements of this response is a technology diffusion strategy to improve Canadian firm use and management of technology and to acquire it from domestic and foreign sources. In addition, consideration is being given to consolidate the numerous technology assistance programmes into a single-window mechanism so as to simplify access to users. Of equal importance, is a potential re-structuring (and downsizing) of the whole of the federal departmental apparatus flowing from a recent Prime Ministerial committee appointed to review the whole of government organisation.

The National Research Council of Canada (NRC) is the premier science and engineering organisation of the Government of Canada. Over the course of its history, NRC has spun out such agencies as Atomic Energy of Canada Limited, the Medical Research Council, the Defence Research Board, the Natural Sciences and Engineering Research Council, and the Canadian Space Agency. In 1992–93, the NRC had an annual budget of roughly $500 million and more than 3000 employees working in sixteen research institutes and a variety of programmes and facilities across the country. The NRC accounts for about nine per cent of federal expenditures on s&t, and its research and technical services provide technical advice and assistance to more than 500,000 Canadians annually. NRC is one of the Government of Canada's most important instruments for the support of science and technology. The NRC reports to Parliament through the Minister for Science. A Long Range Plan (1990–1995) has recently been completed. It's three strategic objectives are (a) to strengthen the NRC's world class research, (b) build partnerships with industry and other Canadian research organisations, and (c) concentrate on areas of research that can raise Canada's competitiveness. (For more on the NRC, see Chapter Four).

Until 1978, the NRC carried out the nation's university research grants and scholarships function. In that year, however, a tripartite research council system effectively came into being with the NRC activities in this area forming the Natural Sciences and Engineering Research Council (NSERC). (The other research councils were the Social Sciences and Humanities Research Council, created in 1978 out of the Canada Council, and the Medical Research Council, created in 1969 out of the NRC. In an interesting playback of *déjà vu*, SSHRC has been returned to its home, the Canada Council, as a result of a February 1992 federal budget decision). SSHRC has a number of programmes that support s&t from the wider perspective of their impact on society, and also jointly funds (with NSERC) university chairs in the management of technology. (See Chapter Twelve).

Reporting to Parliament through the Minister for Science, NSERC is Canada's largest granting agency, channelling over $500 million into university research and the training of new scientists and engineers. The functions of NSERC are to promote and assist research in the natural sciences and engineering, and to advise the Minister in respect of such matters relating to research as the Minister may refer to the Council for its consideration. The role of NSERC is to build a healthy research base in universities and, as a corollary, to secure a sound balance between this diversified research base and the more targeted research programs; to secure an adequate supply of highly qualified personnel who have been well educated in basic science as well as trained with state-of-the-art facilities; and to facilitate collaboration between r&d performing sectors in Canada.(For more details on NSERC, see Chapter Six).

In addition to NSERC, there is the Medical Research Council (MRC) which has a budget of $256 million and which reports to Parliament through the Minister of Health and Welfare, and the Social Sciences and Humanities Research Council, which reports to Parliament through the Secretary of State. It has a budget of $103 million.

As already mentioned, the Science Council of Canada (SCC), which was established in 1966 and abolished in 1992, was Canada's principal public science policy agency responsible for the analysis of s&t policy issues; the recommendation of policy directions to government; the 'policy watch' of important s&t issues to Canadians; and stimulating action on s&t issues amongst governments, the private sector, and academia. At it's closure, it had a budget of $3 million, and a total staff of 29. This was down from it's 1984 budget of more than $6.5 million and a staff complement of nearly 60.

During its life, the Science Council reported to Parliament through the Minister for Science. As a Crown agency, the Council was independent of government and could establish its own research program. In its early years, the Council conducted 'inventory studies' in an attempt to canvas the broad spectrum of science activities in Canada. Thus, it published studies on physics, psychology, water resources, chemistry, and scientific and technical information in an attempt to establish the backdrop for understanding the policy issues affecting these activities. As its information base grew, and as the level of sophistication mounted in the policy analysis (supported by the growth of similar bodies abroad), the Council took on various thematic studies, exploring, for example, the question of a 'conserver society', technological sovereignty and the need for domestic industrial r&d capability, and basic research. All of its work was public and it served as a lightning rod for the technical and scientific community in presenting thorny recommendations to the government of the day. On occasion, the government would request a special study (this was done, for example, in the case of the optimal funding required for the Canadian space program), but, on the whole, the Council gradually lost favour with the government (particularly within a bureaucracy that grew frustrated over its inability to control the agenda of this arm's length agency) and was eventually squeezed out by the creation of other bodies closer to the government, such as NABST. In the provinces, every government eventually established its own advisory group for s&t, thus limiting the Science Council's national impact.[3] The Council had undertaken studies and released reports on such topics as science education, big science, 'innovation in a cold climate', provincial research organisations, threshold firms, sustainable agriculture, technology and local community development, polar science, and water resources. It had recently released its first (and last) national report on the health of s&t in Canada. This 1992 report included an assessment of the competitiveness debate in Canada, and chronicled the ongoing discussions on new frontiers in science, education, training and literacy, science infrastructures, industrial innovation, and the environment.

Parliament and Science

Unlike other parliamentary democracies, Canada's parliament has traditionally paid little attention to s&t; this, despite the fact that the Senate Special Committee Report on Science Policy remains a seminal work. As part of major procedural reform within the House of Commons, the Committee for Research, Science and Technology (CRST) was established in March 1986. While committees had existed for years to deal with topics like transportation, health, and energy, s&t policy had no permanent home in the parliamentary committee structure. Following a revision of the parliamentary committee system in 1990, the CRST was re-configured – becoming the Standing Committee on Industry, Science, Technology, Regional and Northern Development. As part of this restructuring, the number of members was reduced from fourteen to eight. As a result, the attention paid to s&t has been even less focused. Like other parliamentary committees, this committee has representatives from all three political parties. The Library of Parliament's Research Branch provides the secretariat to the committee. Among it's tasks, the Committee is required to examine the spending estimates of the major departments and agencies responsible for the funding of s&t in Canada. On a regular basis, therefore, witnesses from these organisations are called before the Committee to defend their annual budget statements. As the Committee's structure and composition is controlled by the governing Party, it is difficult for the Committee to probe for potentially embarassing policy questions, with the result that its reports have little teeth, and are often accompanied by a dissenting opinion from the opposition.

The Committee's most recent inquiry has been an examination of the new directions, management and funding of the NRC's Industrial Research Assistance Program. In December 1990, the Committee produced a report that attempted to prod the government into action on a number of s&t fronts. Their report, *Canada Must Compete*, covered a vast range of policy concerns that had been raised by witnesses to the Committee. Unfortunately, because of its structure, the reports' thirty-one recommendations received little play both within the community from which it had received testimony and the government. The government's response (requested by the House of Commons) and released in May 1991 remains probably the most lucid and comprehensive assessment of the rationale behind the federal government's support of science and technology. Summarised in this document are the government's official positions on the inappropriateness of a national r&d target; ongoing support mechanisms of a fiscal and non-fiscal nature, assistance to the granting agencies and university research, procurement policy, technology transfer and diffusion, community initiatives, s&t assistance in remote and northern communities, regulatory obstacles to innovation, intellectual property rights, financing and venture capital. A year later, the government had initiated a large-scale exercise to examine these and other issues in its Prosperity Initiative.

Another body organised to find a common bridge between parliamentarians and the science community has been the Committee of Parliamentarians, Scientists and Engineers (COPSE) which meets infrequently under the aegis of the Royal Society of Canada. COPSE has its origins rooted in the World War II era when one of Canada's first science and society organisations was formed, the Canadian Association of Scientific Workers. (More description of various science and society associations is found in Chapter Twelve).

National Science Policy

Upon the signing of Canada's first National Science and Technology Policy (NSTP) in 1987, a Council of Science and Technology Ministers (CSTM) was formed. This council, which is supported by a secretariat housed at ISTC, has the following terms of reference;

● To promote the optimization of existing s&t policies and programs of both orders of government
● To encourage more effective consultation on s&t policies and programs being brought forward by individual governments so as to promote compatibility between government actions in the s&t area
● To identify s&t policies and program actions that might be pursued cooperatively by the federal, provincial and territorial governments, and to review the s&t actions undertaken cooperatively
● To suggest appropriate amendments to the National Science and Technology Policy over time, as changing circumstances dictate
● To encourage a better linkage between s&t and the goals of the federal, provincial and territorial programs related to the national and regional economic development of the country
● To create, where appropriate, working groups composed of federal, provincial and territorial officials, and others representing appropriate non-government sectors, to study and report on areas of concern identified by Council Members

The CSTM meets periodically and has published a 1991 framework for action, which is now under review by all levels of government. This action plan includes a host of proposals targeted at enhancing r&d, stimulating innovation, developing human resources, promoting science literacy.

Along with the NSTP, 1987 saw the announcement of the federal government's Science and Technology Strategy. Called 'InnovAction', its goals were to;

● Increase industrial innovation and technology transfer
● Develop and promote strategic technologies
● Manage federal s&t resources more effectively
● Ensure the supply and development of human resources required for science and technology
● Promote public education in s&t and a more science-oriented culture

The original programs and initiatives of InnovAction ($1.3 billion over five years) included;

● Science and Technology Programs (Strategic Technologies, Sector Competitiveness, Services to Business) – $466 million
● the Canadian Space Program – $256 million.

- Funding for 15 Networks of Centres of Excellence – $240 million
- Increases to the Granting Councils – $200 million
- a Canada Scholarships Program for undergraduate students - $80 million

The principal thrusts of InnovAction have recently been refinanced for a further five years, to the tune of $1.5 billion. Most of this new funding is targeted at increasing the base funding levels of the national granting councils, which had been eroding as a result of inflation and the increased cost of research infrastructure. In addition, other funding support was provided for education (scholarships), industrial technology diffusion, and directed research projects in international programs.

In addition to these agencies, numerous federal line departments have extensive resources and laboratories across the country. These are described in Chapter Four.

The role of central agencies, such as the Department of Finance, Privy Council Office and Treasury Board are clearly critical to the decision-making system for s&t, as budgets and other levers for r&d (such as tax credits, intellectual property regimes, procurement and foreign investment) are major determinants of the growth of science and technology in the country. Science and technology is not a separately-defined line item in the federal budget; it has no statutory condition; rather it is considered as a discretionary item in the overall federal expenditure envelope. Departmental budgets, including science and technology components, are negotiated directly with Treasury Board.

To assist the overall coordination of s&t expenditures, ISTC manages a decision framework process. This requires all major science-based departments and agencies to prepare a strategic plan and assessment of their respective major s&t activities for the current and following three years by outlining the purpose of major S&T programs, expected results from them, and their impact. These plans also include financial and resource information for major activities. While the allocation of resources to science and technology based activity is the responsiblity of individual departments, various consultative and coordinating mechanisms exist to ensure that national objectives are kept in view. For example, an interdepartmental steering committee of Assistant Deputy Ministers responsible for science meets regularly to review pressing issues and policy directions for science and technology in the federal government. Other consultative mechanisms among government departments exist, for instance, in international s&t relations, space, and biotechnology. On university research funding, the three granting councils had an Inter-Council Coordinating Committee to review policy issues of common concern. Despite all of these mechanisms, there is only one arbiter for decisions leading to new expenditure in s&t. Proposals from all federal Ministers with economic and scientific and technological implications are considered by one Cabinet committee, now called the Economic and Trade Policy Committee. This is to ensure that an effective overview is given at the political level to new s&t proposals.

Provincial science policy

As we have noted, Canada's s&t policy apparatus is not conditioned solely by the federal government. Indeed, Canadian governments have spent considerable resources in s&t targetted at regional development. Apart from gradually decentralising the laboratory research infrastructure (which has been increasing over the past decade), a number of federal-provincial initiatives were designed in the mid-1980s to assist in the s&t development of the provinces. A series of Economic and Regional Development Agreements signed in 1984 and beyond have significant s&t components to them. Agreements with the provinces of Nova Scotia, Quebec, Saskatchewan, British Columbia, and New Brunswick all had joint federal-provincial funds committed to technology and innovation. In addition, these and several other provinces had sub-agreements on forest resource development, mineral development, agriculture, and fisheries and ocean-related with s&t components to them. From 1984 to 1989, these financial commitments (from both levels of government) amounted to close to $700 million. The tradition of regional development support for s&t has continued with the establishment by the federal government of the Atlantic Canada Opportunities Agency and the Department of Western Economic Diversification, both of which sponsor and support r&d projects of strategic importance to their respective regions.

Complementing the national science policy process and initiatives, all ten provinces and two territories have, at one time or another, developed respective technology development strategies. The provincial s&t initiatives have evolved over time and have taken on different approaches to reflect regional and sectoral specificities of the economies.

British Columbia

In British Columbia, a major research organization, B.C. Research, was established in 1944. One of the first policy statements on research was issued in 1976 with an overview provided of the research funding in the province. A Science Council was established in 1978 to provide advice to the government as well as to award grants and scholarships in the field of science, engineering and research. A Discovery Foundation, and Discovery Parks (a form of science parks), were established in 1977, followed by the creation of the B.C. Innovation Office.

Along with the three other Western provinces, British Columbia's innovation policy was examined in 1988 by the OECD. Among the key issues the OECD identified as strategic to the provincial economic base were promotion of microelectronics research, development of biotechnology, upgrading resource industries through new technologies, aquaculture, and strengthening basic science, including the particle physics installation managed by four Canadian universities, known as the Tri-University Meson Facility (TRIUMF).

British Columbia formed the Premier's Advisory Council on Science and Technology in 1987. Its mandate is to advise the Premier of the province on s&t policy development and implementation; to advise the premier on specific issues and initiatives of particular importance to the province; and to promote awareness of s&t. The Council, composed of fifteen senior representatives from all sectors, provides advice through the evaluation of the province's $420 million Science and Technology Fund. The Council recommended a s&t policy in 1987 that was adopted by the provincial government. It's nine-point desiderata guides public policy for investments in s&t, and the Science and Technology Fund was put in place to assist the development of innovation in the province. The Council has also undertaken examinations of a wide range of issues, including the KAON factory proposal as an upgrade of the TRIUMF facility, hydrogen as an alternative fuel and pharmaceuticals.

The Technology Fund, established in 1990, is designed to assist in the province's economic diversification and to enhance productivity and industrial competiveness in new technologies and traditional industries. The Fund provides support in the areas of public awareness of science, infrastructure, r&d, human resource development and special projects. The first annual review of the directions for the Fund took place in early 1992.

Alberta

Alberta is home of the oldest provincial research organisation in Canada, the Alberta Research Council (est. 1921). Among the province's other major research institutions are the Alberta Heritage Foundation for Medical Research, the Centre for Frontier Engineering Research and the Alberta Oil Sands Technology and Research Authority. This latter organisation has been an important catalytic agent in encouraging energy research at Alberta universities and in making efforts to access technology worldwide (through a number of international agreements).

Alberta has produced a Science and Technology Policy. The overriding philosophy of Alberta's approach has been a dual one;
- to develop and apply advanced technology for upgrading the province's traditional resource-based industries
- to engage in high technology opportunities that are identified by the private sector and encouraged by partnerships between the province's stakeholders

Like British Columbia, Alberta has a Premier's Council on Science and Technology. Formed in 1990, the Council has twenty-four members appointed by the Premier. It's mandate is to advise the government on s&t as they relate to economic and social development and to the ability of the province to compete effectively in the global marketplace. It also recommends guidelines for the allocation of government resources devoted to s&t. The Council has examined various policy issues, including women in s&t, total quality management and research consortia.

Saskatchewan

Saskatchewan created a Research Council in 1947. The Council provides technical and managerial support for the province's small and medium-sized businesses. In 1975, the Saskatchewan Science Council was created. In 1983, the province released a high technology strategy aimed at creating value-added in the resource sector, transportation and in biotechnology and microelectronics/computer applications.

In 1989, the province established an Advisory Council on Science, Technology and Innovation renamed in 1990 as the Saskatchewan Economic Diversification Council. It's mandate is to advise the government on the development and diversification of the provincial economy. This Council has clearly moved beyond solely s&t issues and is examining the role of taxation policy, environmental regulations and federal-provincial relations. It is advancing an economic blueprint for the province to the year 2000.

Manitoba

Manitoba was later than the other Western provinces in establishing its own research organisation. Only in 1963 did it create the Manitoba Research Council which manages two major research facilities; one in particular, the National Agri-food Technology Centre, has worked extensively with the agri-food industry in the province. In the mid-1980s, Manitoba's technological development strategy had two major components: emphasis on technology adaptation and on security, preservation and creation of employment.

Among its' key industrial r&d and technology support programs are the Health Industries Development Initiative, the Manitoba Aerospace Technology Program and the Environmental Industries Development Program.

In 1991, the provincial government restructured the Manitoba Research Council into the Economic and Innovation Technology Council. It appointed 29 members to the Council who represented the academic, business and labour sectors. The Council's mandate is to provide advice to the province's Economic Development Cabinet Committee and assist it in linking innovation to economic growth.

The four Western provinces have also been active in combining their common efforts in s&t. In the late 1970s, a Western Canada committee of officials for s&t was established, with government representatives from the four provinces meeting regularly to discuss common policy concerns. Recent activities on this common front include potential projects in space research and high energy particle physics research. Because their economies have similar weaknesses (*e.g.*, diversification of a natural-resource based economy), British Columbia, Alberta, Saskatchewan and Manitoba became the target of an OECD review of their innovation policies in 1988. Each province spent considerable effort during the course of this examination to identify its' own unique characteristics in s&t. Much of the OECD report was devoted to suggesting a stronger innovation culture in the provinces, with attention paid to the need to diversify from a natural resource-based economy. One could therefore argue that the OECD report served, in part, to trigger some of the developments that occured in the late 1980s.

New Brunswick

New Brunswick established a Research and Productivity Council in 1962 and has devoted a number of efforts, over the past several decades, to elaborate a science and technology policy. Like some of the other provinces, New Brunswick worked closely with the federal government in the mid-1980s in defining it's own approach to technology transfer and innovation. It's Legislature created a special committee on Science, Technology and the Future in 1984 and a jointly-financed (federal-provincial) study was commissioned in 1985 to develop a s&t strategy for the province. The overall goal of this strategy was to accelerate the development and application of technology to increase and enhance employment opportunity in the province.

In 1987, the province established the Minister's Advisory Board on Science and Technology to advise the government on the strategy. At present, this board consists of seven members, appointed by the Minister of Economic Development and Tourism. The Board has recently examined a study on a s&t policy for the province.

Nova Scotia

Nova Scotia, probably more than any other of the Atlantic Provinces, has undergone numerous explorations of directions for s&t in the province. It has an extensive university research base and, in 1946, set up the Nova Scotia Research Foundation Corporation. Following a series of task force reports and White Papers on the economy and innovation, the province instituted a Council of Applied Science and Technology in 1987.

The Council, which reports to a Minister of Industry, Trade and Technology, is active in advising the government on the application of s&t to economic development. Among its' previous examinations were the preparation of a comprehensive program to raise awareness of s&t to the small and medium-sized companies based in the province; preparation of a report on a regional approach to technological innovation and diffusion; and advice on community colleges and technology transfer to rural areas. More recently, it has conducted background work on environmental issues and on the regulatory environment affecting innovation.

Support programs for Nova Scotia research and technology transfer include the Technology Advancement Program (assistance for acquisition of technology-based capital equipment), the Technology Collaboration Program (assistance to business in using technological expertise from provincial research institutions) and the Technical People in Industry Program.

Newfoundland and Labrador

Newfoundland and Labrador, the country's most recent member of the Confederation, has a relatively younger and more specialised s&t infrastructure than other Atlantic provinces. The province began its exploration of an s&t policy in 1981 with a White Paper. This paper suggested that the province follow a 'leap-frog' strategy of technological development that focussed on its' areas of comparative strength, including marine sciences and cold oceans engineering. Newfoundland has an extensive research complex in these areas with a Centre for Cold Oceans Resource Engineering and the Institute for Marine Dynamics as major research facilities.

In May 1988, the Newfoundland and Labrador Science and Technology Advisory Council was formed. With fourteen members appointed by the Premier of the province, the Council has an extensive remit covering all domains of s&t policy. It's major focus for the past three years has been in industrial activity, education and science awareness. In the first of these areas, work has been developed on a sector-by-sector strategic plan to establish targets for s&t-based industrial innovation. Among the sectors examined have been fish and food processing and information technologies.

Prine Edward Island

Prince Edward Island, the country's smallest province, has also used the federal-provincial regional development agreements to help position its science and technology strategy. A 1984 strategy called 'Blending Tradition and Innovation' served to establish the elements of a technology strategy. This was later re-examined in 1986.

In 1988, the Prince Edward Island Advisory Council on Science and Technology was established. The Council, composed of ten members appointed by Cabinet, advises the Minister of Industry on directions for s&t. Among other accomplishments, the Council has presented a brief on s&t to the Cabinet, sponsored a campaign to educate people of all ages on the importance of science and technology; and recommended the establishment of the province's innovation programs.

Like the Western provincial grouping, the Atlantic provinces have also long-standing inter-provincial relations in s&t. The Council of Maritime Premiers (CMP), for example, has had considerable involvement in preparing work on the science and technology activities of the Maritime provinces. As early as 1979, the CMP had produced a discussion paper on the r&d of the region. In 1981, it produced a report, *Technological Innovation: An Industrial Imperative*, designed to recommend to the three Maritime provinces how best to tackle r&d and innovation for industrial competitiveness. In 1991, CMP, as part of it's exercise to initiate an economic union for the region, asked the three Maritime science advisory councils to develop a regional approach to technological innovation and diffusion. The report has served to position the three provinces in formulating an enhanced economic union.

Yukon and the Northwest Territories

The two territories, Yukon and the Northwest Territories, have also been active in developing s&t policy for economic and social objectives. Both territories were signatories of the 1987 National Science and Technology Policy and both have created quasi-advisory councils. The Science Institute of the Northwest Territories was established in 1985 to provide scientific, engineering and technological perspectives to the Legislative Assembly on the social and economic needs of the people. It has four complementary programs to accomplish this;

- science advisory services (which issues licenses for research conducted within the territories)
- scientific services (which offers logistical support to researchers through three research centres)
- information/education (which assists in developing a science culture within the northern society)
- technology development (which brings viable technology to northern businesses and assists northern industry to develop unique technologies)

The Institute's Executive Director also sits on a number of national and international committees dealing with Arctic research.

The Yukon Science Institute was established in 1985 to foster s&t to meet the social and economic needs of the Yukon people. Like its sister organisation in the Northwest Territories, the Yukon Institute is actively involved in Arctic research activities. It's major work covers public awareness programs and research assistance (such as deliver of the National Research Council's Industrial Research Assistance Program).

Ontario and Quebec

We have not commented in this section on the evolution of the s&t policy advisory mechanisms of the country's industrial heartland, Ontario and Quebec. The Premier's Council on Economic Renewal in Ontario, and the Conseil de la science et de la technologie in Quebec have extensive roots and have done much to position their respective provinces in developing and implementing science and innovation policies.

Quebec's Conseil de la science et de la technologie is the oldest of the provincial science advisory mechanisms, having been first established in 1972. It's early years were devoted to working closely with government and key decision-makers on the directions for science policy in the province. In 1983, the Conseil was reformed with a much closer link to the government. It now reports to the Minister of Higher Education and Science. The $1.3 million Conseil consists of fourteen members appointed by the Quebec Cabinet and has a secretariat of almost twenty employees. Over the past eight years, the Conseil has been active in outlining the health of s&t in Quebec through a series of 'state of the union' reports (these have since been discontinued). It has developed an inventory of the s&t infrastructure in all of the province's major regions, working closely with local and regional players. A strategic plan for the city of Montreal, for example, is it's latest remit. The Conseil has also examined a wide range of science policy issues, from biotecology and plant biology to Quebec's participation in federal s&t policy measures.

Ontario has also been active in supporting advisory mechanisms on innovation policy. It first established a Premier's Council on s&t in 1986. This Council commenced an ambitious examination of the province's directions in s&t. Following the publication of four separate volumes that suggested an action plan for the province in competing in a global economy, the Council, under a new government, was re-labelled the Premier's Council on Economic Renewal in 1991. The new $3.7 million Council consists of the Premier, nine Ministers and 37 members and is supported by a secretariat of about twenty professionals. As the secretariat is privy to confidential Cabinet documents, some of it's ongoing work includes reviews of the province's technology and learning policies. It has recently reviewed, for example, the Ontario Technology Fund and provided advice on it's general orientation and prospective new directions. (For more description of both Ontario and Quebec's technology policies, see Chapters Four and Five).

In helping to coordinate the work of all these bodies, a National Forum of Science and Technology Advisory Councils has been established which meets annually to discuss progress on a number of common fronts. Three reports have been issued from this group, all directed at increasing the nation's capacity to upgrade itself with s&t. The fourth meeting (slated for Toronto in the Spring 1993) has as its principal focus Canada's role in international s&t. To prepare for this, the Forum has held a series of roundtables on research collaboration within the North American, Pacific Rim and European contexts.

The federal and provincial governments' policy apparatus depends extensively on the ability to monitor and analyse the performance and funding of s&t. Many of the s&t indicators used in such analyses are collected by Statistics Canada. This organisation has been collecting and disseminating statistics on r&d and science and technology for several decades now, and has produced a credible time series of data used extensively by governments (it collects national data, including that for most provinces) and international organisations. Unfortunately, due to recent fiscal constraints, the organisation's Science and Technology Division has come under the budgetary knife, and has had several of it's well-established surveys excised as a result. The following section briefly highlights some of the most recent data.

The performance of s&t

In 1991, Canada's gross expenditure on r&d were $9.7 billion (with preliminary estimates for 1992 at $10 billion). In that year, industry funded forty-one per cent of all R and D in Canada, while the federal government funded thirty per cent. Provinces and provincial research organizations funded seven per cent, and the higher education sector funded ten per cent.

Fifty-three per cent of all r&d in Canada is carried out by industry. However, out of 3300 companies that reported carrying out r&d in 1989, twenty-five (that is, less than one per cent) accounted for a full half of the total r&d. These twenty-five include such publicly-owned utilities as Hydro-Québec, Ontario Hydro,

AECL, and B.C. Telephone as well as a number of foreign-owned multinationals such as IBM Canada and Imperial Oil. Only fourteen companies in Canada spent more than $50 million in 1989, twenty-seven more than $25 million, and 121 more than $5 million. Of the 3300 r&d-performing firms, thirty-eight per cent were foreign-controlled. Further augmenting this concentration of industrial r&d in Canada is the fact that six major industries have dominated industrial r&d, these being telecommunications equipment, aircraft and parts, engineering and scientific services, business machines, other electronic equipment, and computer and related services. Together, these industries account for more than half of all intramural R and D.

Seventeen per cent of all r&d in Canada is carried out by the federal government. This, in 1991, amounted to $1.6 billion. The bulk of this (roughly $1.1 billion) goes on in-house r&d. The remainder is spent on r&d contracts, r&d grants and contributions, research fellowships, r&d-related capital expenditures, and extramural r&d administration. An additional $1.728 billion was spent in 1991 on related scientific activities (RSA) such as data collection ($873 million), information services, feasibility studies, testing and standards, museum services, and RSA-related capital expenditures. More than 20 federal government departments and agencies are represented in these total expenditures, as are about 35,000 s&t personnel.

Twenty-six per cent of total r&d in Canada is carried out at Canadian universities, amounting to more than $2.5 billion. (There are 147 colleges and more than 70 degree-granting institutions of higher education.) In 1990, more than 130,000 degrees were granted by Canadian universities. Thirty-eight per cent of these

Table 3.1   External Funding of University Research by Source, 1971–72 to 1990–91
('000$)

	Fed. Govt.	Prov. Govt.	Business Enter.	Priv. Non Prof.	Total
1971	134	30	11	34	209
1972	134	37	12	38	221
1973	141	42	13	42	238
1974	147	45	13	47	252
1975	156	50	13	51	270
1976	167	55	15	53	290
1977	192	60	15	57	324
1978	210	70	19	73	372
1979	234	76	51	46	407
1980	238	96	56	49	489
1981	354	114	68	58	594
1982	393	142	68	53	656
1983	457	153	80	66	756
1984	517	169	87	70	843
1985	515	178	102	80	875
1986	523	207	105	81	916
1987	561	217	136	99	1,013
1988	625	261	175	113	1,174
1989	662	277	186	119	1,244
1990	715	299	201	129	1,344

Source: Statistics Canada.

Table 3.2   Science and Technology Expenditures
($000,000)
deflated by CPI 1986

	1963–64*	1973–74*	1983–84*	1991–92*
ISTC-TC – MOSST	80.5	310.5	198.9	186.7
Energy, Mines & Resources (EMR)	146.3	194.7	324.2	286.1
Environment	—	585.8	393.6	449.0
Agriculture	139.0	227.2	319.2	294.5
Fisheries/Oceans	72.8	—	238.9	207.4
National defence	197.2	170.0	183.4	230.6
Atomic Energy	240.2	259.4	155.9	134.6
National Health and Welfare	34.1	121.4	124.1	145.8
NRC	179.7	435.3	464.5	385.9
Communications	—	83.1	96.9	46.4
Forestry	—	—	82.3	58.2
Transport	86.2	75.0	43.6	30.7
National Museums	—	85.0	92.8	97.3
Statistics (census year)	—	221.7	259.5	361.3
Can. Inter. Devlop. Agency	—	—	213.3	286.5
Inter. Develp. Res. Cen.	—	45.8	72.0	89.3
NSERC	—	—	318.8	384.9
MRC	21.1	114.4	159.0	196.5
SSHRC	—	53.6	68.4	78.1
Science Council	—	5.0	5.5	2.8
Canadian/Space Agency	—	—	—	232.3
Total	1244.3	3319.4	4128.9	4509.4

1963–64 First year of science surveys. Covered only natural sciences. Social sciences were surveyed in 1970–71 for the first time.

Source for CPI: Statistics Canada, Canadian Economic Observer, Catalogue # 11-210, Bank of Canada Review, May 1992.

Source Statistics Canada

were in the social sciences, twenty-seven per cent in the natural sciences and engineering (including health), and the remainder were in such fields as management and administration, and humanities. A breakdown of external funding for university research, since 1971, can be seen in Table 3.1.

In terms of r&d expenditures as a percentage of provincial GDP, only Quebec and Ontario exceed the national GERD/GDP ratio of one point four three per cent. Only Nova Scotia exceeds one per cent, while those of Newfoundland, New Brunswick, British Columbia, Alberta, Saskatchewan, Manitoba, and Prince Edward Island fell between half a per cent and one per cent. Moreover, approximately fifty-six per cent of industrial r&d is performed in Ontario, while almost twenty-seven per cent is performed in Quebec.

Both the overall and relative size of the federal s&t population changed little during the past five years. The major changes occurred in the proportion of r&d and RSA personnel. There were declines in almost all categories of r&d, whilst almost all categories of RSA person-years (PYs) saw increases.

Table 3.2 outlines the major shift in s&t expenditure of federal government performers in four time periods, 1963–64, 1973–74, 1983–84, and 1991–92. These serve to demonstrate the changes to the federal s&t policy and funding system

Table 3.3   Science and Technology Personnel

	1975–76	1983–84	1991–92
NRC	3,103	3,424	3,116
ISTC-ITC-MOSST	282	307	305
Energy, Mines & Resources (EMR)	2,373	2,660	2,718
Environment	7,057	3,941	4,557
Agriculture	4,274	4,539	3,986
Fisheries/Oceans	2,158	2,595	2,374
National Defence	2,034	1,947	1,896
Atomic Energy	2,435	2,665	2,528
National Health and Welfare	1,085	1,444	1,620
Communications	509	613	448
Forestry	—	1,097	953
Transport	188	133	109
National Museums	715	1,051	1,123
Statistics (Census year)	5,907	4,671	5,576
Can. Inter. Develop. Agency	50	57	166
Inter. Dev. Research Centre	304	266	293
NSERC	—	111	180
MRC	35	31	64
SSHRC	—	104	101
Science Council	64	67	29
Canadian Space Agency	—	—	329
Total	34,678	35,313	35,621

Source: Statistics Canada, Canada Review, May 1992.

over that four-decade period. They also show the growth and decline of s&t budgets by individual departments and agencies. Table 3.3 describes the changes in s&t personnel within the federal government over the period 1975–1991. Most remarkabe is that in a fifteen year period, s&t personnel in the public service have remained virtually unchanged despite the enormous growth of s&t activity within the federal government itself as it has strived to re-direct intramural resources to current issues. Thus, in both Tables 3.2 and 3.3, one can immediately see the present s&t attention to the environment.

During the 1970s, the turnover rate in the federal s&t community was typically five per cent, although it sometimes reached seven to eight per cent. The current rate is three per cent, or less. One reason for this low turnover is the median age of 45. There are fewer promotional opportunities for younger staff as long-service employees occupy senior positions. As a result, research establishments are unable to hire significant numbers of recent graduates.

Much of this slack is being picked up in the business enterprise sector. Since 1971, there has been a marked shift in r&d funders and performers with the federal government becoming relatively less involved and the business sector growing in significance. Thus, where the federal government performed twenty-nine per cent of r&d and funded forty-three per cent in the 1971–75 period, this has declined to seventeen per cent and thirty per cent respectively in 1991–92. On the other hand, business enterprise, which performed thirty-five per cent of r&d, and funded twenty-nine per cent in 1971–75, has improved on both counts, performing fifty-three per cent and funding forty-one per cent in 1991–92. One would expect such a trend-line to develop (and in fact, it has been replicated in most OECD industrialised nations) as efforts to build up industrial r&d capacity and policies via the transfer of public s&t assets to the private sector have taken hold.

The following chapter describes this trend.

Conclusions

Canada has been witness to extensive surgery to its decision-making apparatus for s&t over the past several decades.Some observers would argue it is cosmetic. Others might argue that s&t have moved, in fact, from the periphery to the centre of the decision-making process. A good number would also hold the view that the constant shifting of approaches to managing s&t in the national interest has done scant little to instil visionary leadership within the scientific and technical community and the business and labour sectors. The result, some would argue, has been a weakened power base for s&t in the country, with the various lobby groups and associations that represent well-defined interests unable to interpret unambiguously the shifts in policy direction. The constant change has also led to cynicism within the very organisations expected to revitalise the country's s&t base. This has proved a difficult challenge to overcome. Nevertheless, in Canada, much progress has been made in bringing s&t policy issues to the front of the political agenda of the day. As s&t become increasingly critical as determinants of social well-being and economic growth, and as the general public assess the pros and cons of the impact of these forms of knowledge on their everyday lives, one can expect that science and technology will assume the mantle of a civic duty for a democratic society.

References

1 Uffen, Robert J. (1972). "How Science Policy is made in Canada," *Science Forum*, December, 3–8.
2 For more on the use of s&t indicators, see the special October 1992 issue of *Science and Public Policy* edited by Adam Holbrook of Industry, Science and Technology Canada.
3 For a more detailed analysis of the reasons for the Science Council's demise, see de la Mothe, John (1992), "A Dollar Short and a Day Late: A Note on the Demise of the Science Council of Canada" *Queen's Quarterly*, Winter.

Chapter 4

The government research infrastructure

Frances Anderson and Charles Davis

The development of government research infrastructure

A public laboratory in Canada typically fulfills six or more primary research roles (SCC, 1975);

1. It may undertake r&d that has implications for national security, making private-sector execution unsuitable
2. It may undertake regulatory r&d if no private institution that is independent of the regulated firms is available
3. It may do r&d that a government agency or department requires in order to accomplish it's mission or mandate, and which is either inappropriate for private industry or unavailable
4. It may undertake r&d in support of domestic or international consumer standards
5. It may undertake r&d to permit intramural monitoring of the state-of-the-art in a particular area, without which the parent department might miss opportunities or mismanage contract research
6. It may undertake r&d to support intramural capital facilities that meet Canadian industry's research and testing needs

Of course, Canadian government labs – as well as their public sector counterparts in other countries – may also undertake research for less obvious reasons related to their core mandates. These include: to enlarge the international pool of knowledge, contribute to the development of national capacity in fields in which universities or the private sector are absent, provide demonstration capability for new, emerging technologies, or develop technologies that Canadian industry is thought to need. Government research labs can offer confidentiality, intellectual independence, continuity, in-house expertise to oversee procurement of scientifically or technically complex inputs, and close contact with the user of research – the Government itself.

Government research has played an important historical role in the development of Canadian r&d capabilities (see Chapter Two). Recent trends, however, suggest the development of modifications of traditional roles for public-sector r&d in Canada. The federal science and technology infrastructure, in particular, is undergoing substantial change, and new activities and institutions are emerging at both the provincial and the metropolitan levels. The six primary roles and the other rationales for government r&d described above have not changed, but greater emphasis is now being laid on questions of performance assessment, accountability, priority determination, and linkage mechanisms with clients – especially industrial – to augment commercialisation possibilities. As the 1981 OECD report, *Science and Technology Policy for the 80s*, described the situation for small open economies like Canada, the degree of difficulty of adapting the government sector to changing priorities depends on the extent to which a nation's s&t resources are concentrated in the government sector. This chapter provides an overview of the primary features, current trends, and major issues of the Canadian public sector research system.

Historical perspectives

The Canadian government's commitment to direct support for r&d originated with the Geological Survey of Canada, established in 1842 to support mineral research and exploration. Forty-four years later, in 1886, the federal government established five experimental farms to undertake agricultural research, and in 1898 it created the predecessor of the Fisheries Research Board of Canada. Direct support for the mineral resource sector continued in 1907 with the establishment of the Canadian Centre for Mineral and Energy Technology (CANMET), which was mandated to undertake r&d on the extraction of Canadian mineral resources.

Federal government involvement in r&d for industrial development began in earnest in 1916, when the Honorary Advisory Council for Scientific and Industrial Research was established to provide policy advice to the government and to supervise the distribution of research grants to universities and industry. The Honorary Advisory Council evolved into the National Research Council of Canada, which opened it's first laboratory in 1932. As described in Chapter Two, the NRC grew slowly in the interwar period, but by the outbreak of World War II it was well positioned to make a major contribution to Canada's military science effort. Most other early government initiatives in s&t followed a pattern of providing public support for so-called 'inventory science' and the exploitation of natural resources.

But two significant research organisations that were spun out of NRC in the post-war period clearly deviated from this pattern: the Defence Research Board (DRB) and Atomic Energy of Canada Limited (AECL). Established in 1947, the DRB took over most of the defence research activities of the NRC and attempted valiantly (but largely unsuccessfully) to maintain and expand the technological leadership in a number of fields, including electronics, aeronautics, computer technology and telecommunications. The DRB has since been disbanded.

Similarly, the National Research Council, which had orginally been set up to promote industrial development through applied research, gradually shifted its emphasis to civilian research of a more fundamental and long-term nature. NRC's role in basic research was justified in large part by insufficient development of university research capacity in Canada prior to the 1950s.

Three provincial governments became involved in the direct funding of r&d activities in the inter-war years. The Alberta Research Council was established in 1921, the Ontario Research Foundation in 1928, and the Saskatchewan Research Council in 1930. These three organisations were formed as a result of a confluence of pressures from the public and private sectors to develop Canadian manufacturing. This pressure was, in part, a result of the increased awareness of the utility of science that emerged from the experience of World War I. Two more provincial research organisations were established in the aftermath of

World War II: the British Colombia Research Council in 1944 and the Nova Scotia Research Council in 1946. Once again, arguments put forward in favor of the formation of these organisations emphasised the importance of provincial government support for scientific activities that would lead to economic and industrial development.

The period from the end of World War II to the mid-1960s was the "golden age" of science and technology in Canada, as it was in many other industrialised countries. Widespread political support for s&t allowed federal and provincial science and technology activities, infrastructure, and expenditure to expand significantly during this period.

The last twenty years, however, have been characterised by a decline in the importance of publicly-executed research relative to the industry and university sectors. In 1965, for example, government laboratories performed a total of thirty-six per cent of Canada's Gross Expenditures on r&d. By 1992, the proportion had declined to seventeen per cent.

With this decline has come a questioning of the role and mission of the government research sector. Until the 1960s, a widely held view in Canada was that the government research sector could be the driving force behind industrial technological development. Indeed, some have referred to these Canadian research institutes as surrogates for industrial research. This view has shifted to one in which a major purpose of government funding of r&d is to build up research and technological development capacity in the industrial sector. In the 1970s and 1980s, the Canadian government increased commitments to the r&d effort in a number of new fields of economic and social importance: in particular, environmental sciences, scientific and engineering research for exploring and exploiting offshore resources, new industrial materials, information technologies, and biotechnology. The government's contracting-out policy of 1972 (known as 'Make-or-Buy') was another example of this attempt to bolster industrial r&d capacity by requiring that government r&d be contracted out to the private sector. National security and public health and safety were exceptions to the general rule. By the mid to late 1970s, the government was actively examining the justification for intramural scientific activities and the extent to which they should be contracted out to the private sector and universities. During the Lamontagne Committee hearings from 1970–77, considerable national debate took place over the future role of government laboratories, with the Committee eventually suggesting that the government's intramural r&d activities serving industry be consolidated in one multi-disciplinary institution known as the Canadian Industrial Laboratories Corporation.

This recommendation was not implemented, but efforts were made to establish new laboratories in response to potential industrial needs. This resulted, for example, in the establishment of the NRC's Industrial Materials Research Institute and NRC's Biotechnology Research Institute, both set up in the Montreal region. The concept of research relevance and the principle that government research establishments should be accountable for their use of public funds in relation to government policy objectives has increasingly taken hold.

In a countervailing trend to that developing on the federal scene, the late 1970s and early 1980s saw increasing provincial commitments to investment in public sector r&d organisations. Provincial governments began to regard r&d as essential to their own industrial development policies. This was particularly true in the larger Canadian provinces – Ontario, Quebec, British Columbia, and Alberta. Provincial governments proceeded to set up a series of consultative s&t policy organisations and ministerial agencies, in addition to new provincial research facilities, in order to support their move into industrial policy.

An overview of federal support for s&t

Federal scientific and technical activities support a constellation of missions and mandates, including national defence and security, protection of the environment, health and safety through regulatory research and testing, development of national safety standards, transportation, communications, aerospace, social and economic information, production of knowledge of natural resources, and international economic and diplomatic relations. The Canadian federal government is expected to spend $5.9 billion on s&t activities in 1992–93, an increase of two point six per cent over the previous year. In 1990, there were over 200 federal scientific establishments distributed throughout the country. They represented more than twenty federal departments and agencies and ranged in size from a few person-years to over 4,000 person-years. Just over 35,600 person-years were allocated to federal s&t activities in 1991–92.

Federal scientific and technological activities are divided into two principal categories: research and development (r&d) and related scientific activities (RSA). The former refers to basic research and applications and technology development as defined by the OECD's *Frascati Manual*. The latter refers to activities that complement or extend R and D by contributing to the generation, dissemination and application of science and technology. In 1992–93, federal r&d expenditures were $3.13 billion. The principal federal agencies and departments supporting r&d were Environment Canada, the National Research Council(NRC), the Natural Sciences and Engineering Research Council(NSERC), Agriculture Canada, National Defense, and Industry, Science and Technology Canada(ISTC).

In 1992–93, RSA expenditures were $2.4 billion. The main supporters of RSA were Environment Canada, Statistics Canada, the Canadian International Development Agency, Energy, Mines and Resources (EMR), and Health and Welfare Canada.

Intramural s&t expenditures include costs incurred for in-house personnel, acquisition of land, buildings, and equipment related to s&t, the administration of scientific activities, and the purchase of goods and services in support of in-house scientific activities. Of the $5.92 billion spent by the federal government on science and technology in 1992–93, $2.95 billion (fifty-nine per cent) was performed intramurally. (Details on these statistical data can be found in the Appendix).

The regional distribution of federal intramural and extramural r&d expenditures is a perpetual political issue in Canada. In 1990–91, the Western provinces received seventeen per cent of federal s&t expenditures, Ontario (including the National Capital Region) fifty-nine point seven per cent, Quebec eleven point seven per cent, and the Atlantic provinces, nine per cent. The regional distribution pattern of federal science and technology expenditure reflects a variety of factors, such as the existing r&d base, the nature of the r&d, the location of departmental clientele, the location of laboratory facilities, and political factors. The historical concentration of federal research establishments in the National Capital Region (which includes Ottawa, Hull and Gatineau) is viewed by less privileged regions as a situation requiring redress. Given the recent constitutional discussions between the federal and provincial governments, questions – and actions – regarding the redistribution of federal government laboratory activities throughout the country will undoubtedly persist.

The federal government laboratory infrastructure circa 1990

Four agencies and departments account for fifty-one per cent of the total intramural activities: the National Research Council, Energy, Mines and Resources, Agriculture Canada, and Environment Canada. Compared to 1984–85, the share of federal funds received by the industrial sector increased from fourteen per cent to seventeen per cent in 1992, in contrast to federal intramural expenditures that decreased from sixty-two per cent to fifty-nine per cent in the same period. The Canadian Space Agency is the largest funder (twenty-nine per cent) of scientific activities in the industry sector followed by Industry, Science and Technology Canada at twenty per cent. The three granting councils for academic research, the Medical Research Council, the Natural Sciences and Engineering Research Council, and the Social Sciences and Humanities Research Council, are responsible for seventy-nine per cent of total federal government expenditures in the university sector. Below we quickly describe the scientific and technical activities carried out by major federal agencies in Canada.[1]

The National Research Council

In 1991, the National Research Council (NRC) celebrated it's 75th anniversary. (For more on the NRC's history and it's key role in the development of Canadian science policy, see Chapters Two and Three). The NRC is the single most important public scientific institution in the country, accounting for about one-twentieth of all r&d expenditures in Canada. The NRC aims to perform, support and promote scientific and industrial research for the economic and social benefit of the country. The agency emphasises its commitment to relevance and excellence in all its' activities: as a partner with industry, as a source of R and D in the national interest, as a contributor to the national s&t infrastructure, and in fostering the development of a highly skilled workforce. In it's Long-Range Plan for 1990–1995, NRC committed itself to pursue it's mandate along three complementary dimensions: to produce world-class research, to engage in partnerships and collaborations with socioeconomic players in Canada, and to contribute to enhancing the competitiveness of industry in Canada.

The National Research Council is one of Canada's scientific agencies that has been hit hard by budgetary compressions. It lost about 700 person-years between 1984–85 and 1992–93. It's approved budget (excluding space research) increased slightly from about $411 million in 1983–4 to about $506 million in 1991–92.

Research is carried out in three sectors: Science, Engineering, and Biotechnology. The NRC also provides support for important s&t infrastructures, and it provides technical 'outreach' services. Of the NRC's 3,000 scientific and technical staff, many work in a central complex in Ottawa. Major research centres are also located in British Columbia, Saskatchewan, Manitoba, Quebec, Nova Scotia, and Newfoundland.

In the biotechnology sector, for example, NRC spent about $64.8 million and had 589 person-years in 1990–91. It operated the Biotechnology Research Institute (BRI) in Montreal, the Institute for Biological Sciences in Ottawa, the Institute for Marine Biotechnology in Halifax, and the Plant Biotechnology Institute (PBI) in Saskatoon. Two major facilities provide services to external clients. BRI's Pilot Plant in Montreal offers a range of fermenters and purification equipment and provides services in bioprocess optimisation, downstream processing, purification, and pilot scale production of recombinant proteins. PBI's Transgenic Plant Centre provides laboratories and greenhouse space for the safe propagation and assessment of transgenic plants.

NRC's activities in the science sector primarily cover the physical sciences. NRC spent about $124 million and had 752 person-years for activities in this sector. Work took place under the auspices of six institutes: the Herzberg Institute of Astrophysics, the Industrial Materials Institute, the Institute for Environmental Chemistry, the Institute for Microstructural Sciences, the Institute for National Measurement Standards, and the Steacie Institute for Molecular Sciences. Most of NRC's major facilities in the science sector are for the pursuit of fundamental research in particle physics and astrophysics. Some are managed externally. Those managed by the NRC include the Herzberg Institute of Astrophysics' two observatories, both in British Columbia. NRC also supports the operation of facilities that are cooperatively funded and externally managed. These include a share of the Canada-France-Hawaii Telescope, a share of the James Clerk Maxwell Telescope (also in Hawaii), the Tri-University Meson Facility (TRIUMF, a joint venture between NRC and four western universities), and the Sudbury Neutrino Observatory, a Canadian-based solar neutrino detector involving the UK and the USA. Also under the auspices of NRC's science sector, but dedicated in part to prototype production, is the Focused Ion Beam Facility, a molecular epitaxy device serving the optoelectronics and microelectronics industry and supported by a novel consortium of public and private organisations in the Ottawa area.

In the engineering sector, a budget of $91 million and 909 person-years (1990–91) supported research that took place within institutes for Industrial Technology, Aerospace Research, Marine Dynamics, Mechanical Engineering, and Research in Construction. All of NRC's major engineering facilities are owned by NRC. They include wind tunnels and facilities for airborne experiments and tests, structural tests, vehicle dynamics, gas dynamics, hydraulics, engine tests, low temperature experiments, and facilities for smoke and fire studies. The Institute for Marine Dynamics in St. John's, Newfoundland, is concerned with naval and offshore engineering in conditions of ice-covered water.

NRC provides two important technical outreach services. One is the Canadian Institute for Scientific and Technical Information (CISTI), Canada's largest scientific and technical information service; CISTI was created in 1974 from the National Science Library. CISTI's library, documentation, and information service cover all areas of science, engineering, technology and medicine. The CISTI collection includes several million books, 50,000 periodicals, and over two million technical reports in microform. The second technical outreach service provided by NRC is the Industrial Research Assistance Program (IRAP), a highly regarded national industrial extension service that was initiated in 1962. IRAP assists Canadian firms in technology transfer and diffusion. About 220 industrial technology advisors across the country provide technical and financial support for r&d and technology acquisiton activities of Canadian companies. Most of the staff are housed in, and employed by, organisations other than NRC, especially the Provincial Research Organisations. IRAP had a budget of about $76 million in 1990–91. A major debate is underway on the future of IRAP. One of the options under consideration would have the IRAP as the node of an expanded national technology extension service to serve the technical needs of the Canadian industrial sector. Indeed, in its December 1992 Economic Statement the federal government announced an expansion of the IRAP along these lines.

Environment Canada

Environment Canada has the mandate to preserve and enhance the quality of the environment for the benefit of present and future generations of Canadians. It is responsible for matters of federal jurisdiction, not assigned to other agencies, relating to the natural environment, including air, water, and soil; renewable resources, including migratory birds; meteorology; environmental matters in Canada-U.S. relations; the coordination of federal environmental policies and programmes; and the protection of natural and historic sites.

Environment Canada reported s&t expenditures rising from $376 million in 1984–85 to an estimated $645 in 1992–93. About ninety-two per cent is spent intramurally, and about sixty per cent of the expenditure is devoted to data collection. The agency's scientific and technical staff stood at about 4,500 person-years. Some of it's work is targetted in support of the Canadian environmental industry sector, a sector valued at $7–10 billion and employing about 150,000 people.

Environment Canada operates laboratories and scientific and technical services in nearly every province. It has atmospheric environmental service facilities in Vancouver, Edmonton, Montreal, and Halifax. The Canadian Meteorological Centre (the central processor of meteorological data) is near Montreal. Canadian Wildlife Service facilities are in British Columbia, Alberta, Ottawa, and Quebec City. The National Wildlife Research Centre, in Ottawa, conducts research on wildlife toxicology and migratory birds. Environmental protection service facilities are in Vancouver, Edmonton, Toronto, and Hull.

Environment Canada operates a cross-country network of water quality facilities that perform analyses and in some cases conduct r&d and undertake technological development. Inland waters directorates and facilities are in Vancouver, Regina, Burlington, Quebec City, Moncton, Dartmouth, and St. John's. The National Hydrology Research Institute in Saskatoon undertakes research on hydrological processes and climatic change, the impact of urban, industrial, and agricultural activity on water systems, and the characteristics of northern water systems.

Environment Canada operates two environmental technology centres. One, the Wastewater Technology Centre in Burlington, has been reorganised as a government-owned, company-operated laboratory (see below). The other is the River Road Environmental Technology Centre in Ottawa, which specializes in air pollution measurement and other analytical services, as well as environmental emergencies technology (e.g., arctic and marine oilspills, chemical hazards).

Environment Canada is deeply involved in delivering the federal 'Green Plan', a six-year, coordinated set of initiatives to achieve environmentally sustainable development. About forty per cent of the Green Plan's $6 billion expenditures will be made by Environment Canada. Many initiatives undertaken within the framework of the Green Plan have a science and technology dimension. Among those in which Environment Canada is involved are: a $50 million fund over six years for training the next generation of environmental scientists, administered through the three Granting Councils; the Canadian Global Change Program, under which the Royal Society of Canada receives $4.5 million from Environment Canada to coordinate research on global change and make findings available for environmental policy; the Environmental Innovation Program, that has $1.7 million available in 1992–93 to support extramural r&d to attain Green Plan goals; and the International Institute for Sustainable Development in Winnipeg, a not-for-profit organization established in 1990 to promote the concept of environmentally sustainable development. Environment Canada is providing more than half of the Institute's $25 million budget over six years.

Environment Canada's achievements in research include the development of the brewer spectrophotometer, a state-of-the-art remote sensing instrument for measuring stratospheric ozone; Canada's first doppler radar for the detection of tornadic storms and severe winds; equipment and reference methods for the measurement of air pollutants and other complex media; an instrument for measuring organic and inorganic contaminants in water and soil; a microwave process for removing oil and other contaminants from soil/sand; and advanced biotreatment technologies.

Agriculture Canada

The objective of Agriculture Canada's s&t work is to increase opportunities for improving the long-term marketability of Canadian agri-food products to promote growth, stability and competitiveness. Research is focused on longer-term, higher-risk projects with economic or environmental significance that are un likely to be undertaken by the private sector alone. The Canadian agri-food sector employs fourteen per cent of the labour force and contributes eight per cent of GDP. Of the total agri-food research conducted in Canada, forty per cent is found in federal installations and forty-three per cent in universities and colleges.

Agriculture Canada possesses one of the larger s&t budgets of federal agencies - $373 million in 1992–93, supporting nearly 3,400 person-years of science and technology activities. The bulk (ninety-six per cent) of Agriculture Canada's science and technology expenditures are intramural. Agriculture Canada is another federal agency that has seen its' s&t actitivities reduced from past levels. In 1984–1985, Agriculture Canada was able to devote over 5,500 person-years to s&t, and spend $393 million. By 1991–92, approximately $274 million and 3330 person-years were required to deliver programmes and services under Agriculture Canada's scientific research and development activity, which encompasses subactivities for r&d on resources, animals, crops, and food. Resource research focuses on assessment of the effects of agricultural production on soil, water, or air, and developing compensatory mechanisms as needed. Animal research focuses as a priority on the development of sustainable agricultural production systems. Crop research endeavors to develop better or alternative crops, including the development of new, innovative, or environmentally acceptable controls for pests. The Food Research group is strengthening it's research to create new food processing techniques, improve the diversification of the Canadian food industry, and increase the competitiveness of Canadian food products.

Agriculture Canada is also delivering the agricultural component of Canada's Green Plan (described earlier). Agriculture Canada's contribution to the Green Plan will support R and D and other scientific and technical activities to limit greenhouse gases produced by agricultural activities. A second component of Agriculture Canada's participation in the Green Plan promotes action to achieve environmental sustainability through research on soil resources, water quality, wildlife habitat, air and climate, energy, pollution and waste management, and genetic resources. A third component addresses issues of bio-diversity, the effects of ozone on vegetation, pest management strategies, and bio-engineered plants.

Agriculture Canada has about twenty-five major research establishments across the country, with at least one in every province. The Central Experimental Farm complex in Ottawa includes the Centre for Land and Biosystematics Resource Research, the Centre for Food and Animal Research, and the Plant Research Centre. Agriculture Canada also operates major inspection and regulation establishments in every region of the country.

With the relative decline in financial and human resources devoted to s&t in Agriculture Canada since the beginning of the decade, the agency has decided to consolidate research activities at fewer, more highly-equipped facilities. Two new multidisciplinary laboratories are under construction, one in Brandon, Manitoba, and another in Normandin, Quebec. The agency has also made collaborative research and technology transfer a priority. It is involved in about 700 collaborative research agreements, and is developing a coordinated approach to the commercialization and transfer of technology.

Energy, Mines and Resources

Energy, Mines and Resources Canada (EMR) is a scientific and economic department concerned primarily with Canada's landmass and with the responsible development of the resources it contains. EMR's priorities are development and international competitiveness of client industries; maintenance and enhancement of environmental quality; health and safety of those associated with the mineral and energy industries; economic equity and fairness; security of supply of mineral and energy commodities; regional development; and Canada's sovereignty. The s&t responsibilities of this department include policies, programmes and activities for s&t in the areas of energy, geological resources, minerals and metals, and the environment. EMR is also deeply involved in mapping and remote sensing of mineral, vegetative, and ocean resources.

EMR spent about $366 million on s&t in 1992–93 (about eighty-seven per cent of which was intramural), and employed about 2,800 scientific and technical personnel. Of these resources, about one-third are devoted to mineral and energy technology activities. EMR coordinates federal energy r&d through the Office of Energy Research and Development (for more detail on it's energy mandate, see Chapter Eight). The Department is also responsible for the Polar Continental Shelf Project, a unique organisation for the logistical support and funding of Arctic-related research.

CANMET (Canada Centre for Mineral and Energy Technology) is the main research and technology development arm of EMR. CANMET has six laboratory groups: Mineral Sciences, Metals Technology, Mining Research, Energy Research, Coal Research, and Energy Diversification Research. According to EMR estimates, forty-five per cent of CANMET's work aims to enhance the competitiveness of the mineral and energy industry in Canada, forty per cent aims to improve environmental health and safety in client industries, and fifteen per cent is in support of policy. CANMET is mandated to enhance the competitiveness of the Canadian minerals, metals and energy industries, to improve and develop energy efficient and alternative energy technologies, to improve health, safety and environmental control in the client industries, and to support government policy initiatives. CANMET works to improve methods to develop and use Canada's mineral and energy resources. In partnership with its' clients, CANMET performs and sponsors predominantly commercial and cost-shared research and development, and technology transfer. CANMET serves the minerals, metals and energy industries across Canada through its' research facilities in Nova Scotia, Quebec, the National Capital Region and Northern Ontario, and it's regional office in British Columbia.

EMR is especially interesting from an s&t policy perspective in two respects.

In the first place, it is one of the science-based federal departments that has begun to explore ways to understand and measure the performance of public s&t units. The performance of CANMET is monitored through a set of key indicators that measure the effectiveness of CANMET's interaction with its' clients in terms of the levering of funds, attracting cost-recovery revenue, and promoting exchange of personnel with the private sector. In 1990–91, CANMET recovered $5.6 million from the sale of services, and levered about $14 million and $29.5 million in task-shared and cost-shared work, respectively.

In the second place, EMR has announced plans to develop and implement a science and technology policy for the department. The policy will establish goals for EMR's science and technology activities, linked to the achievement of departmental and federal objectives relating to the economy, environment, and health and safety.

Fisheries and Oceans

Canada has the world's longest undefended coastline, and the second-largest continental shelf. About seven point five per cent of Canada's surface is covered by freshwater, representing sixteen per cent of the world's total freshwater surface area. The Department of Fisheries and Oceans is the only federal department with resource-management responsibilities with a primary focus on water and the resources it contains. Fisheries and Oceans has responsibility for all matters respecting oceans not assigned by law to any other department. DFO's program aims to undertake policies and programmes in support of Canada's economic, ecological and scientific interests in the oceans and inland waters; to provide for the conservation, development and sustained economic utilisation of Canada's fisheries resources in marine and inland waters; and to coordinate the policies and programmes of the Government of Canada respecting oceans. Fisheries and Ocean's three primary constituencies are the Canadian public, the fish harvesting and processing sectors, and the ocean manufacturing and services industries.

The Department's Science Sector ensures that appropriate scientific information is available for it's own use in developing policies, regulations, and legislation regarding the oceans and aquatic life, and for the use of other government departments, private industry and the public in planning and carrying out aquatic activities that impact on fisheries and fish habitat. The Sector is managed through three sub-activities;

1. the Biological Sciences subactivity is responsible for conducting stock assessment on major fisheries resource species. It also carries out research on marine and freshwater ecosystems and provides technical advice in aquaculture and aquatic habitat management
2. the Physical and Chemical Sciences subactivity conducts the Department's physical oceanographic research program and associated information management services in support of fisheries management, offshore development, climate prediction, marine services, coastal engineering, defence, and shipping
3. The Canadian Hydrographic Service conducts surveys and gathers information on tide and water levels, and publishes navigational charts

Fisheries and Oceans manages the following major research facilities:

– the Northwest Atlantic Fisheries Centre in St. John's, Newfoundland, which conducts multidisciplinary research in the area from the Cabot Strait east to

the Flemish Cap and north to Baffin Island
- the Bedford Institute of Oceanography in Dartmouth, Nova Scotia, a multi-disciplinary institute designed to service a multi-departmental clientele. This institute provides a full range of ocean research and service functions for the east coast and the Arctic
- St. Andrew's Biological Station in New Brunswick, which covers marine vertebrate and invertebrate fisheries, marine ecology, and aquaculture
- the Champlain Centre for Marine Science and Surveys near Quebec City, which is responsible for oceanographic and hydrographic surveys
- the Bayfield Institute in Burlington, Ontario, comprises two departmental components of the Canadian Centre for Inland Waters: the regional Canadian Hydrographic Service, and the Great Lakes Laboratory for Fisheries and Aquatic Sciences
- the Freshwater Institute in Winnipeg, Manitoba, is the major Canadian centre for freshwater and arctic fisheries research
- the Pacific Biological Station in Nanaimo, British Columbia, undertakes research in support of the management and conservation of the range of living marine resources of the Pacific Coast
- the Institute of Ocean Sciences in Sydney, British Columbia, is a centre of oceanographic research and provides hydrographic services. It is active in the Pacific and in the Arctic

Technology development is relevant to Fisheries and Oceans' sub-activities, particularly in the areas of aquaculture and resource development, ocean sciences, hydrography and charting. The Department actively attempts to transfer technologies to the Canadian oceans industry. A Marine Science Strategic Plan, for example, covers a ten to twenty year planning horizon that identifies the nature and extent of the challenges facing marine science, and proposes ways to address these challenges.

Between 1984–5 and 1992–93, the s&t expenditure of Fisheries and Oceans increased marginally from $254 million to $265 million. About ninety-four per cent of this expenditure is intramural. The recent crisis in the east coast cod fisheries resulted in a much publicised increase of about $20 million in scientific expenditures.

National Defence

The Department of National Defence (DND) conducts r&d to improve the capabilties effectiveness of the Canadian Forces. Unlike most other federal agencies, DND has reported increasing s&t expenditures in recent years: from $195 million in 1984–84 to $275 million in 1992–93. However, the federal government has been downsizing its' global military commitments and has cut back on the overall operating budget of the department.

The human resources devoted to s&t remained at about 1,930 person-years in this period. About fifty-six per cent of DND's expenditure 1992–93 was intramural; thirty-nine per cent was conducted in the business sector.

DND has six defence research establishments. The Atlantic facility in Dartmouth, Nova Scotia, conducts anti-submarine research in such areas as acoustics and signal processing. The Defence Research Establishment in Valcartier, Québec, conducts research in the fields of armaments, electro-optical aspects of surveillance and remote sensing, command and control, propellants, explosives, aerospace, lasers and weapon systems analysis. The facility in Ottawa handles research in a variety of areas including military communication, electronic warfare, remote sensing, and defensive aspects of chemical and biological warfare.

The Defence and Civil Institute of Environmental Medicine in Downsview, Ontario performs research on human health and safety in adverse environments (for example space, ocean, flight, and the Arctic). The Defence Research Establishment in Suffield, Alberta, does research on medical and chemical aspects of chemical and biological warfare and the disposal of toxic materials. The establishment in Victoria, British Columbia, is primarily concerned with anti-submarine detection in the North Pacific and Arctic.

In addition to it's intramural research, DND cooperates with Canada's allies on bilateral or multilateral development projects which often require the mobilisation of r&d. Furthermore, it participates in the Defence Industrial Research Programme, a shared-cost program with industry designed to improve Canadian industry's ability to provide high-technology equipment to the defence services.

DND, over the past several years, has spent considerable effort in forecasting technological opportunities and trends and continues to do so through an interdepartmental committee.

Health and Welfare Canada

Health and Welfare Canada is responsible for developing and delivering programmes to meet health and social needs. Research in support of programmes is organised in the following branches: Health Protection, Income Security Programmes, Medical Services Branch, Social Services, Health Services and Promotion, and Fitness and Amateur Sport (for more description of Canada's health-related research, see Chapter Eleven).

In 1992–1993, the Department of National Health and Welfare devoted $176 million to scientific activities, of which seventy-nine per cent was performed intramurally. National Health and Welfare is responsible for developing policies and delivering programmes to meet health and social security needs. Much of the research takes place in the Health Services Branch. It operates twenty-two district offices and five regional office in which scientific, technical and administrative activities take place. It is responsible for assessing risks from food, drugs, radiation-emitting and medical devices, consumer products and environmental contaminants as well as investigating the occurrence and causes of injuries and communicable diseases.

Within the Health Service Branch are a number of research laboratories including the following;

- The Laboratory Centre for Disease Control, with a staff of 173, provides national leadership and co-ordination in the identification, investigation, control, and prevention of human disease
- The Federal Centre for AIDS, with a staff of seventeen, is responsible for planning, directing, and coordinating federal government activities under the National AIDS Program
- The Drug Directorate, with a staff of 476, is responsible for ensuring the safety and efficacy of human and veterinary drugs that are manufactured or offered for sale in Canada
- The Environmental Health Directorate, with a staff of 268, serves to protect Canadians from health hazards associated with natural and man-made environments, the components of which are physical, chemical, and technological
- The Food Directorate, with a staff of 258, is responsible for the protection of the public against deficiencies in the nutritional quality of foods and from microbial and chemical hazards that might cause illness

Atomic Energy of Canada Limited (AECL)

AECL is a Crown Corporation responsible for Canada's nuclear energy programme. It was created from the NRC in 1952. It's broad research programme on
the peaceful uses of nuclear energy has resulted in the development of products
and services in use worldwide. The corporation's world-renowned flagship
product, the CANDU reactor, currently satisfies sixteen per cent of Canadian
electricity requirements and is a key component of the energy programmes on
four continents. According to a recent international evaluation, there were 339
large power reactors operating around the world by the end of 1989, twenty of
which were CANDUs. On the basis of lifelong performance, four of the top ten
were CANDU systems. CANDU enjoys the third largest share of the world nuclear reactor export market. AECL represents the single largest concentration of
electric industry r&d in Canada.

AECL in 1990–1991 employed 4,400 people. With headquarters in Ottawa,
AECL operates world-class research facilities at Chalk River, Ontario, and
Pinawa, Manitoba. It also operates engineering and design offices in Mis-
sissauga, Ontario, and Montreal, Quebec.

AECL scientists conduct basic research programmes in physics to advance
knowledge of the structure of matter. Since 1987, AECL has been responsible for
managing Canada's National Fusion Program which includes the Centre Canadien de la Fusion Magnetique's Tokamak program in Varennes, and the Canadian Fusion Fuel Technology Project which supports work at universities, in
industry and at AECL. AECL has established two commercial enterprises as part
of the federal government's privatisation programme. These enterprises, Nor-
dion International Inc. and Theratronics International Ltd. have respectively offered radiochemical services and medical products. AECL itself is being
re-structured into two operations: AECL Research and CANDU Operations.
The first of these will be jointly financed by the federal government and the
hydro-electric utilities in New Brunswick, Ontario and Quebec. The second operation will facilitate the participation of private and public sector actors in the
commercialisation of CANDU.

Forestry Canada

Forestry Canada'a mission is to promote the sustainable development and competitiveness of the Canadian forest sector for the well-being of present and future generations of Canadians. Forestry Canada was created in 1987–88 from a
section of the Department of Agriculture. It spent $93 million on s&t in 1992–903,
with eighty-six per cent of the expenditure intramural. It had 944 person-years of
science and technology personnel. These are the s&t resources that the Canadian
federal government devoted to a sector whose products' sales value amounted
to $50 billion in 1989. Shipments of forest products generated $685.5 million in
corporate income taxes in 1987 and $9.6 billion in salary and wage payments in
1988, and represented seventeen per cent of Canada's export trade in 1989.

Forestry Canada operates eight major regional forestry research centres. The
Newfoundland Forestry Research Centre in St. John's provides scientific, technical, and economic information and services to the local forest industries. The
Maritimes Forest Research Centre in Fredericton, New Brunswick, addresses
problems of forest renewal, acid rain, forest productivity, tree genetics, and
pests. The Laurentian Forest Research Centre near Quebec City applies systems
analysis to problems of forest ecology, epidemiology, and population dynamics.
The Great Lakes Forestry Research Centre, in Sault Ste. Marie, Ontario, conducts

research on long-term wood supply and forest ecosystems. The Northern Forest Research Centre in Edmonton conducts research into a variety of pest, management, and improvement problems. The Pacific Forestry Centre in Victoria, British Columbia, conducts research and technology transfer activities in silviculture, physiology, hydrology, meteorology, entomology, remote sensing, and biological and fire control. Forestry Canada operates two major facilities in addition to the regional centres. The Petawawa National Forestry Institute in Chalk River, Ontario, is responsible for a forest resource data program and for developing and demonstrating new methods of establishing and managing forests. The Forest Pest Management Institute in Sault Ste. Marie, Ontario, conducts research exclusively on pest control methods and strategies.

Forestry Canada's research objectives in 1991–92 included increasing it's work on environmentally sound forest management practices, strengthening it's work on applications of biotechnology to stock propagation and pest control, stepping up activities for monitoring forest health and demonstrating improved practices, emphasising work on integrated pest management, and developing decision-support systems for forest managers. Forestry Canada also announced it's intention to stimulate increased industrial support of forestry r&d with a Research Partnerships programme, increase support for graduate education in forestry s&t, and, within the framework of the Department's strategic plan, develop a Science and Technology Agenda for the Canadian forest sector.

Communications Canada

In addition to fulfiling a broad mandate having to do with culture, Communications Canada is responsible for formulating telecommunications policies, developing new technologies, providing telecommunications services to federal agencies, and supporting strategic policies with respect to communications technologies. Communications Canada's s&t expenditure declined from $94 million in 1983–84 to an estimated $71 million in 1992–93, and it's s&t personnel declined from 660 person-years to 448 person-years in the same period. In 1989, about $32 million was spent on r&d under the Research and Development subactivity. The department's r&d efforts focused on precompetitive communications technologies and components, satellites, broadcast technologies, and workplace automation technologies. Two major research facilities are the Communications Research Centre in Ottawa and the Canadian Workplace Automation Centre in Montreal. Six Regional Applications Centres were established around the country to provide the nurturing environment needed for dynamic growth in the research and development of specialised applications for communications technologies. For example, a centre in Newfoundland supports r&d for marine communications, telemedicine, and telehealth. (See Chapters Nine and Ten for more description of Canada's space, telecommunications and informatics capability).

Government policies

In addition to these individual activities, the line departments and agencies involved in technology and research have implemented the government's Contracting-Out Policy established in the early 1970s when it was known as the Make-or-Buy Policy. Designed to help develop a technological capability outside the public sector, the policy has had mixed success. One negative consequence of the policy has been to create a population of firms essentially dependent on government contracts. On the positive side, the policy has led to the creation of some technologically successful firms that have benefitted from the government as a purchaser and demonstrator of new technologies. Thus, a number of firms, especially in the space and aerospace sectors have grown and have become internationally competitive.

In 1974, another procurement policy instrument, known as the Unsolicited Proposals Programme, was added to the contracting-out policy. Administered by the Department of Supply and Services, the UP Programme was designed to allow firms to submit unsolicited proposals that built on government research competence. By 1989, the value of the programme reached $52 million annually, and the programme's impact was felt in terms of: creation of new companies, new products and new services; regional benefits; novel technological capability gained by government; and accelerated technological learning on the part of companies. The Programme was discontinued at the end of the 1980s, only to be resurrected after considerable pressure from the user community in December 1991 as the Unsolicited Proposals Brokerage Service. The new programme, in the spirit of the previous one, invites innovative scientific and technological proposals from the private sector which may be of interest to federal governnment departments.

Other recent initiatives by the federal government to better direct it's large intramural science and technology expenditure includes an examination of federal s&t alliances with private and other public interests; a "Partnerships for Success" initiative emphasising that federal s&t labs are seeking to expand their collaboration with businesses through alliances and assistance for technological development and diffusion; and the development of a 'toolbox' for government science-based departments that provides structured assistance to managers in cooperating with industry in a variety of ways. This latter vehicle has been developed by an Interdepartmental Committee on Intellectual Property Management, which examines new approaches to improving technology transfer from government to industry.

Provincial government laboratory infrastructure

Provincial research organisations (PROs) are one of the principal means by which provincial governments have sought to reap the benefits of r&d for provincial industrial development. All provinces, except Newfoundland and Prince Edward Island, have PROs, which generally specialise in applied r&d and technology transfer as described in Chapter Three). Total expenditure on PROs in 1990 totalled $183 million and the total number of employees was roughly 2,200.

PROs carry out a wide range of activities, including scientific research, development, resource surveys, analysis and testing, industrial engineering, feasibility studies, library and technical information services and industrial innovation. In 1990, PROs devoted forty-three per cent of their resources to secondary industry, thirteen per cent to primary industries, thirteen per cent to environment and lesser amounts to the natural resources sector, the construction industry, service industries, and utilities.

In terms of sources of funding, PROs rely for thirty-eight per cent of their budgets on provincial government subsidies, grants and contributions and on contracts from provincial governments (thirteen per cent), the federal government(ten per cent) and Canadian industry(thirty-one per cent). The remaining funds come other Canadian sources(one per cent) and foreign sources(seven per cent).

PROs in Canada have come together to form a technology network called the Association of Provincial Research Organisations (APRO). The network provides the opportunity for members to increase the expertise and equipment base upon which to offer service to clients, to increase business opportunities through contracts obtained via APRO, and to enhance the Association's role in Canadian s&t through participation in a national applied s&t network. APRO is comprised of the above-mentioned PROs plus the Newfoundland Association of Technical Industries, the Yukon College, and the Science Institute of the Northwest Territories. APRO will probably increasingly emphasise it's ability to marshall the network's technology transfer and diffusion capability to meet national and regional needs.

At the provincial level, Ontario and Quebec were each involved in the early 1980s in setting up technology centres focused on industrial sectors. Each were motivated by recent comprehensive reviews of provincial technology policy.

In Quebec, where about twenty-five per cent of the country's r&d is performed, Hydro-Quebec's research arm, IREQ, was established in 1967. The Centre de recherches minérales and the Centre de recherche industrielle du Québec were established in 1969. Following the publication of Quebec's first technology policy statement, the *Virage Technologique* (1982), a new generation of scientific and technological facilities was established in the government as well as in universities and colleges. Government laboratories established by the Quebec government in 1984–85 included the Centre Québécois pour l'informatisation de la production (CQIP), the Centre de recherche sur les applications pédagogiques de l'ordinateur (APO), and the Centre québécois de valorisation de la biomasse (CQVB). In addition, a 1985 Economic and Regional Development Agreement (Science and Technology Sub-Agreement) between the Quebec provincial and Canadian federal governments led to the establishment of the National Optics Institute. Another centre, originally planned by the Quebec government, was eventually set up as a consortium between Montreal universities and companies: the Centre de recherche informatique de Montréal.

Similar developments occurred in Ontario, where over fifty per cent of the country's R and D is conducted. Following a discussion document on technology policy, *The Technology Challenge: Ontario Faces the Future* (1974), an Ontario technology policy was developed which led, in the early 1980s, to the establishment of five 'Technology Centres' devoted to resource machinery, farm equipment and food processing, automobile parts, microelectronics, and CAD/CAM and robotics. The establishment of these centres was intended to complement the facilities offered by the Ontario Research Foundation (ORF), Ontario's provincial research organisation. Like many other Canadian public institutions designed to provide innovation support services to industry, ORF was expected to recover a fraction of its' operating expenses from its' clients. In 1980, ORF received twenty-four per cent of its' operating expenses from the Ontario government and estimated that it had worked with one in three manufacturing industries. Today, the organisation (now called ORTECH International) receives almost two-thirds of it's funding from private and foreign sources.

In other provinces, such as British Columbia and Nova Scotia, funds have been created to strengthen the capability of the industrial r&d infrastructure. These have gone, in part, to establish specialised centres of technology. (Detailed descriptions of some of these provincially-based centres can be found in the Appendix).

Municipal and metropolitan governments

In Canada, cities, metropolitan regions, and rural regions are attempting to develop and implement strategies to enhance technological innovation. There are many 'stakeholders' in local innovation systems, including public laboratories, research-producing public agencies, the public sector as user of s&t, and local public or quasi-public agencies with economic development responsibilities. Most Canadian metropolitan areas are experimenting with institutional development, innovation support mechanisms, consensus-building, and strategic planning for industrial or sectoral innovation. Three institutional models for innovation policy and management exist at the local level in Canada;

- The consultative forum. The Edmonton Council for Advanced Technology (ECAT) encompasses over two hundred individuals in the high technology community, including higher education and the public sector. ECAT's focus is on advanced technology rather than on innovation more generally. Established in 1984, ECAT's objectives are to promote discussion on local issues concerning advanced technology, to improve the local economy's ability to generate and attract technologically advanced firms, and to offer certain services to local players interested in s&t. ECAT functions as a forum or as a network of individuals which can stimulate mobilisation and political communication in some cases.
- The lobby group. A good example is GATIQ, the Groupe d'action pour l'avancement technologique et industriel de la region de Quebec. GATIQ was established in 1983 to promote collaboration between s&t players and the business sector in the Quebec City region.
- The industrial service centre. A third model of local concerted action in support of innovation is represented by initiatives to establish specialised services to industry. Two Canadian examples are the Business Advisory Centre (BAC) of Hamilton-Wentworth, and the Ottawa-Carleton Research Institute (OCRI). Each involves a network of players, including local universities, local authorities, and the region's business community. The BAC offers advice to entrepreneurs and SMEs, and includes a network of community-minded sponsors in large firms that provide services and advice to local SMEs. BAC expects to expand its' services to firms through the establishment of a Greater Hamilton Technology Enterprise Centre (GHTEC) and a Technology Transfer and Diffusion Unit (TTDU). OCRI, created in 1983, performs liaison, promotional, and planning roles, offers many technical seminars and training courses, provides infrastructure for technical organisations such as the Canadian Semiconductor Design Association, and spearheads initiatives to bring specialised infrastructure (such as the focused ion beam device mentioned earlier) to the region.

New trends in government research laboratories policy

Given the importance of federal intramural activities and institutions in the Canadian s&t effort, a recurring theme in Canadian science policy has been to define the appropriate structure and mission for these publicly supported assets. In the past thirty years, five major policy reviews of federal scientific activities have been completed. These are the Glassco Commission (1963), the Lamontagne

Committee (1968), discussed in Chapter Two, the Nielsen Task Force (1984), the Wright Report (1985), and the Lortie Report (1990). A fifth, through the Prime Minister's National Advisory Board on Science and Technology, is now underway, and is reviewing how government can spend smarter; *i.e.*, reallocate existing resources to achieve greater benefits for the nation. Among the issues it is examining, the review will assess how government research organisations allocate their existing levels of financial resources and how well clients are served.

These policy reviews have reached sometimes divergent conclusions regarding the health and performance of the federal intramural s&t system. The Nielsen Task Force (chaired by the then Deputy Prime Minister, Erik Nielsen), established by Canada's present Conservative Government early in its mandate, analysed the quality and usefulness of research in the government and concluded that, with a few exceptions, government research establishments were doing excellent work. The Task Force considered that the research establishments' problems were administrative rather than scientific, and it advocated increased support in certain cases. The report recommended a rationalised government-wide approach to r&d and a restructuring of the NRC.

In contrast, the Lortie Report (chaired by a prominent businessman) advocated far-reaching modifications to the federal laboratory system. The Lortie Report was released in 1990 by the Committee on Federal Science and Technology Expenditures of the National Advisory Board on Science and Technology. This Committee, advising the Prime Minister, concluded that fundamental changes were required in the organisation and design of a number of departments' intramural s&t activities and that a new management regime, one better suited to the unique nature of s&t, should be established. The proposed new regime recommended the following elements;

1. Each department would transfer its' s&t establishments into a single s&t institute
2. Contractual relationships would be established between the Department and its s&t institute and the Department would enter into several contracts with the institute for the scientific and technological services it required
3. Research institutes would become revenue dependant on Departments, and funds would be allocated to Departments that would in turn enter into contracts with research institutes
4. A research institute would have the authority to manage its' contracts with the department, to enter into contracts with other clients, to own intellectual property, enter into contracting and licensing agreements, set fees and retain the earnings generated by this activity, and carry out activities in a manner consistent with the best management practices as they apply to its' s&t responsibilities
5. The board of directors and the chief executive officer of a research institute would have explicit authority and responsibillity to ensure the evaluation of the research activities and personnel. These assessments would be monitored by a National Panel for Quality Evaluation

Privatisation of government r&d facilities is emerging as an attractive alternative to intramural s&t activity in recent years. The AECL example cited above is one example. Another is the case of the Wastewater Technology Centre in Burlington, Ontario, a former laboratory of Environment Canada. The 110 employees voted to be managed by a private consortium. In exchange the employees receive a twenty per cent share of any profits the lab makes selling its wastewater treatment technology. The consortium is permitted to make a profit on anything the lab does over and above the basic services it provides for the government. Aside from the twenty per cent earmarked for employees, the federal government will receive thirty-five per cent of the profits and the consortium will receive forty-five per cent. Yet another model being used as a pilot to test out some of the principles suggested in the Lortie report is the Communications Research Centre of Communications Canada.

The few studies of privatisation of government laboratories in Canada shed some light on the constraints and limits on this strategy to reduce costs and increase the effectiveness and efficiency of publicly required services. Privatisation of some governmental environmental regulatory functions in British Columbia, for example, have led to questions of conflict of interest and have required complex and expensive contract-monitoring arrangements, paradoxically necessitating greater government intervention.

An environment of fiscal restraint inhibits the expectations for growth in the public R and D sector. This impacts adversely on the ability of some agencies to fully meet their mandates. The trend in federal r&d spending is to allocate resources to areas of activity in which commercial benefits seem possible. This has entailed reduction of the relative level of resources allocated to intramural r&d activities. Marketing and technology transfer functions have been increased, as has attention to such matters as patenting, licensing and other intellectual property considerations. Much experimentation has taken place with institutional arrangements to maintain the scientific quality of establishments under conditions of fiscal restraint, while making the establishments more responsive to user needs. The extent to which these goals are mutually attainable remains to be seen. Such a regime poses difficulties for longer-term research in support of industrial innovation and development, and may experience difficulty valuing non-commercial programmes serving to safeguard the public interest in such areas as health and safety, preservation of the environment, and conservation of natural resources.

A second set of issues concerns the ability of federal research establishments to attract and retain highly qualified and highly motivated personnel. Recruitment of younger scientists into federal research positions has been slow, due to the slow-growth regime most federal science and technology activities have experienced. The average age of the federal occupational group of physical and natural scientists increased from 45.9 to 46.7 years between 1987 and 1989. This relatively rapid rate of aging of the principal group of qualified federal research scientists reflects the stagnation in recruitment to federal research positions. The turnover rate in the federal s&t community has declined from seven per cent or eight per cent to three per cent or less. The demographics of federal s&t personnel are such that there are fewer promotional opportunities for younger staff as long-service employees occupy senior positions. As a result, research establishments are unable to hire significant numbers of recently graduated researchers. In addition, political decisions to decentralise federal government facilities away from Ottawa and move them to other regions of the country have not always been accompanied with well thought out incentives for the personnel involved, thus making it difficult for key researchers to adjust to the move. Furthermore, with continued cutbacks affecting the quality of some long-standing public missions, a severe morale problem has emerged in several quarters of the public service. Cynicism is rampant and it has become exceedingly difficult to engender strong leadership within these labs.

A new system of state-industry relations in s&t is emerging in Canada. The contours of the emerging institutional arrangements for federal involvement in science and technology are only faintly visible at present. The new arrangements are meant to be more flexible, more cost-effective, and more directly supportive of industrial competitiveness and environmental sustainabilty. Another likely feature of federal s&t arrangements, if public sector s&t is to be maintained in an era of severe fiscal constraints, is increased collaboration in s&t initiatives with provincial governments, universities, and international partners.

Notes

1 What follows describes the principal federal government laboratories in Canada, but it is by no means comprehensive. For further details on federal r&d organisations, the reader should turn to the Appendix

References

Agriculture Canada 1992 *Agriculture Canada. 1992–93 Estimates. Part III Expenditure Plan* Ottawa Department of Supply and Services

Alexander J 1990 On the Vicissitudes of Knowledge and Power or, Why he Shot the Messenger in Crousse B Alexander J Landry R (eds) *Science and Technology Policy Evaluation. National Experiences* pp 333–364 Québec Les Presses de l'Université Laval

Anderson F 1985 *Répertoire analytique des centres canadiens de recherche qui ont comme mission l'aide à l'industrie* Québec

Anderson F Berseneff O Dufour P 1991 Le Développement des conseils de recherche provinciaux: quelques problématiques historiographiques in Jarrell R A Hull J P (eds) *Science, Technology and Medicine in Canada's Past: Selections from Scientia Canadensis* pp 129–146 Thornhill The Scientia Press

Anderson F Dalpé R 1991 The Evaluation of Public Applied Research Laboratories *Canadian J Program Evaluation* 6(2):107–125

Association of Provincial Research Organizations 1992 *The Canadian Technology Network* Ottawa

Clarke T E Reavley J 1988 Problems Faced by R&D Managers in Canadian Federal Government Laboratories *R&D Management* 18(1):33–44

Communications Canada 1991 *1992–93 Estimates. Part III. Expenditure Plan* Ottawa Minister of Supply and Services

Conseil de la science et de la technologie du Québec 1988 *Science et Technologie: Conjoncture 1988* Quebec CSTQ

Cordell A Gilmour J 1972 *The Role and Function of Government Laboratories and the Transfer of Technology to the Manufacturing Sector* Ottawa Science Council of Canada

Davis C H 1992 L'analyse du 'Virage technologique' au Québec in Dalpé R Landry R (eds) *La politique technologique au Québec.* Les Presses de l'Université de Montréal, forthcoming

Davis C H 1992 Les régions métropolitaines canadiennes et la promotion locale de l'innovation technologique *Revue d'économie métropolitaine et urbaine* forthcoming

Davis C H (ed) 1991 *Local Initiatives to Promote Technological Innovations in Canada: Eight Case Studies* Ottawa Science Council of Canada Manuscript Report

Davis C H Alexander J MacDonald L 1990 The Political Incorporation of Innovation Systems: Collective Action and Canadian Science and Technology Policy *Réseaux* 58–59–60:65–82

Energy, Mines and Resources Canada 1992 *1992–93 Estimates. Part III. Expenditure Plan* Ottawa Minister of Supply and Services

Environment Canada 1992 *1992–93 Estimates. Part III. Expenditure Plan* Ottawa Minister of Supply and Services

Fisheries and Oceans 1992 *1992–93 Estimates. Part III. Expenditure Plan* Ottawa Minister of Supply and Services

Forestry Canada 1991 *1991–92 Estimates. Part III. Expenditure Plan* Ottawa Minister of Supply and Services

Hanna J 1991 NRC: Corporate Handmaiden or Savvy Survivor? *Ottawa Business Magazine* October:23–32.

Harrison K Stanbury W T 1991 Privatization in British Columbia: Lessons from the Sale of Government Laboratories *Canadian Public Administration* 33(2):165–197

Industry, Science and Technology Canada 1990 *B.O.S.S. Directory of Research and Development Laboratories/Facilities in Canada* Ottawa ISTC

Industry, Science and Technology Canada 1990 *Strategic Overview of Science and Technology Activities in the Federal Government* Ottawa ISTC

Industry, Science and Technology Canada 1990 *Technology Networking Guide Canada* Ottawa ISTC

Le Roy D J Dufour P 1983 *Partners in Industrial Strategy: The Special Role of the Provincial Research Organizations* Ottawa Science Council of Canada

National Defence 1992 *1992–93 Estimates. Part III: Expenditure Plan* Ottawa Minister of Supply and Services

National Research Council 1992 *1992–93 Estimates. Part III: Expenditure Plan* Ottawa Minister of Supply and Services

National Research Council 1991 *The Competitive Edge: Long Range Plan 1990–1995* Ottawa NRC

OECD 1987 *Government Research Establishments: Synthesis Report* Paris OECD Ad hoc Group on Scientific and University Research

RockCliffe Research and Technology Inc 1989 *Models and Policies for the Commercialization of Government Science and Technology*. Ottawa ISTC

Science Council of Canada 1992 *Reaching for Tomorrow. Science and Technology Policy in Canada, 1991* Ottawa Science Council of Canada

Science Council of Canada 1990 *Grassroots Initiatives, Global Success: Report of the 1989 National Technology Policy Roundtable* Ottawa Science Council of Canada the Canadian Advanced Technology Association Canadian Chamber of Commerce

Science Council of Canada 1990 *Firing up the Technology Engine: Strategies for Community Economic Development* Ottawa Science Council of Canada and Canadian Advanced Technology Association

Science Council of Canada 1975 *Technology Transfer: Government Laboratories to Manufacturing Industry* Ottawa Science Council of Canada

Statistics Canada 1992 *Science Statistics* Ottawa Statistics Canada

Statistics Canada 1992 *Federal Scientific Activities 1991–92* Ottawa Catalogue 88–204 Statistics Canada

Statistics Canada 1990 *Estimates of Canadian Research and Development Expenditures (GERD)* Ottawa Statistics Canada

Chapter 5

The industrial research infrastructure

Richard Botham and Michel Giguère

Introduction

Public policy encouraging the private sector to increase it's spending on scientif-ic research and experimental development (r&d) is not new to Canada. Since the 1960s, but even more since the early 1980s, the federal and provincial govern-ments in Canada have tried to stimulate industrial r&d and innovation through many different policies, instruments and mechanisms. While the 1920–1970 era was characterised by a consolidation of the public research infrastructure, the last twenty years have seen the focus shift towards a concern about the strength of industrial research. As discussed in Chapter Three, as early as 1970, the Lamontagne Committee on science policy warned the government that too much of the (already poor) research effort in Canada was concentrated in gov-ernment laboratories and universities. Accordingly, and in parallel with a trend that was existent in almost every industrialised country, the state began to exam-ine ways to substantially increase the industrial r&d effort.

The federal government was the first to act significantly on this plan. The fed-eral government is a bigger government, has more financial resources and – through its' responsibilities for external affairs – tends to be more aware of inter-national trends than most provinces, although some provinces, most notably Quebec, Ontario, Alberta and British Columbia, have also had an international presence. At the time it was creating science policy bodies, such as the Science Council in 1966 and the Ministry of State for Science and Technology in 1971, the federal government was starting to elaborate policies and programmes for s&t, and, in particular, to try to correct the noted weakness in industrial r&d. The provinces were to follow soon after, with Quebec leading the way in the 1970s and Ontario, British Columbia, Alberta and other smaller provinces doing the same in the following decade.

As far as provincial initiatives are concerned, this chapter will focus mostly on Quebec, Ontario, British Columbia and Alberta, which represent the four Cana-dian provinces where industrial activity is concentrated in Canada.

Canada's industrial infrastructure

As in most developed economies, services are becoming a more prominent part of Canada's economy. There has been a shift in the proportion of the workforce employed in the service producing sector from the goods producing sector. The Economic Council of Canada reported that employment in the service sector increased at an average annual rate of three point two per cent between 1967 and 1988; employment increased by less than one per cent in the goods sector in the same period. From 1967 to 1988, the service sector's share of employment increased from fifty-nine per cent to seventy-one per cent and the goods sectors' share dropped from forty-one per cent to twenty-nine per cent.

Many analysts argue that the service economy is becoming increasingly important as the barriers that separate manufacturing and services breakdown and as services become an increasingly tradeable commodity[1]

Table shows the importance of the service sector in employment and output to the Canadian economy.

Profile of key manufacturing sectors[2]

Table 5.2 shows the relative weight of selected industries in their contribution to manufacturing output. There is also a brief description of the three largest industries and sectors that are of regional significance.

The transportation sector contributes more than any other sector to Canada's manufacturing output, measured as total value added. The sector is dominated by the production of automotive vehicles and parts for the U.S. market. Most of the automotive vehicle production is centred in southern Ontario and consists of facilities owned by subsidiaries of General Motors, Ford, and Chrysler. It is dominated by a few large final-assembly plants, several captive component plants, and a small and independent parts industry. The automotive assembly, parts and specialty vehicle sector accounted for $46.2 billion in exports and 185,200 jobs in 1989. Competitive advantage of firms in this sector has depended on the importation of state-of-the-art machinery and low labour costs. Proportionally, very few firms perform research and development in Canada; firms in

Table 5.1

Sector	Employment 1991	Output in 1991 (millions of $1986)
Forestry	51,132	$2 926
Mines, Quarries, Oil Wells	146,543	$19 998
Manufacturing	1,571,549	$84 784
Construction	405,228	$31 797
Transportation, Communication, other Utilities	824,920	$60 068
Trade	1,664,707	$55 782
Finance, Insurance, Real Estate	637,338	$84 954
Community, Business and Personal Services	3,441,929	$55 593
Public Administration	713,405	$33 657
*Goods Producing	2,174,454	$140 376
Service Producing	7,282,301	$316 355
Industrial Aggregate/GDP at Factor Cost	9,456,755	$501 863

Includes forestry, hunting, fishing and trapping, mines, quarries and oil wells, manufacturing and construction.

Sources: Statistics Canada and Conference Board of Canada

Table 5.2

Manufacturing Industries	Total Value Added, 1989 (thousands $1986)	Sector's Share of MFG Total Value Added
Transportation	$16,422,202	13.9%
Primary Metal & Fabricated Metal	$16,216,812	13.8%
Forest Products	$15,979,835	13.6%
Food, Beverage and Tobacco	$15,931,110	13.5%
Chemical & Chemical Products	$11,171,421	9.5%
Electrical and Electronic	$9,068,881	7.7%
Printing, Publishing & Allied	$7,373,082	6.3%
Leather and Allied, Textiles and Clothing	$6,380,462	5.4%
Machinery	$4,803,372	4.1%
Non-metallic Mineral, Refined Petroleum & Coal	$3,857,733	3.3%
Rubber and Plastic Products	$3,788,847	3.2%
Furniture and Fixture	$2,155,298	1.8%
Scientific and Professional Equipment	$1,128,863	1.0%

Source: Statistics Canada, 31–203, 1989

this sector had r&d expenditures of zero point two per cent of sales in 1989. In part, this is the result of the high levels of foreign ownership. Canadian-owned automotive-parts manufacturers tend to spend more than foreign-owned firms in the sector.

A strong high-technology sub-sector of transportation is the aerospace industry. In Canada, this industry grew rapidly through the 1980s, so that by 1988 the industry employed 63,650 people, had domestic sales of $2 billion and exports of $4 billion. The principle companies in the aerospace industry are Pratt & Whitney, de Havilland, Bell Helicopter Canada, MBB Helicopter Canada, Garrett Canada, Rolls-Royce Canada, Hawker Siddeley Canada, Bristol Aerospace, McDonnel Douglas Canada, Canadian Aircraft Products, Amherst Aerospace, and Oerlikon Aerospace. With a few notable exceptions, Spar, CAE Electronics, Fleet Aerospace, and Canadair, Canadian owned aerospace firms tend to be smaller firms that supply subcontract products and services.

Primary and fabricated metal constitutes the second largest sectoral contributor to Canada's total value added in manufacturing. There are three main components of this sector: iron and steel, nonferrous-metals, and metal fabrication. Canada was the world's thirteenth largest steel-producing country in 1988, accounting for less than two per cent of the world's crude steel output. Two Ontario-based integrated steel companies, Dofasco/Algoma and Stelco, accounted for about seventy-one per cent of steel production in Canada in 1988, most of the remainder was produced by mini-mills. In the past, Canada's iron and steel industry successfully produced mostly commodity steel with imported technology, relatively inexpensive and high quality raw materials, low energy costs and skilled labour. Lack of spending on research and development is now making the industry increasingly more vulnerable to competitive pressures.

Non-ferrous metals have made a significant contribution to Canada's economy. Canada leads the world in value of mineral exports and ranks fourth among the diversified-minerals producers in non-fuel minerals production. Most of the activity in this sector is in mining (sixty-three per cent), the remainder comes from smelting and refining (twenty-seven per cent), and semi-fabrication (ten per cent). The industry is dominated in Canada by a few large vertically integrated firms: Alcan in aluminum; Inco Limited, Falconbridge, and Sherritt Gordon in nickel, lead and zinc; Cominco in lead, copper and zinc; and Noranda in copper and zinc. Many of these companies have mining, smelting, refining or fabricating operations outside of Canada. The concentration of Canadian firms on mining and exploration has been at the expense of investing in downstream research and development.

The third largest sectoral contribution to Canada's manufacturing output comes from the forest products sector. Firms in this sector provided direct employment for 300,000 people and had exports of $22 billion in 1990. The sector holds a twenty-two per cent share of the world trade in forest products and is the most important net exporting sector. This striking success has been built on access to a vast quantity of high-quality raw material, low power costs, and investment in up-to-date equipment, most of which was purchased from other countries.

The sector can be divided into two types of products: paper and allied products, which account for sixty-three per cent of total shipments; and wood industries, contributing the remaining thirty-seven per cent of the value of the sector's shipments. The sector is under pressure to increase spending on research and development, to improve its environmental record and to move to the production of higher-value added goods. Canadian wood products and paper and allied firms spent point five per cent and point four per cent respectively of their sales on research and development in 1989. The low levels of intramural r&d by firms, about forty per cent of expenditure in this sector, is supplemented by cooperative research organizations which perform seventeen per cent of the forest products r&d, with governments and universities contributing the remaining forty-three per cent.

In a study of the Canadian economy's competitiveness, Rugman and D'Cruz identified the Western Canadian Forest Products cluster as one of ten strategic clusters of industrial activity. This sector is of particular importance in British Columbia and Alberta. In their report, Rugman and D'Cruz point out that the basis for the industry's past success, an inexpensive resource, led to the export of commodity-based products which are now vulnerable to downward pressures on price and demand. They state that lack of rivalry in the forest industry has also blunted the competitiveness of the industry leader, MacMillan Bloedel(owned by Noranda Forest).

Another strategic cluster of regional significance identified in the report, the Alberta Energy cluster, is also dependent on fluctuations in primary products' prices. Leading firms in this industry are Imperial Oil, Shell, Nova and Petro-Canada. Rugman and D'Cruz argue that the boom and bust character of the industry has hampered supporting industries and services from developing deep roots. Another weakness that they identify is the failure of the industry in sponsoring the development of an advanced research and educational infrastructure.

Also of regional significance is the Prairie farming cluster centred in Saskatchewan and Manitoba. Like the previous two clusters this is a commodity-based industry that is vulnerable to downward pressure internationally on prices. The leading firm is Cargill, a U.S.-based corporation. Federal government laboratories have played a vital role in providing r&d for this sector, but Rugman and D'Cruz argue that Cargill and other leading firms should be called upon to play a more visible role in the development and marketing of new grain products.

Rugman and D'Cruz identify two weaknesses which have hindered the move away from commodity-based exports to higher value-added goods. The first is an absence of a strong technological infrastructure of training centres, university departments and research centres. Secondly, they argue that the position of this industry in the production chain, one step removed from the processors and linked only by their main commodity market in Boston, has prevented collaboration with the downstream processors which could have led to the development of new products. The two leading firms in the industry are National Sea Products and Fisheries Products International; other significant players are the offshore foreign fleets. Disadvantages of scale in the inshore fishing fleet, and a weak market in the Atlantic provinces are often seen as reasons for government support of technological infrastructure for training, product development and testing and standards.

This brief introduction to Canada's leading goods industries highlights five significant characteristics: lack of processing of natural resources; a high degree of foreign ownership; an underdeveloped domestic equipment and machinery industry; few internationally competitive, large non-resource based indigenous firms; a lack of private sector research and development.

These persistent weaknesses will be discussed in more detail later.

A persistent characteristic of the Canadian economy has been a dependence on the export of unprocessed and semi-processed natural resources. A study of the competitiveness of the Canadian economy found that Canada's share of world resource-based exports increased from four point nine per cent in 1978 to eight point three per cent in 1989.[3] Canada's share of world exports in non-resource-based exports rose from two point four per cent in 1978 to three point eight per cent in 1985 and then subsequently decreased to two point eight per cent in 1989. Michael Porter's analysis succinctly summarises this tendency.

> Exports of natural resource-intensive goods are by no means undesirable. However, a high proportion of exports of unprocessed resources makes Canada sensitive to commodity price shifts, technological substitution, and the emergence of lower cost competitors. The problem is made more acute by the fact that technological change is reducing resource intensity in advanced economies, and that unprocessed resource industries are especially accessible by developing countries.

Canada's exports are also concentrated geographically. There has been a high, and growing, reliance on the U.S. market. The U.S. was the destination of fifty-six per cent of Canada's exports in 1960. This share rose to seventy-five per cent in 1989. Some of this export concentration can be attributed to high levels of foreign (U.S.) ownership of firms in Canada.

Foreign direct investment provided a channel for the importation of advanced technologies and provided the impetus for the development of skills in Canada in the early part of the twentieth century. But this investment has also had negative consequences. Chief among these has been the tendency of many foreign-owned operations to centre their knowledge-based strategic activities outside of Canada. These activities are increasingly being recognised as essential to the maintenance of high standards of living in advanced economies. A related negative outcome of this behaviour is a disconnectedness with domestic suppliers, customers and institutions. The strong linkages that these firms maintain with their home-base operate at the expense of the development of similar networks and linkages that would drive innovation in Canada. The absence of strong linkages between large foreign-owned firms and their suppliers has lessened demands for the development of a sophisticated domestic machinery and equipment industry.

Table 5.3 illustrates the high degree of foreign ownership in the Canadian economy, and the dominance in manufacturing and mining. The data also show that although foreign controlled firms account for only one point two per cent of manufacturing firms, these firms account for almost one-half of sales and forty-six per cent of assets. A counterpart to the characteristic of a high degree of foreign controlled firms in Canada is the relative absence of strong medium- and large-sized companies capable of competing in world markets. In a 1987 study of Ontario's economy, the lack of indigenous world class firms, outside of the resource-based industries, was identified as a structural weakness of the economy. Provincial and federal government economic development support has been channeled to develop medium-sized companies so that they are better able to compete in the world markets. The next section focusses on industrial research and development in Canada.

Industrial r&d expenditures[4]

Expenditures on research and development in the business enterprise sector in Canada are low relative to other advanced economies. Expenditures in the business sector in Canada were point seven two per cent of the gross domestic product (GDP) in 1989. Although this proportion is comparable to expenditures in Italy, the average expenditure on research and development in the business sector for G-7 countries in 1989 was one point five per cent of GDP.

Compared to other OECD countries, Canada's business enterprise sector also performs a relatively small proportion of the gross expenditure on research and development (GERD). In 1989, Canada's business sector accounted for fifty-four per cent of the GERD. This proportion is projected to remain constant through 1991. The average proportion of GERD performed by the business enterprise sector in the other G-7 countries in the same year was sixty-three per cent. Although lower than in many other advanced economies, the proportion of the GERD accounted for by the business enterprise sector has risen steadily from thirty-three per cent in 1971. It is the largest performing sector of research and development.

Government financing of expenditures on research and development in Canada's business sector is relatively low compared to other G-7 countries. In 1989, government financed ten per cent of research and development in this sector. Most of this funding, ninety per cent, came from the federal government and ten per cent came from governments at the subnational or provincial level. Japan was the only country that had a lower proportion of financing by government, and the average for all G-7 countries was fifteen per cent.

Table 5.3
Share of Canadian Non-Financial Industries Under Foreign Control (1987)

Sector	% of all firms	% of total assets	% of total sales	% of total equity
Mines	1.8	30.6	40.1	32.5
Manufacturing	1.2	45.6	48.6	49.9
Construction	0.1	5.7	5.4	7.6
Transportation	0.3	3.2	5.4	4.3
Trade	0.5	23.5	20.1	29.1
Business, Personal and Commercial Services	0.1	12.9	9.9	16.2
Total Non-financial	0.4	24.6	27.4	31.2

Government funding of intramural research and development varies widely between industries. The federal government funded more than one-quarter of intramural research and development in the wood, aircraft and parts, and scientific and professional equipment industries in 1989. It was the source of less than one per cent of funds for intramural research and development in the telecommunication equipment and pharmaceutical and medicine industries in the same year.

Industry financed seventy-three per cent of research and development expenditure in the business enterprise sector in Canada in 1989. This proportion is comparable to most other G-7 countries with the exception of Japan, where industry spending accounted for almost ninety-nine per cent of expenditure. However, the compound growth of industry financed expenditure in Canada was lower than in all other G-7 countries except the UK through the last half of the 1980s.

A substantial proportion of research and development expenditure in the business enterprise sector in Canada is accounted for by foreign sources. In 1989, seventeen per cent of funding came from abroad. The average for other G-7 countries (excluding the U.S.A.) was six point five per cent. The predominant source of foreign funds in Canada, more than three-quarters, was related companies. The share of financing by foreign sources of intramural research and development varied widely between industries, but it accounted for the largest proportion of funds for intramural research and development in the business machines industry (sixty-eight per cent in 1989).

Intramural research and development activity is not distributed evenly in each of Canada's ten provinces. It is heavily concentrated in two provinces, Ontario and Quebec, which accounted for eighty-four per cent of total intramural expenditure in 1989. Ontario alone accounted for fifty-seven per cent of total intramural research and development expenditures in 1989, and was the location of forty-seven per cent of industrial research and development units. Intramural expenditure in two other provinces, Alberta and British Columbia, accounted for twelve per cent of total expenditure in Canada in 1989. These two provinces were the location of twenty-three per cent of research and development units in Canada in 1989.

The heavy concentration of expenditure is also evident at the metropolitan area level. Almost half of the total Canadian expenditure on intramural research and development in 1989 occurred in Montreal and Toronto.

The concentration of intramural research and development activity is more pronounced in a few key industries. Activity in Ontario and Quebec accounted for ninety-seven per cent of Canadian expenditure in the aircraft and parts industry. Eighty-five per cent of intramural research and development expenditure in the telecommunication equipment and business machines industries, and sixty-seven per cent of expenditure in the computer and related services industry took place in Ontario in 1989.

In 1989, 3,311 firms in Canada reported performing research and development activities to Statistics Canada. Half of the industrial research and development expenditure in 1989 was made by twenty-five companies, each spending more than $25 million. The top fourteen spenders each reported research and development expenditures greater than $50 million. The top 100 spenders on research and development accounted for sixty-nine per cent of expenditure. Each of these companies spent more than $5 million on research and development in 1989. The average expenditure for all reporting companies was $1.4 million.

Not surprisingly, the largest companies reported the highest total intramural research and development expenditure in 1989. Companies with sales of at least

$400 million had an average expenditure of $21.4 million. There were 112 companies in this category. At the other end of the scale, 1,297 companies reported sales lower than $1 million and an average research and development expenditure of $200,000. Spending by these companies accounted for less than half of one per cent of total expenditure on research and development.

Companies with relatively lower sales accounted for a small share of the total expenditure, but each of these companies spent, on average, a greater proportion of their sales on research and development. Firms with less than $1 million in sales reported spending forty-three point five per cent of their company sales on intramural research and development. This proportion drops sharply to eleven point one per cent for companies in the next reported size category, from $1 million to $10 million in sales. Companies in the largest size category, more than $400 million in sales, spent an average of one per cent of their sales on intramural research and development.

The country of origin of the ownership of a company also affected the proportion of sales spent on research and development. In 1989, Canadian controlled companies spent an average of one point six per cent of their sales on current intramural research and development, compared to an average of one point two per cent for companies controlled in the USA, and one point eight per cent for companies controlled by other foreign countries.

Foreign controlled companies made up thirteen per cent of the 3,311 companies reporting research and development activities in 1989. These companies accounted for thirty-eight per cent of the total expenditure on intramural research and development. The foreign controlled companies' share of total intramural research and development expenditure varied widely between industry groups, from a low of two per cent in utilities to a high of seventy per cent in trade. Foreign controlled companies accounted for forty-five per cent of manufacturing intramural research and development expenditure.

The sources of funds for intramural research and development differed according to the size of a company's research and development program. Companies reporting less than $1 million of expenditure received an average of eleven per cent of their funds from the federal government and four per cent from foreign sources. Companies with research and development expenditure greater than $1 million received eight per cent of their funds from the federal government and nineteen per cent from foreign sources. Research and development performers spending below the $1 million threshold financed an average of seventy-six per cent of their intramural programs, compared to an average of sixty-one per cent for companies above the threshold.

Research and development expenditure is also concentrated by industry. Six industries – telecommunication equipment, aircraft and parts, engineering and scientific services, business machines, and other electronic equipment and computer and related services – accounted for half of the $4,667 million in total intramural research and development expenditure in 1989. The shares of total expenditure for these six industries has been constant from 1985 and is projected to continue through 1991. The telecommunication equipment industry recorded the largest share of industrial research and development spending, fifteen per cent, in 1989, companies in the aircraft and parts industry accounted for nine per cent, and the share for engineering and scientific services was eight per cent.

A private Canadian consulting company which monitors developments in research and development published a list of Canada's top 100 performers of industrial research and development. Although the list is not a complete survey of Canadian companies, it gives a good indication of the top performers. Northern

Table 5.4
Total Intramural R&D Expenditures, by Industry, 1979 and 1989
(in millions of $ 1986)

	Expenditures	
Industry	*1979*	*1989*
Telecommunication Equipment	$234	$623
Aircraft and Parts	$231	$361
Utilities	$167	$342
Engineering and Scientific Services	$74	$305
Other Electronic Equipment	$61	$266
Business Machines	$56	$261
Computer and Related Services	$12	$222
Other Chemical Products	$100	$164
Pharmaceutical and Medicine	$50	$151
Forest Products	$80	$145
Primary Metals	$118	$139
Refined Petroleum and Coal	$203	$114
Other Transportation Equipment	$70	$99
Mining and Oil Wells	$138	$82
Food, Beverages and Tobacco	$61	$60
Total Manufacturing	**$1,518**	**$2,817**
Total All Industries	**$1,921**	**$4,050**

Telecom reported the highest expenditure of all companies in 1990, at $902 million. All of the top ten companies reported spending more than $100 million. Two provincial utilities, Ontario Hydro and Hydro Quebec, and the federal crown corporation AECL are included in the top ten. The list also includes Pratt & Whitney Canada Inc, IBM Canada Ltd, Alcan Aluminum Ltd, CAE Industries Ltd, Bombardier Inc, and Bell Canada. The list reinforces Statistics Canada's data showing the dominance of the telecommunications and aircraft and parts industries in research and development. The headquarters of five of these companies are in Ontario, the others are in Quebec. Six of the ten companies are Canadian controlled.

Of all companies reporting research and development activity to Statistics Canada, thirty-nine per cent undertook software development. Expenditure on software development accounted for twenty-four per cent of total research and development spending in 1989. The greatest proportion of this spending, sixty-four per cent, was in the manufacturing sector. Almost all of this, eighty-nine per cent, occurred in the telecommunication and other equipment and business machines industries. Two other large performers of software development were the computer and related services industry which accounted for seventeen per cent of the total expenditures, and engineering and scientific services which had a seven per cent share.

Manufacturing companies with fewer than 100 employees performed only four per cent of that sector's software development. In some industries, however, small firms accounted for substantially more of an industry's software development. Firms with fewer than 50 employees accounted for forty per cent of spending by the scientific and professional equipment industry, and seventeen per cent of expenditure by the machinery industry. Companies with fewer than 100 employees in the service industry accounted for fifty-six per cent of software research and development expenditure.

In Statistics Canada's survey of industrial research and development, respondents were asked to record expenditure and receipts for research and development to and from foreign sources. From 1963 to 1982, payments for technological services by research and development performers exceeded receipts. This deficit peaked in 1979, and a payment surplus was recorded in 1982. Receipts have exceeded payments through the 1980s, growing from $7 million in 1963 to $789 million in 1989. Payments grew from $29 million in 1963 to $426 million in 1989.

The surplus in payments received for research and development has been offset by a deficit in payments made for other technology acquired through patents, licences, and technical 'know-how'. The deficit in other technology rose from $19 million in 1963 to a peak of $464 million in 1985, and was $427 million in 1989. Canada had a positive balance for foreign payments made or received for technological services, including both research and development and other technology, for the first time in 1988.

The balance for technological services was negative for both manufacturing and services in 1989. However, the telecommunication equipment industry had a surplus of $224 million in 1989 and received the most payments for research and development and other technology, $333 million, of the manufacturing industries. The greatest deficits were recorded in the pharmaceutical and medicine, other chemical products and rubber products industries. Computer and related services and engineering and scientific services recorded positive balances of $21 million and $38 million respectively.

Tax, financial and other incentives to support industrial r&d and innovation

In this section, we will focus on federal initiatives and on the policies of the four largest provinces to support industrial r&d, moving quickly through some past initiatives to give as comprehensive a picture as possible of the current situation.

Tax measures

While much of the debate on the relative virtues of direct (subsidies) and indirect (through the tax system) support to r&d was going on in the world, Canada was experimenting with a mix of initiatives of both types. We will start by describing some fiscal mechanisms that were put in place to encourage Canadian firms to modify their traditional behaviour of under-investing in research and development. Government initiatives at the federal and provincial levels have combined to effectively reduce the after-tax cost of R and D performed in Canada below the U.S. cost, according to two tax specialists (Meneely and Glover, 1992).

1—The federal measures

Since the early 1960s, a firm's r&d expenditure has been fully deductible from it's taxable income. Changes were made over the following two decades to the time frame in which a firm may deduct the expenses and to the maximum deduction it could claim.

In 1977, an additional r&d investment tax credit was introduced. The credit was five per cent of the expenditure, except in economically disadvantaged regions where it was raised to ten per cent. The following year, the credit was doubled and a supplementary credit of twenty-five per cent was added for small business.

By 1983, Canada already had one of the most generous r&d tax incentives in the world, trailing only Singapore. Despite that fact, in it's April 1983 Budget, the federal government announced it's intention to improve the fiscal treatment of r&d. In October, a Bill was introduced to create the Scientific Research Tax Credit (SRTC). The SRTC was to become one of the biggest fiascos in the history of s&t policy, and that episode deserves a short description.

By 1981, accountants and tax specialists realised that the r&d tax credit was only useful if a firm could claim it against a certain amount of tax to be paid, which, in turn, only happened if the firm had made a profit. Many firms performing r&d were in no position to take advantage of the credit, and tax specialists started to engineer financial deals where firms were selling their credit to another firm in need of a way to reduce their tax bill.

Late in 1982, Revenue Canada restricted such practices but the 1983 SRTC in effect officially authorised this type of arrangement. The SRTC was designed to be a flexible instrument to encourage investment in r&d, but turned out to be simply ill-conceived and inadequate. The Credit made it very easy for large firms to buy a r&d tax credit from a small firm before research was actually conducted. Many small firms which sold their credits were unable to perform the amount of r&d they had anticipated. However, for several reasons, Revenue Canada was unable to recover most of the money these small firms then owed the government for not performing the research for which a credit had been allowed.

The program lasted only ten months. In Linda McQuaig's analysis of the SRTC (1987), she estimates that $2.8 billion was drained from the federal coffers. It is impossible to know how much of this money went for legitimate r&d projects, but it is almost certain that one to two billion dollars were lost by the federal government in the sense that the money was not used for the purpose the Credit was intended for.

The current situation
Warda, in 1990, updated the 1983 McFetridge and Warda study of r&d incentives and came to the same conclusion:

> 'the Canadian corporate tax system provides greater overall incentive for companies to engage in r&d than does the tax system of any of the other major industrial countries discussed.'

Currently, any firm may deduct from it's taxable income all of it's r&d expenditure. This expenditure may be deducted in the year it occurred or in any subsequent year. Current expenditure on r&d (such as salaries, benefits, materials,...) is wholly deductable, while the capital expenditure deduction includes only the goods acquired for use in r&d (such as machinery and equipment). Buildings used for r&d are only deductible through the cost of depreciation.

In addition, r&d performers are entitled to an investment tax credit (ITC) on the qualifying expenditure incurred in Canada. The credit is now twenty per cent of the eligible expenditure, except in the Atlantic provinces where it is thirty per cent. A firm may claim it's credit up to seventy-five per cent of the tax payable. However, there is a three year carry-back provision and a seven year carry-forward provision. This credit is taxable: any credit claimed reduces the total amount of r&d eligible for the deduction. According to the estimate of two tax specialists, in effect the Canadian federal income tax system will pay about fifty-five per cent of each r&d dollar spent in Canada. The federal government itself estimates that through the tax credits, it provides $750 million to r&d performing firms annually. Furthermore, analysts claim that changes announced in the 1992 Budget will ease the rules so that an additional $230 million will be claimed for r&d expenditures over the 1993–1997 period. In addition the federal government's 1992 Economic Statement announced improvements to the scientific research and experimental development tax incentives. Among them were: a new and simpler method of allocating overheads to these activities; partial investment tax credits for equipment that is used primarily (more than 50 per cent) but not exclusively in scientific and experimental activities; and clarification of activities qualifying for tax incentives.

2-The provincial measures

In addition to the federal r&d tax incentives, some provincial governments have designed their own fiscal instruments to encourage local firms to perform research and entice foreign research-intensive companies to locate in their jurisdiction. Quebec was the first province to use tax incentives to stimulate r&d. Ontario followed with it's own tax incentive in 1988. The only other provincial government having such a measure is Nova Scotia, which enacted a ten per cent r&d tax credit to reduce a firm's payable provincial taxes. Quebec and Ontario deserve a closer look because they constitute the industrial heart of Canada.

Quebec Incentives
Since 1978, Quebec has had r&d tax incentives that have complemented those offered by the Federal government. In the 1987 Budget, and in most of the subsequent annual budgets, Quebec has significantly improved it's r&d fiscal measures to become, arguably, the jurisdiction in North America where the after tax cost of r&d is the lowest.[5] As of November 1991, a r&d tax specialist described Quebec as a 'fiscal paradise' as far as r&d funding was concerned.

The current situation involves a variety of creative tax mechanisms that are available to reduce the cost of research and development: a one hundred per cent deduction on current and capital expenditures; a twenty per cent fully refundable credit for wages related to r&d. This credit is increased to forty per cent if the firm is a small or medium sized enterprise (SME); the credit is replaced by a forty per cent credit on all r&d expenditures if the research is judged "precompetitive" (a visa recognising the precompetitive character of the research has to be requested from the government before the r&d is conducted). The credit is also forty per cent if the research is performed in collaboration with a university or a recognised public research organisation. Special treatment is also given to large scale R and D projects accepted through the Technological Development Fund (see below); the non taxation of the federal tax credit; a two-year personal income tax holiday for foreign researchers working in Quebec.

In addition, to help firms raise external funding for r&d, a company may decide to relinquish it's right to part or all the r&d tax credit and pass it along to individual investors. Firms getting external financing from either a Quebec Business Investment Company (QBIC – a private corporation which holds capital stock in SMEs) or from the Quebec Stock Savings Plan (QSSP – a vehicle through which individuals can invest in a publicly traded company and obtain a tax reduction) may use part or all the funds for R and D. If the firm decides to surrender it's tax credit, the investors benefit from an additional deduction to the basic deduction they are entitled to for their investment in a QBIC or QSSP.

Another vehicle is a r&d venture capital company (VCC) whose sole purpose must be to finance r&d in firms whose assets are less than $1 billion. Again, if the firm surrenders it's tax credit, an investor gets an additional fifty or one hundred per cent deduction on top of the basic one hundred per cent deduction.[6]

In 1989, the provincial government was forecasting that the combination of all the r&d tax incentives would be worth $760 million over five years.

These tax incentives have led to the launch of some innovative strategies in Quebec to attract investors to r&d projects. For example, the Caisse de Dépôt et de Placement du Québec (the manager of the province's pension fund), Hydro-Québec and the Fonds de Solidarité des Travailleurs du Québec (a $315 million investment fund launched by the Quebec Federation of Labour) teamed up with other partners such as Bombardier, Bell Canada and the Quebec Order of Engineers to create Capitecq, injecting initial capital of $21 million in r&d projects. Capitecq is now looking for individuals to invest an additional $5 million in the fund.

Another initiative set up as a direct result of the tax measures is a series of third-party R and D financing schemes created by Montreal-based Corporation de financement Helycorp. Using the forty per cent refundable tax credit for university research, Helycorp gathered $24 million in individual investment from doctors, dentists and optometrists to finance five research projects in various universities in Quebec. Helycorp was also the first to create a r&d venture capital corporation, trying to raise up to $25.6 million to finance six university-based research projects.

More recently, the brokerage firm Lévesque Beaubien Geoffrion has started to sell shares to raise $400 million for a dozen r&d projects, mostly in medical and pharmaceutical research.

It is still early to evaluate the impact of the tax measures introduced in Quebec after 1987 to encourage the investment of private venture capital in r&d projects. As of November 1991, it was estimated that $100 million had been raised through QBICs and VCCs for R and D projects. At that time, following rumours that the measures were used for purposes other than the ones they were intended for, a problem that was not unlike the one that affected the federal SRTC earlier, Quebec introduced rule changes that were designed to halt the development of new projects. Grandfather clauses were introduced to allow projects to proceed that were at least in a preliminary stage. If all the projects that were tabled at that time were to go ahead, a total of $591 million could be invested in 39 r&d projects in the following years. The viability of many of these projects is in doubt. Despite the abrupt end of these measures, it is clear that Quebec had put together a set of very innovative mechanisms that deserve a serious assessment to see if the model is worth implementing in other jurisdictions.

Clearly, in recent years, the provincial government in Quebec has privileged the tax system as it's favoured instrument to influence the r&d investment behaviour of industry. Several studies have questioned the validity of these mechanisms, for example Switzer (1989) who writes 'it is fairly clear that direct government spending, as opposed to tax credits, is a more cost efficient way of stimulating industrial r&d'. At the Quebec Summit on Technology held in 1988, Premier Bourassa explained the four reasons why his government thought that an approach based on tax-related instruments was more appropriate to stimulate r&d: Tax credits in Quebec are refundable, so they have the same advantages as subsidies without the administrative burden associated with grants; companies prefer the simplicity and confidentiality of tax measures; tax measures minimise the need for bureaucratic control; tax credits have no associated administrative costs, which means that public money is allocated more efficiently.

Although one can argue with the reasoning, especially with the redundancy of these four arguments, they illustrate that, at least at the time this speech was made, Quebec was firmly committed to using the tax system as an easier and more effective way to encourage investment in r&d.

Ontario Incentives
In the 1988 Budget, the Government of Ontario introduced what was termed an 'R and D Superallowance'. It consists of a deduction of twenty-five per cent of eligible R and D expenditure incurred in the year, the deduction being thirty-five per cent for Canadian-controlled private corporations. On top of this deduction, incremental r&d expenditure (calculated as the year's expenditure minus the average of the three previous years) are eligible for a thirty-seven point five per cent deduction, or fifty-two point five per cent for Canadian controlled private corporations. In contrast to Quebec, the federal R and D investment tax credit is treated as taxable income.

It is estimated that the value of the r&d superallowance is about $50 million annually.

Direct support for industrial r&d

We include in the direct assistance category support given to industry through subsidies, grants, loans and loan guarantees as well as contracts. As mentioned above, several studies have argued that this type of support is, in fact, more efficient than indirect support through tax incentives.

Federal initiatives
Many programs are available at the federal level that provide more direct support for industrial r&d. In 1990–91, federal expenditure for industrial r&d was estimated at $666 million in total. What follows is an examination of the most important programs that were in place in the early 1990s.[8]

The Industrial Research Assistance Program (IRAP)
IRAP was created in 1962 and is managed by the National Research Council (NRC), one of the principal public organisations performing scientific research in Canada. (See Chapters Three and Four). IRAP is among the oldest policy instruments currently in place and has been rated highly by many sources.

The basic idea of IRAP is to make technological expertise available to Canadian firms so they can solve their technical problems rapidly and in a cost-effective manner. IRAP has a national network of Industrial Technology Advisors (ITAs) who combine technical expertise and industrial experience. At it's peak in 1989, the network included 300 ITAs.

The ITAs come from various backgrounds: about one-third come from the NRC itself, one-quarter come from one of the provincial research organisations, the balance being contracted from other research institutes, universities and colleges.

Once a firm's problems have been examined, an ITA tries to locate a technology source that can help solve them. This source could include the National Research Council's laboratories, other public laboratories, universities, and private companies. To locate the best source, ITAs rely on their own expertise and that of their colleagues throughout the network.

If a financially viable solution is found, IRAP then provides grants to help a firm acquire the needed technology or to hire staff to implement the technical solution. Five different cost-sharing instruments were available before a recent restructuring of the program, depending on the type and magnitude of the financial support required.

Although one of the most important and successful public policy instruments for science and technology in Canada, IRAP's budget has declined in real terms since 1985. In constant 1985 dollars, the programme's total expenditure has declined twenty-two per cent from $82 million in 1984–85 to $64 million in 1991–92.

Late in 1991, the program went through a major restructuring accompanied by a significant crisis over it's management and it's orientation. Following the NRC's governing Councils adoption of a new 'strategic plan for the 1990s' in September 1991, the programme now offers only two funding instruments, is more decentralised and provides smaller grants. An industry-led IRAP advisory board was dismissed by NRC over disagreements on the orientation of the programme.

One of the funding mechanisms, the *Technology Acquisition and Information Exchange* is for small-scale projects. Financial support is limited to $15,000 (this limit was to be raised to $30,000 eventually). The other instrument, *Research, Development and Adaptation* funds r&d projects to a maximum contribution of $350,000.

While restructuring the programme, NRC has been lobbying the federal government for a restoration of IRAP's budget to the 1985 level in real terms. Although minimal additional public resources have been given to this very popular program,[9] IRAP has been exempted from a new federal policy that replaces most grants over $100,000 with loans. In December 1992, the government announced its intention to strengthen IRAP as well as expand its scope by having IRAP responsible for the delivery of some technology diffusion currently provided by ISTC.

The Defense Industry Productivity Programme (DIPP)

DIPP was created in 1959 by the federal government to develop a defence-related industry. The programme was substantially modified in 1968 and is run by the Department of Industry, Science and Technology (ISTC). It supports the pre-production development stage of industrial r&d in the defence industry, including aerospace and defence electronics. DIPP's budget experienced significant growth between 1984 and 1988 but it's allocation has decreased since then. For the fiscal year 1992, it's budget dropped $30 million from 1990 to $220 million.

DIPP's fate has been closely associated with Canada's participation in international military projects. It was designed to help Canadian firms compete in the lucrative international market with competitors who also receive significant support from their respective national governments. The end of the Cold War has coincided with a levelling-off of the Canadian defence budget and a reduction in DIPP's budget.

DIPP has apparently been a successful programme. According to a programme official, external audits have been positive and eighty per cent of the products developed as a consequence of DIPP funding have reached export markets. The programme provides conditionally repayable loans for larger firms, and grants for smaller ones. It offers four types of support, including support for r&d, which accounted for one half to two-thirds of the total budget in the late 1980s.

DIPP's allocations have traditionally been awarded disproportionately, in terms of population or share of GDP, to firms in Quebec and Ontario. In 1988 for example, the two provinces accounted for ninety per cent of the total awards.[10]

The Strategic Technologies Programme

In early 1988, the federal government announced its intent to provide funding, under the Strategic Technologies Programme, for the development of 'enabling technologies' linked to three broad generic technological streams: information technology, biotechnology and advanced industrial materials. The programme's goal is to encourage firms to establish vertical and horizontal alliances. It is divided into two major sub-programmes: industry-led pre-competitive r&d and technology application alliances. The first is aimed at long-term collaborative r&d projects between firms and other public research institutions. The second applies further downstream in the innovation process. It funds technology feasibility trials, demonstration projects, international market development, standards development and research to meet regulatory requirements.

The rules of the programme are flexible, allowing the programme delivery officers to develop funding arrangements that are project-specific. The programme provides non-repayable grants that must be matched by other funds from non-government sources. Eligible costs include feasibility studies (maximum $50,000), salaries for professionals, a share of management and overhead costs, testing and prototype evaluation, travel and communication expenses, subcontract fees, tuition fees, and special training fees. Materials and apparatus costs of up to twenty-five per cent of the total project budget are eligible for inclusion.

Over four years, the programme was to provide $30 million to each of the three technologies identified as strategic. A typical project was to be scheduled on a three to five-year period, include four to eight partners, and be worth between $2 million and $10 million. As of late 1991, one research project had been funded in information technology, sixteen in biotechnology and ten in advanced industrial materials.

The Artificial Intelligence r&d Procurement Fund
The government created a $10 million fund in 1989 to finance r&d in computer-based expert systems. It is a procurement-driven instrument, for projects supported by other federal ministries.

The Sector Campaigns
The Sector Campaigns allow the government to spend some money on r&d and other activities (such as marketing, export promotion and diversification initiatives) in targeted industrial sectors. Campaigns have been announced to support advanced manufacturing and the fisheries, forestry, software, semiconductors/ microelectronics, automotive, environmental and medical devices industries. In 1990–91, Industry, Science and Technology Canada was to spend $7.8 million on these various projects. The forestry strategy alone included $18 million over three years for r&d and innovation. Funding is available to match private investment in a project.

The Microelectronics and Systems Development Program
Announced in 1987 but effectively in operation since 1989, the Microelectronics and Systems Development Program (MSDP) supports the development of advanced microelectronics and information technology systems that can be used in Canadian manufacturing or service industries. It has a $60 million budget over four years.

There are two sections to the programme: one for r&d in microelectronics components and one for r&d in systems based on advanced microelectronics and information technologies. The program covers fifty per cent of the eligible costs, to a maximum of $5 million. The funding is repayable when it exceeds $500,000.

Eligible costs include materials, prototypes, travel, patent searches, subcontracting and salaries for scientists, engineers and technicians.

The Technology Outreach Programme
This programme has a budget of $17 million and provides finanacial assistance to technology centres to promote the acquisition, development and diffusion of technology and critical management skills, particularly in small Canadian firms. An additional component of TOP ($4 million) is dedicated to the Advanced Industrial Materials sector. The programme supports over twenty-three centres in such fields as microelectronics, biotechnology and industrial innovation. (See Appendix for descriptions of some of these centres).

The Regional Development Agencies
Regional development functions are now assumed by decentralised agencies in Canada. The Atlantic provinces are served by the Atlantic Canada Opportunities Agency, western provinces by the Department of Western Economic Diversification. These agencies provide some support to industrial r&d. In 1989, the Atlantic Canada Opportunities Agency said that innovation and technology transfer were to be among it's top priorities within it's new $274 million "Cooperation" programme. The two agencies provided $24 million to industry for r&d in 1990–91.

The Green Plan
In 1991, Canada announced a $3,000 million six-year Green Plan for investment in environmental initiatives. The Plan included a Science and Technology Action Plan worth a total of $250 million. From this envelope, $100 million was to be set aside for industrial technology development and commercialisation and $20 million was targetted for unsolicited r&d proposals from the private sector. Combined, this represents an additional $120 million in support for industrial s&t.

Other Departments' Contract r&d Budgets
In addition to all of the programmes specifically targetted at r&d, several Departments have their own programmes to contract out r&d or to provide r&d grants. In 1990–91, Energy, Mines and Resources Canada, Atomic Energy of Canada Limited, Transport Canada and the National Research Council each awarded about $20 million in r&d contracts. The Department of Communications also provides significant funding for industrial r&d. (See Chapter Four for more details).

The provincial initiatives

Industrial r&d Programs in Quebec
Quebec was the first provincial government with a significant involvement in science and technology policy in the early 1980s. In the late 1980s, following major injections of money into s&t by the neighbouring Ontario provincial government (the Ontario Technology Fund), Quebec responded by developing a set of programs to complement the federal and provincial fiscal incentives for industrial r&d outlined above.

Quebec's Technological Development Fund
Announced in the 1989 provincial budget, the Technological Development Fund is an umbrella program that covers several distinct initiatives to stimulate tech-

nological development in Quebec, especially in the private sector and through alliances of industrial, academic and/or governmental partners.

It's initial five-year, $300 million budget was increased, in subsequent budgets, to $350 million in 1990 and then to $370 million in 1991. As it stands now in 1992, the fund includes five components that are all relevant for industrial r&d:

- the Major Opportunities to Stimulate Technology (MOST), which will be described below
- the Government Opportunities, which serves as discretionary funding for government priorities in s&t such as the national Radarsat project (see Chapter Nine)
- an Environmental Research and Technological Development Fund, which will pump $50 million into exploratory research and technology development related to waste management, pollution control and restoration, and sustainable development
- a $32 million 'Synergy' component that encourages joint university-industry collaborative r&d projects
- a $20 million fund for r&d projects presented by small and medium sized enterprises

The Major Opportunities to Stimulate Technology element is aimed at funding a limited number of relatively large projects of commercially-oriented r&d. For example, the first two projects to receive funding were worth $38 million and $20 million respectively. The first was a software development project that included sixteen partners, the second involved five partners in a joint project to develop advanced electronic sub-systems for subway cars. Subsequently, other projects were announced in the fields of telerobotics, information technology for environmental management, forest engineering applications for remote sensing and geographic information systems, molecular endocrinology, and powder metallurgy. A project on flight simulators for water bombers is being studied.

When a project is approved, it automatically qualifies for the forty per cent r&d tax credit on all r&d expenditures (see above). In addition, the government finances fifty per cent of the costs associated with pre- and post-r&d activities. In the first example, given above, Quebec's direct contribution was to be $10.4 million while $5.7 million was to be injected in the second project.

The Société de développement industriel: r&d risk sharing loans
The Société de développement industriel (SDI) is the financial arm of the provincial Ministry of Industry, Commerce and Technology. It is the agency in charge of the core support for industry. Created in 1971, it has supported more than 11,000 projects submitted by Quebec firms, representing a total investment of $4,000 million.

R&D is a priority for SDI, within it's general mandate of fostering economic development in the province. Industrial r&d activities are assisted through different programs such as PARIQ (Programme d'aide aux activités de recherche et d'innovation) which may award a firm a risk-sharing loan of up to seventy-five per cent of eligible expenditures, to a maximum of $1.5 million for a particular project.

For new start-up firms, PARIED (Programme d'aide à l'innovation technologique pour les entreprises en démarrage) provides loans up to ninety per cent of eligible R and D costs.

In 1989, the provincial government estimated that SDI's support for r&d would account for $120 million over the following five years. This is not a net cost, however, because successful projects produce reimbursement of the loan plus a payment in the form of royalties.

The Industrial Development Fund

This recently announced $300 million programme will provide funding to firms investing in r&d facilities and equipment. The fund will offer repayable loans to large industrial r&d projects involving a minimum capital investment of $5 million, to a maximum of twenty-five per cent of the eligible costs. Priority will be given to firms in the following targetted industries: transportation equipment, plastics processing, aeronautics and aerospace, high-energy consuming materials, pharmaceutical, biotechnology, and electric and electronic products.

Innovatech Grand Montréal

To support the region of Montreal, which was experiencing a severe economic decline, the provincial government in 1992 created a regional economic development fund called Innovatech Grand Montréal. Innovatech will have a total budget of $300 million over five years to support economic development projects in the Greater Montreal region. Within the broader programme, a share of the fund will be reserved for r&d projects. This regional technology fund will function according to the same basic principles that guide the province-wide Technological Development Fund.

Hydro-Québec

Hydro-Québec produces and transmits electricity through the province. It is not a government department, but rather an agency with an arm's length relationship with its only shareholder, the provincial government. Hydro-Québec has a significant budget for technological development and has an important impact on industrial r&d in electricity generation, transmission and distribution. In 1990, Hydro-Québec published a Development Plan that committed the corporation to spend $467 million in various technological activities between 1990 and 1992, the biggest share of which was to be invested in r&d. Hydro-Québec has it's own research institute, the Institut de Recherche en Électricité du Québec (IREQ) and it spends about two point five per cent of it's annual revenues on r&d.

In the 1989–1990 Budget, the Quebec government presented a synthesis of all the support it gives to industrial r&d. A sizeable place in this presentation was occupied by Hydro-Québec's procurement policy. It was argued that sixty per cent of Hydro-Québec's r&d expenditure over the 1989–1993 period ($450 million) would be channelled to the private sector through a r&d strategic procurement policy.

Ontario's Initiatives

Ontario's funding for industrial r&d has been concentrated since 1987 under the Ontario Technology Fund umbrella. Announced as a $1,000 million fund over ten years, the share allocated to direct support was decreased when the Government decided to deduct the value of the tax deductions claimed as r&d Superallowances (see above) from the total budget. The estimated cost of the tax measures to the provincial Treasury is $400 million in the ten year period, lowering to $600 million the amount available for direct support of R and D.

For purely industrial r&d, the most important component, outside of the tax credit, is the *Industry Research Program* (IRP). This program encourages collaborative r&d projects in the private sector. Over the 1987–1997 period, it is estimated that $125 million will be awarded to industrial projects. The funding formula provides grants up to a maximum of fifty per cent of the eligible costs.

Another component of the Fund that may be relevant to industrial r&d is the *Technical Personnel Program* that subsidises the salary of engineers, scientists or technicians hired by small firms in Ontario.

The Technology Fund has also provided a framework for the signature of agreements to enhance collaborative r&d efforts between companies in Ontario and firms in the Rhône-Alpes region of France, Baden-Wurttemberg in Germany, Catalonia in Spain and Lombardy in Italy.

The Technology Fund has been used to support specialised research centres and centres of excellence to encourage collaborative research between universities and industries (see below). Ontario has also allocated the money to pay for it's share of the national Radarsat project (see Chapter Nine).

Environmental Technologies Program
Ontario has set aside $30 million for environmental r&d matching grants in several areas of specialisation such as technologies which eliminate or reduce emissions within industrial processes, waste reduction, recycling and composting, waste management, and wastewater treatment.

Innovation Ontario
Innovation Ontario is a crown corporation that was created in 1986 for a five-year term. It's mandate was renewed in 1991 for a subsequent five-year term until 1996. It's mandate is to assist the growth of small technology-oriented firms that are at the start-up stage. Innovation Ontario provides venture capital and recovers its investment through divestment of its equity in a company.

By 1992, Innovation Ontario had invested more than $30 million in almost 200 firms. It's budget was increased from $11.8 million to $20.8 million in the 1991–92 Ontario Budget. A further ten per cent increase was scheduled for the 1992–93 fiscal year.

Support for r&d in British Columbia
The provincial government in British Columbia has been very active in promoting s&t in the late 1980s as a way to enhance future economic development. In 1987, the Premier created an Advisory Council on Science and Technology and in 1988, a provincial Science and Technology Policy was approved by the Cabinet. In 1990, a Five-Year Plan for s&t was adopted and the process culminated in the 1990 Provincial Budget, with the announcement of a $420 million, five-year Science and Technology Fund, later made official by a special legislative Act.

The Science and Technology Fund incorporates a broad set of sub-programmes that support infrastructure, r&d, human resource development, public awareness and special projects. The programmes are delivered by several public agencies and departments, but the Fund is ultimately the responsibility of the ministry of Advanced Education, Training and Technology. Each of the five components was given a share of the total budget: r&d accounts for twenty-seven to thirty-five per cent, infrastructure for twenty-five to thirty-two per cent, special projects for twenty-five to thirty per cent, human resource development for seven to ten per cent, and public awareness projects for two to five per cent. The total support for r&d over the five-year period should be between $113 million and $147 million.

The r&d envelope encourages innovation and technology application. Support is given to projects from the research stage to pre-production marketing assistance. Eligible research activities include market research, prototypes, commercial design and marketing.

There are two levels and types of assistance: industry-based r&d is funded at fifty per cent of the cost by the government through repayable loans; strategic research projects undertaken by partnerships of higher education institutions, firms and not-for-profit foundations may be funded totally. The *Product Development Fund*, for example, supports the final stages of technical work needed to bring products to a commercial stage. It funds up to seventy-five per cent of the costs, to a maximum of $100,000, repayable as royalties in sales revenue. The *Technology Assistance Programme* assists small and medium-sized firms in implementing productivity-increasing technological solutions. It funds up to fifty per cent of the costs to a maximum of $20,000. *Technology B.C.*, provides grants to fund r&d projects. In late 1991, this sub-programme announced $5.7 million in funding for 88 projects. In it's first three years of operation, it provided $23 million in support of 380 projects.

Provincial support for industrial r&d in Alberta
Alberta has limited programme activity providing direct support to industrial R and D, though it has a significant number of technology centres in support of industrial competitiveness. The *Technology Commercialization Program* supports companies in bringing innovations into production. It is managed by the Department of Technology, Research and Telecommunications and it has a $2.5 million annual budget. Other major industrial technology centres include the Laser Institute, the Alberta Environmental Centre, the Alberta Telecommunications Research Centre, the Centre for Frontier Engineering Research and the Alberta Microelectronics Centre. In addition, the province is the home of the oldest and largest provincial research organisation in Canada – the Alberta Research Council. ARC, like its sister organisations in other provinces, provides a wide range of services and technical assistance to the province's small and medium-sized businesses.(For more information on s&t activity in other provinces, see Chapter Three).

Other support for the industrial r&d infrastructure

Apart from the usual programmes that provide financial support for firms to perform r&d, the federal and provincial governments have developed several innovative ways to fund institutes, associations, foundations, agencies and research centres that provide technical support for firms or perform research on behalf of industrial partners.

The Provincial Research Organisations

Eight out of ten provinces have created a provincial research organisation. Despite the variation in the models used by each province, the basic idea is the same in each of them: to provide technical expertise and r&d to firms that are too small to have their own significant r&d capabilities, at a rate that is cheaper than the market rate. These centres are not specialised: they each have a broad mandate that permits them to work on issues where the needs are the most pressing.

The research centre with the biggest budget, $56 million in 1990, is the Alberta Research Council. Ranking second is Quebec's CRIQ (Centre de recherche industrielle du Québec) at $35 million, followed by Ontario's ORTECH International at $32 million.

Government grants and subsidies account for a different share of each centre's revenues. In effect, this is the rate by which the services rendered to the client are discounted. The British Columbia Research Corporation receives no grants at all from the provincial government, while other provincial research organisations receive between nine and fifty-eight per cent of their funding as grants from their respective provincial governments. The most heavily subsidised organisation is CRIQ, which gets fifty-eight per cent of it's funding in grants from the Quebec government, followed by the Manitoba Research Council, which receives fifty-five per cent of it's budget from the province. The Alberta Research Council gets almost fifty per cent from the provincial Treasury, the Saskatchewan Research Council forty-one per cent and the Nova Scotia Research Foundation Corporation gets forty per cent. At the other end of the spectrum, Ontario's ORTECH receives only fifteen per cent in base provincial funding and the New Brunswick Research and Productivity Council receives only nine per cent. (See Chapter Four for more on the PROs.)

The Specialized Research Centres
Many provinces have also invested in more specialised research centres, networks or institutes to serve the needs of a particular industry or to develop a particular technology. Most of these organisations have at least partial funding from industrial partners that hope to benefit from the results of the research. Almost all have participation from university-affiliated researchers. Some of these centres have base funding from both the provincial and the federal governments. Many have fixed-term funding (typically for three to five years) with the implicit or explicit assumption that if successful, the venture should be increasingly self-financed and eventually function as a completely autonomous entity.

In Quebec alone, at least eighteen such research centres exist. They perform r&d on the following subjects:

- automation applied to mine exploitation
- use of computers in organizations
- materials characterization
- biomass
- food processing
- computer science
- industrial materials
- ocean studies
- optics
- mines
- electricity
- biotechnology
- biology
- pharmaceuticals
- magnesium technology
- electrochemical technology
- micro-electronics
- numerical computation

As well, formal entities are emerging out of the projects funded by the Technological Development Fund. An example is the Volvox Research Centre, a non-profit institution that coordinates several r&d projects undertaken by partners in the consortium financed by the Fund emerged out of an r&d project on applications of information technology to environmental management.

Seven Centres of Excellence were created in Ontario following a competition in 1987. Their first five-year mandate was extended for another five-year term.

They all involve researchers in universities and industrial partners. Their funding is significant: over the ten-year period, they are slated to receive a total of $450 million from the province's Technology Fund. When renewed for the second mandate, the centres were given the task to orient their activities more squarely towards industrial applications and the 'creation of wealth'. The seven centres work on:

- Groundwater technology
- Information technology
- Materials research
- Advanced manufacturing technology
- Space and terrestrial science
- Laser and lightwave
- Telecommunications

In addition, Ontario's provincial government helped to create the Industrial Research and Development Institute, an applied r&d centre for the tool, die and molding industry.

Ontario also provided support, through it's Ontario Development Corporation, to industrial r&d infrastructure by awarding $6.2 million to the Ottawa Life Sciences Technology Park. The Park is intended to become an incubator for life sciences companies, providing them with access to common general services and to local universities.

Among other initiatives, British Columbia has supplied $387,000 in seed funding to support an industry-led Software Productivity Centre. The provincial government is involved in setting up the Hydrogen Fuels Technology Agency and the Wireless Communications Research Centre (which eventually became a national foundation with federal support). It also provides support to the SPIRIT Subsea Systems Corp, a consortium of eight firms that apply their expertise in robotics, submersibles, advanced systems and machine vision to joint R and D on subsea technology applied to underwater vehicles.

In Alberta, funding has been given by the provincial government to the Alberta Oil Sands Technology & Research Authority (AOSTRA). Over a sixteen year period following the creation of AOSTRA in 1975, the agency, with industrial partners, has been funded with more than $1,000 million in r&d. Alberta also helped the creation of Westaim, an industry-led venture in advanced industrial materials.

The federal government, especially through it's Sector Campaigns and the TOP programme (see above) has also contributed to support the creation of similar institutions. The Microelectronics Sector Campaign, for example, included the creation of the Strategic Microelectronics Consortium, a cooperative r&d venture with eight industrial founding partners. The fisheries sector campaign created an opportunity for the birth of a Canadian Fisheries Products Technology Consortium, which is expected to initiate $10 million worth of collaborative r&d between industrial partners. The Strategic Technologies Program helped the creation of a National Optics Institute, a r&d institute involving nine industrial members. TOP supported the creation of technology centres in welding, management of technology, advanced materials engineering, plastics, applied microelectronics, medical devices, mining, textile, biotechnology, and gas technology, among others.

In addition, the federal government helped create an Institute for Research on Trucking Issues, to address the needs of the trucking industry and a Regional Nuclear Magnetic Resonance facility in Manitoba, shared by researchers from universities and the industry.

The federal government also sponsored a competition to create fifteen national networks of centres of excellence (NCE) which are mostly university-oriented but involve some industrial participation. One of the main objectives of the NCE is the commercial exploitation of technologies to promote Canada's industrial competitiveness. Five provinces (British Columbia, Quebec and to a lesser extent, Alberta, Saskatchewan and Manitoba) contribute to the overhead costs of the networks. The centres are described more fully in Chapter Six). (More descripions of these and other centres are found in the Appendix).

Recent initiatives to encourage industrial r&d

Data Processing and Transmission

Information processing and transmission capacities are becoming more important items in a nation's industrial research infrastructure. There are six supercomputers in Canada, but the most widely accessible facility is the Ontario Centre for Large Scale Computation.

The Centre is funded by the University of Toronto and the Ontario government. Although used primarily by the university research community, the Centre makes the complex modelling techniques of a supercomputer available to industry. The availability of a supercomputer can help industry reduce time and costs in product development and testing. However, the equipment at the Centre is generations behind available systems capable of processing information much more rapidly and the Centre is seeking to upgrade its present facilities.

Proposals are also being made to upgrade Canada's information transportation networks. The role of high speed digital communications networks in stimulating collaborative research and development has been recognised by other advanced economies. There are networks in continental Europe, Great Britain, Japan, Australia and New Zealand.

Canada has a series of regionally based research networks that are linked with each other and with the NSFNET in the United States by CA*net. The regional networks operate in British Columbia, Alberta, Manitoba, Ontario, Quebec, New Brunswick, Prince Edward Island and Nova Scotia. The capacity of the current Canadian network, 56 Kbit/s, is relatively slow and there have been proposals for a joint industry-government financing to upgrade to a 1.5 megabit per second system. (See Chapter Ten).

Better Protection of Intellectual Property

The regulation of intellectual property can also be used to encourage industrial research and development expenditures. Changes to Canadian law have been made to encourage increased spending by the pharmaceutical and software industries.

Under Canadian law, patents are normally protected for a period of seventeen years. In the early 1980s, the government reviewed legislation passed in 1969 to stimulate price competition between drug manufacturers and decided that it was acting as a disincentive to industry research. The 1969 legislation, amended section 41(4) of the Patent Act so that all patentees of drugs were required to provide licences to anyone wishing to import, market or manufacture patented drugs. In return the patentee would receive a royalty fee of four per cent on the licensee's selling price.

Bill C-22, passed in 1987, amended the federal Patent Act again. It increased the protection afforded to brand-name drugs to seven years; protection extends to ten years if a drug's active ingredient is imported into Canada. In return for this increased protection, the Pharmaceutical Manufacturers Association of Canada agreed to increase their industry's research and development expenditures to eight per cent of sales by 1996. Research and development expenditures averaged four per cent before the patent changes.

Canada's *Copyright Act* remained substantially unrevised from it's proclamation in 1924 through the mid 1980s. Technological changes caused concerns that the spirit of protection offered by the Act no longer provided adequate protection for recent innovations such as computer software. Under the Act, the term of copyright was generally the lifetime of the author and for fifty years thereafter. Although there were precedents in case law to suggest that software would be treated in the same manner as literary works, there was concern to establish this protection in legislation. Under 1987 amendments to the *Copyright Act*, computer programs were defined as literary works and received comparable protection. Exceptions were made to allow the alteration of software for personal needs and the reproduction of software for the purpose of creating back-up copies. The federal government also increased the severity of penalties for the sale, distribution, or importation of an infringing copy of a work.

The use of regulatory and licensing power to influence the r&d behaviour of Canadian firms

In late 1991, the federal Department of Communications announced its intent to possibly set r&d spending targets for radiocommunication firms as a requirement to be met before an airwave license is granted. Although still only a project under discussion, it could have a very significant impact if other regulatory and licensing agencies were to adopt the same principle. This would constitute a breakthrough, since r&d investment has rarely been mentioned as a factor in judging applications for licensing. Although the communications industry is a technology-intensive sector in which such a rule could make sense, it is not unreasonable to think that other fields affected by regulation and compulsory licensing would see this principle enacted.

Proposals have been made to link the screening and authorisation of the acquisition of Canadian firms through foreign takeovers, to guarantees for r&d spending. The provincial government of Ontario in particular has been very active in trying to convince Investment Canada, the federal agency that authorises such take-overs, to pay special attention to the acquisition of Canadian high technology firms. It has argued that any foreign take-over of such a firm should be authorised only following very stringent commitments by the foreign firm to perform a certain level of r&d and to pursue technology development in a way that Canada can still reaps some benefits. It remains to be seen, however, if Investment Canada will incorporate such criteria in its review of take over proposals.

Industry r&d alliances

An increasingly common phenomenon in industrial research and development in Canada is the formation of consortia. This section presents a sampling rather than a comprehensive list of consortia, and indicates the range of partnerships that have been formed to undertake research and development in advanced technology and resource-based industries.

Consortia are a means of distributing the costs and risks of research and development among partners. In most cases, financial commitments made by companies have been matched with funding from governments at the federal and provincial levels. The rationale for government participation in pre-competitive research projects is that benefits are not captured solely by project participants. The traditional practice in Canada has been for government to match industry funding of collaborative research and development projects. An organisation that has argued strongly for the establishment of norms in the negotiation of joint industry-government funding agreements is PRECARN Associates Ltd. More than 30 organisations participate in this research consortium which investigates intelligent systems.

PRECARN's position has been that industry partners should have a one-third share of the costs of pre-competitive research, the balance borne by federal and provincial governments. In the absence of a recognised norm for sharing research costs, PRECARN's projects have been delayed pending the outcome of separate negotiations with governments in different jurisdictions. PRECARN members are able to use the results of the organisation's research without paying any royalty fees.

PRECARN has approved five projects with budgets that total about $39 million. The project's partners include corporations, universities, government institutions, and a provincial research corporation. In each project the federal government is committed to providing most of the funding, with the balance coming from companies and one provincial government per project. PRECARN has had difficulty in attracting the support of more than one sub-national government in a project. PRECARN's projects are APACS (applications or expert systems to advanced process analysis and control), IGI (an intelligent graphic interface for real-time monitoring and control), ARK (active vision of a mobile robot in a know environment), TDS (telerobotic development system) and ARTEMIS (advanced real-time expert system).

Another consortium operating in the high tech sector is the Earth Environment Space Initiative (EESI). It was formed in 1990 and is funded by four aerospace companies, four provincial governments and the federal government. The four industry participants – SED Systems Inc., Bristol Aerospace Ltd, MacDonald Dettwiler and Associates Ltd, and Intera Technologies Ltd – each contributed $70,000 to the project's first phase budget of $1.4 million. The project's goal is to develop new space-based sensors, instruments and data processing systems for both domestic and international earth observation satellites.

Four companies in the micro-electronics industry, Newbridge Microsystems, Gennum Corporation, Mitel Semiconductor and Mosaid Microelectronics Corporation, established the Canadian Semiconductor Design Association in 1985. The consortium's goal was to develop technologies for manufacturing, designing and testing integrated circuits. In one research project for silicon chip design initiated in 1988, the industrial members were joined by six Ontario universities. About half of the funds came from the Ontario government.

In the resource sector a broadly-based consortium of companies involved in the production, storage, transportation and use of hydrogen joined with universities in Canada and Europe, six provincial governments and the federal government to form the Hydrogen Industry Council. The Council's activities include pre-competitive research, technology development and the establishment of a research centre and a national university research network. The total budget of the project is reported to be $85 million.

Another example of a resource-based consortium is a five-year agreement established between the Saskatchewan Potash Producers Association, and the federal and Saskatchewan governments. The agreement covers research to secure supplies of the resource, develop better process methods and product quality, and improve the industry's methods for disposing of mine tailings.

Three other resource-based consortia have been formed in the forestry industry: FERIC, Forintek, and PAPRICAN. All three organisations draw industry and institutional members from across Canada and each has laboratories in Quebec and British Columbia. FERIC, the Forest Engineering Research Institute of Canada, was formed in 1975 to conduct research and development to improve the efficiency of the growing, harvesting and transportation of trees. Forintek, created from the privatisation of a federal laboratory in 1979, undertakes process and product r&d for the solid wood products industry. The Pulp and Paper Research Institute of Canada (PAPRICAN) was founded in 1925. PAPRICAN uses it's strong links with university-based laboratories in Canada to provide data and develop technology for its member companies.

Action by the federal government has also been instrumental in nurturing the development of a consortium for the fisheries industry. The Canadian Fishery Products Technology Consortium was established in 1989 by the federal government to lever industry spending on areas including vessel and on-board equipment design, quality assurance and control, product development and marketing, plant processing equipment development, management systems and aquaculture engineering.

A snapshot provided by survey data indicates that participation in collaborative research and development is the exception rather than the rule. A survey of advanced technology firms in Canada reported that fifty-eight per cent of respondents have never entered in a strategic alliance or partnership with an arm's length foreign or domestic enterprise. Of those firms that had entered into an alliance in the three year period preceding the survey, only thirty per cent had participated in a research and development consortia.

Less than half of firms with fewer than 100 employees (those which could benefit the most from pooling their resources) indicated that they anticipated making future alliances. Surprisingly, only twenty-five per cent of all the responding firms said that governments had a role to play in assisting the formation of strategic alliances - this included a role in subsidising costs. The survey also showed that few advanced technology firms make use of the government research infrastructure. Only thirty-four per cent of the firms reported that they had ever used a government laboratory. Three-quarters of the firms which had availed themselves of the public infrastructure reported using the facilities of the National Research Council. Use of provincial research organisations was highest in Alberta (Alberta Research Council) and Quebec (CRIQ).

More firms, forty-nine per cent, reported having interactions with universities in the year preceding the survey. This rate increased from forty-one per cent for firms with fewer than 100 employees to sixty-three per cent for firms with more than 1,000 employees. The type of interaction also varied with the size of firm. More than half, fifty-one per cent, of the large firms reported using university research and development equipment compared to only thirteen per cent of the small firms.

Respondent firms demonstrated a poor knowledge of government programmes to assist industry-university collaborative research. The Natural Sciences and Engineering Research Council's University-Industry Research Program was recognised by only forty per cent of the firms. Only ten per cent of firms reported having used the programme.

Two-thirds of the firms which recognised the existence of this programme did not know or had not used any other government programme which sponsored industry-university collaborative research. Positive responses came from fourteen per cent of firms for the National Research Council's (NRC) Industrial Research Assistance Program, nine per cent for NSERC, eight per cent for the NRC, and three per cent mentioned Ontario's University Research Incentive Fund URIF.

Conclusion

In the past ten years, the federal and some of the provincial governments in Canada have steadily increased their support for industrial r&d, paying unprecedented attention to it as an issue crucial for future economic competitiveness. New fiscal measures, direct subsidies for individual and collaborative r&d, and the creation of specialised industrial research institutions have strengthened Canada's industrial infrastructure. However, the relatively recent time frame of most of these initiatives makes it difficult to assess whether these enhancements will lever significant incremental increases in private sector r&d spending and if this will, in turn, lead to improved economic performance.

There is increasing support among governments in Canada for a new generation of policies that widen the focus of s&t policy from support for r&d to support for a broader range of activities that may increase value added in the production process. In recent months, the federal government and the provincial governments in Quebec and Ontario have issued policy documents adopting such a perspective. Typical of this new trend is Ontario's *A Framework for Industrial Policy* released in August 1992. R and D still has a significant place in the document but other 'value-adding activities' such as training, innovation (in a broader sense, not just technological innovation), workplace organisation, and networks and linkages among firms are given a higher prominence.

The common point in these documents is that Canadian governments at all levels are recognising changes in the international economic environment and are trying to react to them. In a world characterised by the liberalisation of trade, the deregulation of strategic industries, the emergence of major trading blocks in North America, Europe and Asia, and the new stiff competition offered by newly industrialised countries, public policy has a different yet important role to play in a small open economy such as Canada.

For example, more than ever, Canadian firms are encouraged to export and to take advantage of the new markets that are opening throughout the world. Government support for export-oriented activities is likely to increase in the future. In particular, any activity that will help the Canadian economy to reduce it's over-reliance on a single export market (the USA) will be viewed favourably by governments.

Likewise, the small size of Canadian firms relative to their international competitors is a weakness that public policy may be able to address. Even though it is practically impossible to force firms to merge or consolidate their activities, governments can provide greater incentives for firms to regroup and undertake at least some activities cooperatively. Efficient training programmes, for example, may be economically unfeasible for small individual companies but the government can put in place mechanisms that support sector-wide initiatives whose costs could be shared by the stakeholders, providing service to a whole group of firms.

Precompetitive collaborative r&d is also an example of linkages that will likely get increased support in the future. Such programmes spread the cost and the risk of r&d among several firms, with possible government funding, making it possible for small firms to invest in r&d projects that may otherwise be too risky or costly for an individual firm. In addition, international agreements with partners in other countries may give small Canadian firms access to technology that they would not be able to obtain on their own. Such arrangements will also be encouraged in the future with attention to technology diffusion.

Policies and programmes will increasingly be targetted at particular groups of firms (such as a whole sector or 'cluster'), with governments varying in the degree to which they actually 'pick' winning sectors. Already Quebec has announced that it will encourage fifteen priority clusters, while Ontario's position is that governments do not have the ability to pick winning sectors and that all sectors should be eligible for support if they present collaborative projects that make sense along the 'higher value-added' route.

Alliances will also be encouraged even more between private and public organisations. Increasingly, it is felt that small economies cannot afford duplication of effort and that the public and private sector should help each other by sharing expertise. While public r&d is not likely to dissapear completely, it's role will be more modified to make it more 'relevant' (to economic development, that is) and fundamental, non-oriented research will face increasing difficulties in securing public funding. New public research infrastructure will tend to be more and more 'industry-oriented', with the transfer of knowledge and technology to the private sector being a prime concern. (See Chapter Four).

Training will also become a highly visible priority for governments. The upgrading of professional skills will become more important as governments recognise that Canada's industry trails very significantly *vis a vis* it's competitors in terms of investment in training. While many advanced economies have been very active in designing training programmes for laid-off workers for example, Canada's policy for the unemployed has been 'training-neutral'. As a result, in the early 1990s, Canada experienced high rates of structural unemployment caused by a mismatch of abilities of the unemployed people and the skills required for the available jobs.

Another focus of governments will be to encourage the formation of a genuinely Canadian venture capital industry. Firms have complained in recent years about the difficulty of raising venture capital in Canada, few financial institutions are willing to risk the financing of R and D or technology acquisition projects. It is regularly stated that Canada's venture capital community lacks the sophistication to fully understand the long-term value of complex undertakings or highly technical functions. Moving towards a sophisticated venture capital industry will be an important objective.

Innovation will remain a focus of future policy but in a broader sense than before. Technological innovation was the focus of the 1980s, but innovation in general will be the new focus of the 1990s. For example, organisational innovations
· which concern new ways of organising the production in a workplace, will be valued. Trying to take advantage of best practices elaborated elsewhere in the world, especially in Japan, innovations that tend to give more power and more responsibility to workers will be supported. Innovations that make the production work more efficient will be recognised as important and will receive public support. Government rhetoric and policy is emphasizing the need for 'total quality', especially in the context of the national Prosperity Initiative (see Chapter Thirteen).

Faced with newly industrialised countries that use their low cost of labour as their main competitive advantage, Canada will have to act differently to maintain it's high standard of living. Canadian governments will be arguing that only through moving to higher value added production can this standard of living be sustained.

Typical of this type of position is the following quote from one of the last publications of the Science Council of Canada (1992);

> 'many successful Canadian firms are also beginning to realize that the new, long-term competition is increasingly based on intangibles such as research, innovation, and people – that it has to do with nonprice factors such as quality, corporate strategies and attitudes, and organizational patterns. These – along with staying near the forefront of research and innovation – are the underlying determinants of competition, not simply productivity or costs per se.'

Notes

1 These arguments are made in *Reaching for Tomorrow: Science and Technology Policy in Canada* 1992 Ottawa Science Council of Canada, and by Alan Rugman and Josph D'Cruz 1992 in *Fast Forward: Improving Canada's International Competitiveness*

2 *Unless another source is cited, the anlysis in this section is based on the Sectoral Technology Strategy Series* reports issued by the Science Council of Canada in 1992. These reports analysed fifteen Canadian sectors: telecommunications, automotive parts, iron and steel, automotive vehicles, oil and gas, petrochemicals and resins, banking, nonferrous-metals, forest-products, machinery, food and beverage, electric power, consulting engineering, electronics, and computer software and services.

3 The Business Council on National Issues and the Government of Canada cosponsored a study by Michael Porter and Monitor Company of Canada Limited. The report published from this study is *Canada at the crossroads: the reality of a new competitive environment* 1991. The first chapter of the report discusses weaknesses in Canadian economic performance.

4 Unless otherwise specified, international data in this section is drawn form *Main Science and Technology Indicators*, OECD, 1992. Data on Canada comes from *Industrial Research and Development Statistics 1989*, Statistics Canada, catalogue 88–202.

5 After-tax cost of r&d in different jurisdictions is very difficult to compare, since many complex technical issues have to be resolved before a firm's fiscal situation is deemed to be comparable to one in the other jurisdiction. The comparisons are also rapidly obsolete because governments constantly modify their tax systems. However, several studies point to the low after-tax cost of r&d in Quebec, even when compared with American states such as New York, Michigan or Massachussets. For on example of a methodology of calculating the after-tax cost of r&d, see Bernstein (1986). For a comparison of the after-tax cost of r&d in Quebec, Ontario, Massachussets, Michigan and New York, see the 1989–1990 Quebec Budget, Appendix E. For comparisons between Ontario and Quebec, see Murray (1990) and Samson-Bélair (1988).

6 The Quebec measures are summarised in a government document that is available both in French and English (MICT, 1991a and 1991b).

7 See also Bernstein (1986b) and Longo (1984)

8 When reading the following section, one should be aware that changes are made relatively rapidly to programmes. For example, many programmes announced with a specific allocation have had their funding diminished in subsequent government budgets. Other programmes' spending has been delayed. The amounts given here, although accurate at one point in time, should not be treated as fixed but rather as indicators of the magnitude of the different programmes.

9 The new orientation of IRAP is described in NRC (1991). A thorough parliamentary inquiry into IRAP has also published a report analysing the programme (Canada, House of Commons, 1991). The new additional funding corresponds to $7 million a year for four years, which is much lower than that which NRC was seeking

10 The Department of National Defence also has a matching grants programme for

military r&d. It's budget was $15 million in 1990–91. In addition, it annually spends close to $150 million on r&d contracts.

References

Bernstein J I 1986a *Research and Development, Tax Incentives, and the Structure of Production and Financing* Ontario Economic Council Research Studies, University of Toronto Press

Bernstein J I 1986b The Effect of Tax Expenditures on the Canadian Industrial R&D Expenditures *Canadian Public Policy* **12**(3)

Boismenu G Ducatenzeiler G Anderson F 1985 L'aide directe fédérale à l'innovation industrielle *Politique* **8** Automne

Canada, House of Commons 1991 *IRAP - An Inquiry into the Industrial Research Assistance Program* Standing Committee on Industry, Science and Technology, Regional and Northern Development, December

Conseil de la science et de la technologie 1989 *Séminaire sur les mesures fiscales d'incitation à la R-D*

Ernst and Young 1989 *Directions 89/90 - A Survey of Advanced Technology Companies in Canada*

Industry, Science and Technology Canada 1992 *Industry Profiles* Government of Canada

Leblanc M 1989 *L'approche québécoise pour favoriser la r&d effectuée par les entreprises, in Conseil de la science et de la technologie* pp. 35–53

Longo F 1984 *Industrial R&D and Productivity in Canada* Science Council of Canada

McFetridge D G Warda J P 1983 *Canadian R&D Tax Incentives: Their Adequacy and Impact* Toronto Canadian Tax Foundation

McQuaig L 1987 *Behind Closed Doors* Markham Viking

Meneely K Glover P 1992 Reducing tax with scientific research *CMA Magazine* **65**(10):8

MICT 1991a *Les mesures fiscales pour favoriser la R-D*

MICT 1991b *Fiscal Measures to Foster R & D*

Murray K J 1990 *Recent Developments in Research and Development Tax Incentives* Miméo Montreal Canadian Tax Foundation Annual Conference

NRC 1991 *The Industrial Research Assistance Program - A Strategic Plan for the 1990s*

- OECD 1992 *Main Science and Technology Indicators*

Ostry S 1990 *Governments & Corporations in a Shrinking World* New York Council on Foreign Relations Press

Plourde M 1991 R-D au Québec - Le temps ou jamais d'investir *Info-Log Magazine* **6**(5)

Porter M 1990 *The Competitive Advantage of Nations* New York Free Press

Porter M Monitor Group 1991 *Canada at the Crossroads: The Reality of a New Competitive Environment* Business Council on National Issues and the Minister of Supply and Service Canada Report of the Premier's Council 1987 *Competing in the New Global Economy* vol 1 The Government of Ontario

Re$earch Money (various editions) Ottawa Evert Communications

Roy R 1989 Programme d'encouragement à la recherche scientifique et au développement expérimental in Conseil de la science et de la technologie pp15–33

Rugman A D'Cruz J 1991 Fast Forward: Improving Canada's International Competitiveness Commissioned by Kodak Canada Inc.

Samson-Bélair 1988 *Les avantages fiscaux à la R & D suite à la réforme fiscale et au budget du Québec du 12 mai 1988* Étude préparée pour le Ministére de l'Industrie, du Commerce et de la Technologie du Québec, août

Science Council of Canada 1992 *Reaching for Tomorrow* Minister of Supply and Services

Science Council of Canada 1992 *Sectoral Technology Strategy Series* Minister of Supply and Services

Statistics Canada 1989 *Industrial Research and Development Statistics* Catalogue 88–202

Statistics Canada 1990 *Federal Scientific Activities 1990–91* Catalogue 88–204

Statistics Canada 1990 *Annual Report of the Minister of Supply and Services Canada under the Corporations and Labour Unions Returns Act. Part 1, Corporations* Catalogue 61–210

Statistics Canada 1992 The Provincial Research Organizations, 1990 Service Bulletin, *Science Statistics*, Catalogue 88–001, January

Switzer L 1989 Are Government Fiscal Incentives for R&D Desirable? An Assessment of Rationale, Effectiveness, and Costs in *Conseil de la science et de la technologie* pp145–167

Warda J 1990 *International Competitiveness of Canadian R&D Tax Incentives: An Update* The Conference Board of Canada, International Business Research Centre, Report 55–90

Chapter 6

Educational and scientific infrastructure

Robert J. Kavanagh

Structure of Science and Engineering Education[1]

The Education Continuum

The universities in Canada represent a final phase of an education continuum which begins in the elementary schools. The elementary school phase usually lasts for six years, after which children enter a secondary school at the age of about twelve years. The secondary school phase most commonly involves a further six years of study. The nature of this phase differs to some extent from province to province. Typically there is an opportunity for students to choose between an academic or a vocational stream during the final years of secondary school. Students in a vocational stream may then proceed to a trade school or post-secondary college, enter an apprenticeship, or join the labour force. Those in an academic stream may then enter a university, a post-secondary college, or the labour market. In each case, students normally complete their secondary level of education at about the age of eighteen. In the province of Québec the period of secondary education is one year shorter and at the conclusion of this phase students may then proceed to a *Collège d'enseignement général et professionel (CEGEP)*. These institutions offer a choice of vocational programmes or a two-year programme leading to university entrance.

To illustrate the relative sizes of the various flows through this continuum, it may be noted that about ten per cent of students in secondary schools are enrolled in vocational programmes, full-time enrolment of students at non-degree granting colleges in 1990–91 was about 325,000, and the comparable figure for universities was about 532,000.

The scientific and engineering enterprise in Canada depends upon the universities for the education of professional scientists and engineers and upon the post-secondary colleges and CEGEPs for the education of technologists, technicians and skilled tradespersons. Ultimately, the whole enterprise depends upon the quality of the education imparted to children in the primary and secondary school systems. A recent report published by the Economic Council of Canada has drawn attention to a number of concerns about the Canadian educational system. These include: the relatively poor achievement of Canadian secondary school students in science and mathematics in comparison with students in many other countries; the low esteem in which vocational programs in secondary schools are held; the limited scope of the apprenticeship system; falling post-secondary college enrolments in engineering and applied science fields.

As far as the university system is concerned, the most important issues relating to the school system are the quality of the science and mathematics education, the tendency by students to opt out of science or mathematics courses in the latter years of secondary school, and the relatively low proportion of female students who are motivated to enter university science or engineering degree programmes.

Undergraduate education

Canadian universities have well-developed programmes for the education of scientists and engineers. Some fifty-three universities award bachelor's degrees in the sciences and thirty-two grant bachelor's degrees in engineering. Despite differences in programme structure, regional orientation, language, and size, the Canadian university system is a reliable source of well-educated scientists and engineers. The quality of the education imparted does not vary markedly from one university to another even though the characteristics of that education may be different. By contrast with the USA, Canada has very few degree-granting institutions which are entirely supported from private sources, and those that are, are small entities with religious affiliations. Perhaps as a result, there are no mediocre Canadian universities but, by the same token, there are no Harvards or Stanfords. The university system, as a whole, currently produces about 14,000 new scientists and 7,000 new engineers per year.

It is worth noting, however, that there are some significant differences between the universities in the structure of their undergraduate science and engineering programmes. These differences result, in part, from the fact that jurisdictional responsibility for education rests with the provinces. Consequently, variations in school level education have an effect upon university level education. In those provinces in which there are twelve years of school education (excluding kindergarten) the honours science degree and the engineering degree normally require four years of study. In the province of Québec, as mentioned earlier, there are eleven years of school education followed by two years of intermediate level education at a CEGEP. Québec universities, therefore, require only three years of study for science and engineering degrees. There are other variations which may be noted. In engineering especially, a number of universities offer a cooperative degree programme. In such programmes, the student intersperses periods of academic study with periods of related work experience. A bachelor's degree is typically completed in four and two-thirds years, comprised altogether of six terms of work experience and eight terms of academic study. In Nova Scotia, engineering students spend two years at one of the several universities in that Province and then transfer to the Technical University of Nova Scotia for three years of specialised engineering study. At some universities, it is possible to obtain a bachelor's degree in science after three years of study (following twelve years of school) but this degree is not a specialist degree and will not suffice to gain entry to a postgraduate degree programme in science. These graduates usually have the option of spending one additional year at the university and obtaining an honour's degree in science.

Postgraduate Education

Postgraduate education in science and engineering has two distinct roles in Canada. It may be a period of advanced study, with emphasis on research training, which equips the graduate for a career in research and/or university teaching. It may also be a period with emphasis on advanced courses, which equips the graduate for a career in industry or the public sector in science and engineering work other than r&d. Two degrees are offered at this level: the master's degree and the doctorate. The normal pattern of study is that the student first obtains a master's degree and then enters a programme of doctoral studies and research. The master's degree is frequently a terminal degree and, in practice, only a minority of master's graduates continue their education at the doctoral level. Twenty-six Canadian universities offer master's and doctoral degrees in the sciences and/or engineering. An additional twelve universities, all small, offer only master's degrees in one or more of the sciences. As is apparent from these numbers, there is a more discernible difference between Canadian universities at the postgraduate level than at the undergraduate level. Thus, one can identify three tiers of universities: those which offer only undergraduate science degrees; those which offer undergraduate and master's degrees in science; those which offer all degrees including the doctorate in science and/or engineering. The latter group of twenty-six universities are the more research intensive universities in Canada. The most research intensive among these, with the best facilities and with faculty members having international reputations, tend to attract the best postgraduate students.

Perhaps because of the way in which research activity has developed at Canadian universities over the years (see below), the master's degree can require an unusually long period of study. The time taken by a full-time student to obtain a master's degree can be as little as nine months but is more commonly two years and can be as much as four years. Some master's degrees in engineering or science do not involve research training. The requirements for such degrees are completion of course work and, sometimes, a project. These degrees can be completed in a relatively short time, *i.e.* not more than about eighteen months of full-time work. However, most master's degrees in engineering and the natural sciences involve a significant research training element. In addition to the completion of course work, these master's degrees require the completion of a research investigation and the defence of a thesis. The research component may take a year or more of full-time work. Consequently the research master's degree generally takes a minimum of eighteen months to complete.

The research master's degree plays two roles. It serves as a terminal degree for students seeking employment in research or other technically demanding careers, especially in engineering. Many industrial employers consider the research training component of the engineering master's degree to be important because the student receives some training in research methods and has succeeded in carrying out a substantial investigation and bringing it to a successful conclusion. The research master's degree also serves as a stepping-stone to the doctorate. Master's students with a strong interest in research may enter a doctoral programme which, of course, involves a very substantial research element.

While the significant research component of the master's degree gives that qualification a high reputation, there is increasing concern in Canada about the lengthy period often needed to complete such degrees. There is a feeling that the research investigations required of master's degree candidates are sometimes excessively demanding. Some master's theses are barely distinguishable from doctoral theses. One of the consequences of this type of situation is that a master's degree graduate who then becomes a candidate for a doctoral degree may typically spend seven years between the bachelor's degree and completion of the doctorate. The problem is further aggravated by the tendency, in many science fields, of expecting doctoral graduates to spend a further period (often two years) as a postdoctoral fellow before entering their research career. By this time they will be in their thirties. Efforts are now being made by some universities and in some disciplines to move towards more realistic master's degree research investigations. In addition, the main provider of scholarships for postgraduate students, the Natural Sciences and Engineering Research Council of Canada (NSERC), has recently modified it's regulations in such a way as to give a significant incentive to research students to complete their master's degrees in not more than two years.

The doctoral degree has been well-established in Canada for many years as the principal mechanism for research training. As indicated earlier, the normal entrance qualification to a doctoral programme is the possession of a relevant master's degree. However, there are other possibilities. In some disciplines at some universities, students may be directly admitted from the bachelor's degree level (as in the UK). But this practice is not common. An increasingly common route involves the transfer of a master's degree candidate to a doctoral program after, say, one year without the necessity to complete the master's thesis and degree. Because this procedure can effectively reduce the total time needed to obtain the doctoral degree it is seen as a valuable means for avoiding the excessively lengthy master's degree programmes referred to earlier. This mechanism enables the best students, not seeking a terminal master's degree, to take a short-cut *en route* to the doctoral degree.

The requirements for the award of a doctoral degree vary, to some extent, between universities and between disciplines. A universal requirement is the successful completion of a substantial research investigation. It is expected that the research will result in a significant advancement of knowledge in the field. It is also a normal expectation that publications in refereed journals will result, although in engineering it is accepted that there are alternative methods for disseminating the results of the research. The doctoral thesis must be defended by the candidate in an oral examination. The examining committee will include experts from several cognate departments as well as an examiner from outside the university.

In addition to the research requirement, doctoral candidates must usually take advanced level course work in their own and related fields. The extent of this requirement varies from minimal to quite substantial, depending upon the policies of the department or of the university. Where the U.S. model predominates, the course requirements are significant, whereas if the European influence is strong these requirements can be minor.

It is also common, but not universal, for doctoral candidates to have to pass a 'comprehensive' or 'qualifying' examination within the first year or two of candidacy. This examination is intended to ensure that the candidate has a thorough understanding of the fundamental principles which govern the discipline or research field.

Reference has already been made to the matter of the total duration of the research training process in Canada. While the master's programme duration is of concern, there continues to be active discussion about the time taken to complete doctoral degrees and about the fact that many candidates never complete their degrees. This subject is of concern because of the various ramifications of the present situation: waste of human and financial resources; inefficient utilisation of university space and facilities; graduates who have spent their most productive years as students.

There are two distinct problems: the failure of many candidates to complete their degrees; and the excessive time taken to complete their degrees by those who actually finish. A recent report for the universities in Ontario reveals some interesting statistical evidence concerning these problems. By studying the experiences of all students who commenced their doctoral programmes in 1980 it has been determined that, by 1990, the proportions of these students who had completed their degrees were as follows:

Agricultural and biological sciences	85%
Engineering and applied science	60%
Mathematical and physical sciences	75%

For those students who actually completed their degrees, the median time to completion was as follows:

Agricultural and biological sciences	4.33 years
Engineering and applied science	4.33 years
Mathematical and physical sciences	4.00 years

Another figure of interest is the median cohort time to completion, *i.e.* the time by which fifty per cent of all students commencing their candidacy in 1980 have completed their degrees. These times are as follows:

Agricultural and biological sciences	4.66 years
Engineering and applied science	6.00 years
Mathematical and physical sciences	4.66 years

These data reveal two matters for concern. The median time to completion, which is almost the same for the three discipline groups, shows that half of those completing their degrees take longer than four years to do so. Reduction of this figure is a current objective. Factors involved include the effectiveness of the supervisor, availability of adequate financial support for the student, and excessive involvement of the student in activities, such as teaching, at the expense of research time.

The second problem, which is particularly acute in the engineering disciplines, is the low completion rate. In fact, a further analysis of the engineering figures by citizenship reveals the interesting fact that the proportion of Canadian students who had completed their degrees by 1990 was only about fifty per cent, whereas for foreign students the proportion was about eighty per cent. This striking difference illustrates the influence of the job market upon student behaviour. There has been a strong demand for engineers in Canada, particularly in the latter half of the 1980s. One result has been that engineering doctoral candidates have been tempted, by attractive offers of employment, to discontinue their studies or to try to complete their degrees on a part-time basis. This has contributed to the low completion rates for Canadian engineering students. Foreign students, on the other hand, are not usually allowed to accept employment and have incentives to complete their studies rapidly.

Most full-time postgraduate science and engineering students can expect to receive financial support during their period of study and research. The main sources of this financial support are scholarships and fellowships provided by the federal government. However, some provinces also make similar awards. The most noteworthy example of the latter is the program of awards of the *Fonds pour la Formation de Chercheurs et l'Aide à la Recherche in Québec*. Significant programmes also exist in British Columbia and Ontario. Other sources of support for students are part-time employment as research assistants, using research funds provided to university faculty members by the federal government and by industry, and part-time employment by the university as teaching assistants. Overall, the dominant funding source is the federal government, reflecting the importance placed by successive governments on the of training an adequate number of highly qualified scientists and engineers. The most promising students tend to receive NSERC postgraduate scholarships through a rigorous national competition. While the stipends paid to these scholarship holders are adequate to cover normal living costs, they are not competitive with starting salaries offered to new bachelor's graduates in fields such as engineering and computer science. Many universities supplement NSERC scholarships from their own sources in order to make the total level of financial support more attractive. Even so, a significant number of students in high demand fields decline scholarships offered to them and opt to accept positions in industry.

Postdoctoral studies and research

In the sciences, it is common for a doctoral graduate to spend a period of two or more years as a postdoctoral fellow before seeking a permanent position. This period of further study and specialised research is frequently undertaken in a foreign laboratory. The scientist, therefore, has an opportunity to be exposed to new ideas, to follow up promising avenues of research identified during the doctoral investigation, and often to work with a leading authority in the field. Postdoctoral experience of this type is usually a condition, in Canada, for appointment to a tenure-track position in the sciences at a university. The ambition for many Canadian doctoral graduates in the sciences is to enter an academic career and this fact explains why the postdoctoral fellowship is so commonplace.

In contrast to the sciences, it is not at all common for doctoral graduates in engineering to apply for postdoctoral fellowships. This difference between engineering and the sciences is a consequence of the nature of the job market and of the availability of doctoral graduates in these fields. In engineering, the number of Canadians receiving doctoral degrees has been much smaller than in the sciences. The demand for engineers with doctoral degrees has come from industry as well as from the universities. Demand, in many areas of engineering, has outstripped the supply and most employers are not willing to wait for an engineer to complete a period of postdoctoral study and research.

As an indication of the strength of the tradition for postdoctoral study in the sciences, it may be noted that the Natural Sciences and Engineering Research Council of Canada awards about 250 new postdoctoral fellowships each year. Of these, about ninety per cent are awarded to scientists. In 1990, about 500 doctoral degrees were awarded by Canadian universities to Canadians in the science disciplines. Taking into account the fact that other postdoctoral fellowships for science graduates are awarded by the Medical Research Council of Canada, by the universities themselves, and by foreign organisations, it is evident that the majority of doctoral graduates in the sciences are following the path of postdoctoral study before entering upon their career.

Given that the number of academic positions available each year in the sciences is limited, and given that there is a great need for Canada to increase it's investment in industrial R and D, the wisdom of the above practice may be questioned. In fact, most doctoral graduates in the sciences are likely to obtain regular employment outside the academic sector. Most industrial employers do not require, or place any premium, on a period as a postdoctoral fellow. In fact, many industrial employers do not even consider the possibility of hiring doctoral graduates. As will be described later in this chapter, this state of affairs has prompted the federal government, through NSERC, to offer an alternative to the conventional postdoctoral fellowship: the industrial research fellowship.

Role of the universities in research

Historical background

The earliest scientific research in Canada was carried out as an element of the exploration of the new colonies by British and French military or religious men. The chief emphasis in this activity was on the cataloguing of natural resources by botanists, geologists, and zoologists. Eventually, government support of research evolved, notably with the establishment of the Geological Survey of Canada under the leadership of William Logan in 1842. Until the middle of the nineteenth century, the universities played only a very small role in scientific research.

One of the earliest appointments of a research-oriented academic at a Canadian university was that of Henry Holmes Croft in 1842, as Professor of Chemistry at the University of Toronto. Subsequent early appointments which may also be mentioned include the geologist J. William Dawson as Professor of Geology and Paleontology and as Principal of McGill University in 1854, Loring Woart Bailey, geologist and biologist, at the University of New Brunswick in 1860, and Ernest Rutherford as Professor of Physics at McGill University in 1898. The latter's Nobel Prize, awarded in 1908, was in recognition of the research he carried out at McGill.

Despite these and other appointments, scientific research at Canadian universities was slow to develop. In his 1902 presidential address to the Royal Society of Canada, President James Loudon of the University of Toronto contrasted the research activities of the universities of Germany, the UK and the USA with those of Canadian universities and stated: 'Organized research activities in Canadian universities can scarcely be said to exist as yet, although within the last decade certain beginnings have been made which indicate a movement in that direction.'

The influence of the National Research Council

Although some progress was made in the early years of the twentieth century in the establishment of scientific research capabilities at a small number of universities, Canada found itself, at the beginning of the First World War, with a minimal scientific and technical capacity. The state of industrial research was no better than that of university research. It was estimated that, in 1917, the whole of Canada possessed less than fifty persons engaged in fundamental scientific research.

In 1916, the Canadian government created the National Research Council (NRC) which was, initially, an advisory body. It's mandate was the promotion of researches which would develop industrial production and the utilisation of the natural resources of Canada. NRC acquired it's own research laboratories in 1932, but before that one of it's first actions was to launch a programme of financial support for scientific research at Canadian universities. This turned out to be a most significant step in the development of Canadian universities as centres of scientific and engineering research. (For the NRC's role in the development of science policy, see Chapter Two).

In the first year (1917) of it's programme of grants to universities, NRC awarded a total of $5,000 in the form of scholarships to nine postgraduate students and $8,000 in research grants to university faculty members. Despite this very modest beginning, NRC's decision to establish such a programme was the result of a conviction that Canada could never achieve a strong industrial research capability without having universities in which researchers could be trained in an environment of high quality research activity. From 1917 until 1978, NRC played a pivotal role in the growth of university scientific research in Canada. Without exaggeration, it can be stated that the great majority of academic research in science and engineering during that period was funded by NRC and a similar majority of Canadian researchers owed their postgraduate training to direct or indirect financial support from NRC. The evolution of this support by NRC is illustrated by the following table of expenditures.

In 1978, the function of providing university grants and scholarships was removed from NRC and a new body, the Natural Sciences and Engineering Research Council (NSERC), was created to take over this responsibility.

The development of graduate studies

Postgraduate studies in the sciences at Canadian universities originated in the early years of the twentieth century at the University of Toronto and at McGill University. The first earned PhD. degree was awarded in 1902 at the University of Toronto. McGill University conferred its first PhD. degree in 1909. In 1910, there were twenty-four doctoral students and twenty-two master's degree students enrolled in the science disciplines at the University of Toronto. In that same year, McGill University had thirty-six master's students (including ten in engineering) and eight doctoral students (including one in engineering).

The development of graduate studies, especially at the doctoral level, was closely related to the establishment of a scientific research capability at Canadian universities. Despite NRC's efforts to encourage the latter, the spread of this capability beyond Toronto and McGill was slow to occur. By 1945, only five universities offered doctoral degrees. At the end of World War II, a critical shortage of highly qualified scientific and technical personnel was considered to be a limiting factor in the development of Canadian industrial research. NRC undertook a major expansion of it's university grants programme. Added emphasis to this expansion occurred with the appointment of Dr E.W.R. Steacie as President of NRC in 1952. Dr Steacie believed that the universities were 'the keystone of the

Table 6.1
Expenditures on Scholarships and Grants by the National Research Council

Fiscal Year	Total Expenditure ($ Millions)
1917–18	0.013
1927–28	0.154
1937–38	0.214
1947–48	0.957
1957–58	3.6
1967–68	66.3
1977–78	97.7

whole structure' and was an unceasing campaigner for increased support for university science and engineering research.

By 1953, eleven Canadian universities had awarded doctoral degrees in science but, at most universities, the master's degree was the highest qualification awarded. The master's degree assumed special significance at those universities and, as a result, high standards were set for the master's thesis. A period of great expansion of graduate studies took place during the 1960s and 1970s, and resulted in the current total of twenty-six universities which award doctoral degrees in science and/or in engineering. The number of doctoral students in the sciences and in engineering has increased rapidly, as has the number of degrees awarded as shown in Table 6.2.

The Natural Sciences and Engineering Research Council

Although the National Research Council's role in promoting research and research training in the universities was widely appreciated, this activity was seen by some as being in conflict of interest with NRC's other role, the operation of the country's largest research establishment. Beginning with the construction of it's first laboratories in 1932, NRC had expanded it's activities as a performer of research, especially during and after World War II. It became the principal research organisation for the federal government and evolved into a centre of research excellence with an international reputation. To it's main role as a resource for the promotion of industrial research was added a capability for fundamental scientific research. With the recruitment of excellent researchers, notably Gerhard Herzberg, NRC rivalled the universities in some scientific fields. By some, NRC was suspected of not promoting it's university granting function to the desired extent because of a wish to further it's own ambitions as a research organisation.

Eventually, the Canadian government took action to spin off the university grants activity of NRC by creating a new granting organisation called the Natural Sciences and Engineering Research Council. At the same time, it split off the grants and fellowships activities in the social sciences and humanities disciplines from the Canada Council to form a second granting organization called the Social Sciences and Humanities Research Council (SSHRC)[2]. At an earlier date (1969), the Medical Research Council had also been established out of a former unit within NRC. Thus, three similar granting councils were created as a means

Table 6.2
Number of Ph.D. Degrees Awarded in Science and Engineering

Year	Number of Ph.D. Degrees
1910	0
1920	24*
1930	46*
1940	75*
1950	202*
1960	177
1970	885
1980	735
1990	1,236

*All disciplines. It is estimated that about half of these degrees were conferred in the natural and applied sciences.

for providing federal government financial support for university research. In fact, the functions of NSERC are to:

- promote and assist research in the natural sciences and engineering, other than the health sciences
- advise the Minister in respect of such matters relating to such research as the Minister may refer to Council for its consideration

It's mandate, therefore, is not limited to the support of university research but the role Council has identified for itself is mainly to fund university researchers and students.

NSERC has, to a great extent, continued the general approach established by NRC in it's granting function, but has been innovative in establishing new programmes and has broadened the scope of its activities to meet its objectives:

- to secure a healthy research base in universities and, as a corollary, to secure a sound balance between this diversified research base and the more targeted research programs
- to secure an adequate supply of highly qualified personnel who have been well educated in basic science as well as trained with state-of-the-art facilities
- to facilitate collaboration between r&d performing sectors in Canada

The federal government has been relatively generous in funding NSERC since its creation, reflecting the belief (as was the case in 1916) that national prosperity is intimately linked to the need for internationally competitive industries and that this condition requires the availability of an adequate number of highly qualified scientists and engineers trained in university research laboratories. The growth in total expenditures by NSERC on grants and scholarships since 1978 is illustrated in Table 6.3.

Perhaps the most noteworthy influence by NSERC upon university research in science and engineering has been the increasing emphasis placed upon research in fields of national priority and upon research collaboration between the universities and the industrial sector. In fact, the initial moves in this direction were made by the National Research Council in the 1970s. In particular, a new programme of strategic grants was established by NRC in 1977. These grants, for university researchers, were intended to support high calibre research in certain specific areas (*e.g.* energy and oceanography) which were considered to be of national importance. The creation of this programme provoked substantial opposition from many academic researchers who saw this new activity as draining funds away from fundamental research. Indeed, Dr Steacie had always championed the importance of academic researchers being free to seek knowledge for its own sake rather than to respond to external pressures. NRC's decision to launch a programme of strategic grants was a bold step which recognised that

Table 6.3
Expenditures on Scholarships and Grants by NSERC

Fiscal Year	Total Expenditure ($ Millions)
1977–78	109.7
1983–84	274.6
1988–89	350.3
1992–93	445.5*

*budget (excludes funding for Networks of Centres of Excellence)

increased funding for university research would no longer be readily forthcoming unless academic researchers paid some attention to the needs of society in the orientation of their research. The controversy over the strategic grants programme continues to this day. However, NSERC has continued and expanded the programme. Many distinguished researchers who carry out excellent fundamental research also now devote some of their efforts to research having a specific practical objective by means of a strategic grant from NSERC.

The role of a granting council in steering the direction of university research is a controversial one. Canada does not differ from most other industrialised countries in this respect. Until 1977, there was no discernible steering of scientific research in the universities by NRC. Perhaps because of the great emphasis upon the value of high quality fundamental research, Canadian academic research in engineering and in the applied sciences developed more slowly than was the case for the pure sciences. Graduate education in engineering lagged behind that in the sciences. Much of the research which was carried out in faculties of engineering was barely distinguishable in character from research in the sciences. There was little evidence of the fundamental distinction between engineering research and science research which ought to exist. Since the establishment of NSERC, a considerable effort has gone into the encouragement of engineering research in the universities as a distinct entity. An important aspect of this effort has been the strong encouragement given by NSERC to collaborative research with industry (see later in this Chapter for a discussion of this activity). This activity has been of particular interest to engineering researchers. However, despite efforts to promote engineering research in the universities, there remains a concern that the research conducted in many faculties of engineering contributes more to the furtherance of scientific knowledge than to the solution of engineering problems. To some extent, the criteria used by the universities and by research funding agencies in evaluating research accomplishments and research proposals are still based upon the criteria used in the sciences.

Provincial support of university research

Funding of university research by provincial governments has tended to increase during the past decade. In the case of research in the natural sciences and engineering, the proportion of total funding contributed by the provinces has increased from about fifteen per cent in 1981 to about twenty per cent in 1990. By comparison, federal government funding of university research in these fields made up forty-five per cent of the total in 1981 and forty-six point five per cent in 1990. (The other main funders are the universities themselves and the private sector, each with fifteen per cent in 1990.)

Provincial funding of university research is largely concentrated in Québec and Ontario with some significant support also in British Columbia and Alberta. The province of Québec has been particularly active in the support of university research and the training of researchers. The Québec government organism which has the responsibility for this activity is *Fonds pour la formation de chercheurs et l'aide à la recherche*(Fonds FCAR). In the fiscal year ending May 31, 1991, Fonds FCAR awarded a total of over $44 million in the form of research grants and scholarships. These funds were spent in support of five basic programmes: establishment of new researchers; funding for research teams; research centres; scientific publications; postgraduate and postdoctoral fellowships; and also in support of strategic research.

Ontario's support of university research includes a programme of postgraduate scholarships (Ontario Graduate Scholarships), the University Research Incentive Fund and the Ontario Centres of Excellence. The latter two initiatives are mentioned later in this chapter in connection with university-industry collaboration.

The current status of university research in science and engineering

Scientific and engineering research in Canadian universities is now well-developed and plays an important part in the country's s&t activity. In 1990, the higher education sector (which consists almost entirely of the universities) accounted for twelve point one per cent of the gross expenditure on research and development in the natural sciences and engineering. This fraction compares with sixty-six point six per cent by the business enterprise sector and eighteen per cent by the federal government. Research of the highest quality is carried out in many universities. In the sciences, one can refer to the work of Raymond Lemieux and John Polanyi in chemistry, Tuzo Wilson and William Fyfe in the earth sciences, Scott Tremaine in astrophysics, Pierre Deslongchamps in organic chemistry, H.S.M. Coxeter in mathematics, Tak Mak in biophysics, and the computer science research groups at Waterloo and Toronto as examples of the calibre of the researchers active in Canadian universities. In engineering, the work of Edward Davison, Peter Nikiforuk, and others in the field of automatic control, Alan Davenport and Norbert Morgenstern in civil engineering, Keith Brimacombe in metallurgical engineering, and Simon Haykin in communications engineering are examples of university research which is of international standard.

In order to monitor the state of Canadian research, particularly in the universities, NSERC has developed a process which periodically produces *Health of the Discipline* reports. These reports, which are prepared by NSERC's disciplinary grant selection committees, treat such matters as: the current strength and weaknesses of research in the discipline in Canada; emerging trends over the next five to ten years; important new initiatives and projects in the disciplines; opportunities for Canadian participation in international activities; changes in research funding needs; predictions for graduate student enrolments in the discipline; and other aspects of research in the discipline which relate to maintaining and improving excellence and international competitiveness.

The transformation of many universities during the past 40 years, from organisations devoted almost entirely to teaching into research intensive institutions, has generated a concern in some quarters as to whether the teaching function has suffered from the increased emphasis upon research. In the sciences, in particular, the evaluation of the performance of faculty members tends to give most weight to research accomplishments. New faculty are hired principally upon the basis of their research potential rather than their ability to teach. The possession of research grants from NSERC tends to assume an extraordinary importance for some faculty members and not to be awarded such a grant is often regarded as a stigma. Thus, there are powerful incentives which encourage research activity. On the other hand, most educators accept that the quality of undergraduate teaching in the sciences and the capability of Canadian universities to educate future scientists at the postgraduate level have been greatly improved as the universities assumed their additional role as scientific research centres. Debate about this issue has been stimulated by the recent report of the Commission of Inquiry on Canadian University Education which concluded that the universities are 'fundamentally healthy' but expresses a concern over the reward system in Canadian universities which, in practice, places greater weight upon research publication than upon teaching excellence. The report's conclusions and recommendations have elicited a vigorous negative reaction from university scientists but a more positive response from those segments of the universities in which research and scholarship are not as fundamental a part of the academic profession.

Highly qualified scientists and engineers

Need for persons trained in research

The need for an adequate supply of well-trained scientists and engineers to support a strong industrial research capability was an important factor leading to the establishment of the National Research Council in 1916. Now, some seventy-seven years later, this requirement remains a key feature of federal government policy in Canada. In the interim, much has changed. As has been described earlier, the capacity of the universities to educate scientists and engineers by and for research has expanded enormously. The national need for such persons has also increased dramatically. At certain times the supply has outstripped the need and on other occasions the opposite has been true.

The most recent figures available from the Organisation for Economic Cooperation and Development show that, in the case of research scientists and engineers, Canada ranks behind most of the leading industrialised countries on the basis of the number of such persons per thousand labour force, as shown in Table 6.4.

The distribution of Canada's research scientists and engineers between the three sectors: business enterprise, higher education and government, is different from that of most of the other countries listed in Table 6.4. The fraction of such persons employed in the higher education and government sectors in Canada is relatively high, with the consequence that employment in the business enterprise sector is relatively low in comparison with most other countries, as shown in Table 6.5.

As of 1989, the percentages of Canadian research scientists and engineers employed in the higher education and the government sectors were forty-one per cent and eleven point nine per cent respectively. (It may be noted that, in accordance with OECD definitions, the higher education figure includes postgraduate students. Hence the proportion of research scientists and engineers in regular employment in the higher education sector is significantly less than is suggested by the OECD figure.)

The future need for research scientists and engineers in Canada cannot, of course, be predicted with any precision. A common view is that, provided the national economy experiences healthy growth, the biggest demand for highly

Table 6.4
Total Research Scientists and Engineers per Thousand Labour Force

Canada (1989)	4.6
France (1989)	5.0
Germany (West) (1987)	5.6
Italy (1989)	3.1
Japan (1989)	8.9
Sweden (1987)	5.1
United Kingdom (1988)	4.6
United States (1988)	7.7

qualified scientists and engineers will come from the business enterprise sector, followed by the higher education sector and then by the government sector. No attempt has been made to forecast actual numbers of scientists and engineers likely to be required in each sector by, say, the year 2000. The use of mathematical models for this purpose has not been considered to be a worthwhile activity. Nevertheless, taking into account expected future trends in the requirements of the three sectors and the anticipated output of such persons from the universities, there is concern that a shortage of highly qualified scientists and engineers may occur by the end of the century. Fields such as engineering, chemistry, and computer science are among those considered likely to experience shortages.

The Supply of Scientists and Engineers

Canada has two principal sources of scientists and engineers: graduates of Canadian universities and immigration. (It also experiences a significant loss of these persons through emigration, mainly to the USA) In the years following World War II, large numbers of scientists and engineers came to Canada from Europe and elsewhere. Since that time, Canadian universities have greatly increased their capacity to educate scientists and engineers and immigration is now a less significant, but by no means negligible, source.

The data contained in Table 6.2 show that there has been a great increase in the number of doctoral degrees awarded in science and engineering since 1950. (The same is true of the granting of master's degrees.) In fact, the production of PhD.s peaked in the early 1970s and then generally declined until the early 1980s, in part because of a widespread belief during the 1970s that there was an oversupply of doctoral graduates. Since the early 1980s, however, the number of PhD. graduates has increased significantly. Table 6.6 shows the numbers of persons receiving doctoral degrees from Canadian universities since 1980 in the three discipline groups defined by Statistics Canada: Agriculture and Biological Sciences (ABS); Engineering and Applied Sciences (EAS); Mathematics and Physical Sciences. The numbers of master's degrees awarded have also increased since 1980 (2,493 were awarded in 1980 and 3,644 in 1990). In 1990, doctoral degrees in science and engineering accounted for forty-six per cent of all doctoral degrees awarded by Canadian universities. Master's degrees in science and engineering made up only twenty-one per cent of all degrees at that level.

In assessing the potential supply of scientists and engineers with postgraduate degrees, it is necessary to take into account the substantial number of foreign students receiving degrees in science and engineering from Canadian universi-

Table 6.5
Business Enterprise Research Scientists and Engineers as Percentage of Total and per Thousand Labour Force

	%	per thousand
Canada (1989)	45.8	2.11
France (1989)	45.0	2.25
Germany (West) (1987)	64.7	3.62
Italy (1989)	40.1	1.24
Japan (1989)	56.0	4.98
Sweden (1987)	53.9	2.75
United Kingdom (1988)	68.5	3.15
United States (1988)	75.4	5.81

ties. For many years, Canadian universities have educated students from other countries, especially the developing countries. Many of these students seek postgraduate degrees in science or engineering. Foreign students at Canadian universities are required to leave Canada after completion of their studies and are not permitted to accept employment in Canada. Should they wish to apply to become an immigrant they must apply from outside Canada. As a consequence of these regulations, only a small fraction of the foreign students at Canadian universities will eventually become part of the work-force. This means that the effective supply of scientists and engineers with postgraduate degrees is significantly smaller than the figures in Table 6.6 would imply.

Statistical information about student enrolments and degrees awarded at Canadian universities is collected by Statistics Canada through a data collection system known as the University Student Information System. This system collects information about the citizenship of students enrolled at universities but, at present, does not collect complete information about the citizenship of degree recipients. Therefore, it is necessary to estimate the numbers of foreign students who obtain degrees using information about enrolments. For the year 1990, it is estimated that the percentages of doctoral degree recipients who were foreign students are about fifty per cent in Engineering and Applied Sciences, thirty-five per cent in Mathematics and Physical Sciences, and thirty per cent in Agriculture and Biological Sciences. (At the master's degree level the percentages are about thirty-five, thirty and twenty per cent respectively.) The implication of these percentages is that at the doctoral level, the numbers of 1990 graduates available for employment in Canada were about 205 in EAS, 315 in MPS and 235 in ABS. It is expected that these numbers may increase until the middle of the decade and that a levelling off, or even a decrease, may take place as a consequence of undergraduate enrolments which, in the early 1990s, have either levelled off or fallen in most disciplines.

Actions to increase the supply of scientists and engineers

In recent years, the availability of adequate numbers of scientists and engineers has been an explicit objective of the federal government as well as of some provincial governments. The federal government's 'InnovAction' strategy for s&t, announced in 1987, was intended to 'bring s&t to the centre of the government's agenda' and among other objectives proposed was 'to help Canada attract, de-

Table 6.6
Doctoral Degrees Awarded by Canadian Universities

Year	ABS	EAS	MPS	Total
1980	210	191	334	735
1981	220	215	341	776
1982	221	183	318	722
1983	247	220	340	807
1984	237	188	373	798
1985	248	278	386	912
1986	281	296	410	987
1987	343	299	477	1119
1988	323	350	465	1138
1989	312	328	500	1140
1990	338	413	485	1236

velop and keep world-class scientists and engineers'. The main motive for this strategy was to enable Canadian industries to compete in the global economy. (See Chapter Three).

A number of new or expanded initiatives have arisen as part of the InnovAction strategy. These include several which are intended to increase the number of university students in science and engineering programmes and to improve the capability of universities to undertake research of the highest calibre and to educate researchers. Among these initiatives are the Canada Scholarships Programme and the Networks of Centres of Excellence (see later), as well as increased funding for the Natural Sciences and Engineering Research Council. The Canada Scholarships Programme was introduced in 1988 as a means to encourage more high school graduates to consider university studies in the natural sciences and engineering. Through this programme, at least 2,500 scholarships are awarded each year to students entering undergraduate programs in the natural sciences, engineering and related disciplines. Each award is renewable for up to four years, provided that the student maintains a high standard of academic performance and continues to be enrolled in an eligible field. An interesting feature of this programme is that at least fifty per cent of these scholarships must be awarded to women. In 1992, the programme was expanded to include additional awards for students enrolling in eligible technology programs at community colleges, institutes of technology and CEGEPs.

In Canada, women are seriously under-represented in most science and engineering fields. The situation is most extreme in Engineering and Applied Sciences, where only thirteen per cent of bachelor's degrees, thirteen per cent of master's degrees, and six per cent of doctoral degrees awarded in 1989 were received by women. The percentages are only slightly higher in Mathematics and Physical Sciences. Only in some of the Biological Sciences do women receive as many degrees as do men. Full-time faculty members at Canadian universities are predominantly male. For example, in Mathematics and Physical Sciences only six per cent of the full-time faculty in 1987–88 were female. In 1988, only three per cent of the registered professional engineers in Canada were female.

The under-representation of women in most science and engineering fields is considered to be unacceptable for two reasons. In the first place, the current situation implies an under-utilised pool of talent. The potential for increasing the number of scientists and engineers would be greatly enhanced should more women enter careers in these fields. Secondly, by avoiding science and engineering careers, large numbers of capable women are limiting their employment opportunities to traditional occupations which are frequently less interesting and less well paid.

The Natural Sciences and Engineering Research Council has taken several actions to encourage greater participation of women in science and engineering. These include undergraduate level research experience awards for women, a programme to encourage the universities to recruit great numbers of highly qualified women into faculty positions, the co-funding with Northern Telecom of a Chair of Women in Engineering, and a programme to encourage qualified women to return to research careers after having spent several years away from science because of family responsibilities. A program somewhat similar to the latter is also offered by FCAR in Québec (*bourses de réintégration à la recherche*). In addition to these initiatives, there are a range of other activities sponsored by professional associations and societies designed to improve the participation of women in the sciences, engineering and mathematics.

The availability of academic staff

The current complement of full-time faculty members in science and engineering at Canadian universities is characterised by an abnormal age distribution and a high median age. This situation has arisen because of the rapid expansion of the university system in the 1960s and early 1970s, followed by a period of very limited hiring of new staff since that time. In 1988, the median age of full-time academic staff at the rank of assistant professor and above was forty-seven in the ABS fields, forty-nine in the EAS fields, and forty-eight in the MPS fields. These figures have increased by about three years in the past six years as the population aged, relatively few senior faculty left the universities, and relatively few young faculty were hired.

Because of the current age distribution of faculty members in science and engineering, it is clear that substantial numbers of these persons will retire, die, or otherwise leave the academic work-force within the next ten years. It is likely that most of those departing will need to be replaced, if not immediately, at least within a few years because of demographic data which lead to an expectation of increased overall university enrolments for the foreseeable future and because of the expectation that enrolments in science and engineering in particular will increase. An important concern, therefore, is whether there will be enough suitably qualified Canadians to fill the academic vacancies which will occur. In this connection it is pertinent to note that, in general, Canadian universities may not recruit non-Canadians into faculty positions unless they can demonstrate, by means of an initial search, that there are no suitably qualified Canadians available. Only in a limited number of fields (*e.g.* some areas of engineering and computer science) are the universities free to hire foreign academics.

Canadian universities now, almost universally, require the possession of a doctoral degree for appointment at the usual entry rank of assistant professor. Therefore, in most fields, recruitment will come largely from among recent PhD. graduates of Canadian universities (and among Canadian PhD. graduates from foreign universities). While the numbers of such graduates are expected to exceed the numbers of new recruits to the faculty ranks, this does not imply that the supply of PhD.s will be adequate. For one thing, by no means are all PhD. graduates qualified for, or interested in, an academic career. Young PhD.s who observe their supervisors' never-ending search for adequate research funds, to say nothing of demanding teaching and administrative duties, are not always inspired to enter the academic profession. But, more importantly, there is a growing demand for PhD. graduates in industry. In many fields, only a minority of PhD. graduates will enter academic careers. This will be increasingly the case in fields such as engineering, chemistry, biotechnology and computer science.

Canadian universities face severe competition in some fields from U.S. universities which can, and do, recruit significant numbers of Canadian doctoral graduates. Attractive salaries, excellent research facilities, and the lure of world-famous laboratories have always persuaded many Canadian scientists and engineers to move to the USA. There is a concern that the present manageable flow of Canadians to the south may become an unacceptable flood if predictions about severe shortages of academics in the USA by the end of the century turn out to be correct. Because of concern about the 'brain-drain' to the USA, NSERC has established a computer database of Canadian scientists and engineers studying in other countries at the postgraduate and postdoctoral level. Some 800 individuals are listed, voluntarily, in this register. Information about these persons is made freely available to any Canadian employer in the academic, industrial or government sector seeking to fill positions requiring an advanced degree in science or engineering.

Research collaboration involving universities

The engineering and science faculties of Canadian universities no longer can be justifiably accused of being isolated ivory towers. Slowly at first, but more recently at an increasing rate, university researchers are engaging in collaboration with colleagues in other universities, in industry and in government laboratories, as well as with colleagues in other countries. Four types of research collaboration may be identified: inter- and intra-university, inter-disciplinary, inter-sectorial, international. University researchers have been motivated to engage in collaborative research by the benefits to be gained. These include: the intrinsic value of a collaboration whereby colleagues with similar interests are able to achieve more by working together than by separate activities; the contribution to the national well-being which results from the utilisation by industry of ideas conceived within universities; an improvement in the student training process resulting from a research environment which includes researchers from industry, from other disciplines, or from other countries; an enhanced capability to investigate complex research problems requiring expertise from several disciplines.

University-industry collaboration

It is the realm of Canadian university-industry collaboration which has experienced the greatest change, both in attitudes as well as in achievements, during recent decades. For many years, little contact existed between the universities and the industrial sector. Industry saw the universities as being useful only as suppliers of scientists and engineers. University researchers were frequently viewed as being out of touch with the real world of industry. University researchers, for their part, tended to believe that challenging scientific or engineering problems did not occur in industry. Their best research students were encouraged to seek academic careers and only the less brilliant were thought suitable for an industrial job.

The overcoming of prejudices on both sides and the mutual discovery that the two sectors could benefit from scientific and technical collaboration has been a lengthy process which is not yet complete. The real beginning of this process in Canada occurred in the 1960s. During this decade, many universities had created offices of research administration whose mandates usually included the management of contract research with industry and governments. The particular case of the University of Waterloo, a university of recent foundation (1959) is worth singling out for special mention. From the beginning, the mission of this university was based upon the importance of the university-industry linkage. This importance is manifest in it's large cooperative undergraduate teaching programme, in the high level of joint research undertaken with industry, and in the large number of companies which have had their origin in the university.

During the years since 1980, many initiatives have been taken by universities, by industry and by governments to promote university-industry collaboration, especially in science and engineering research. Only a few of the more interesting examples can be mentioned here.

Bell Northern Research, the research arm of Northern Telecom, and the leading industrial research organisation in Canada, has been very aggressive in building links to engineering, computer science and science departments in the universities. This is accomplished by the funding of research projects in the universities, by membership in research consortia, by the funding of university chairs, and by the general promotion of engineering and science education. The benefits to the company lie in its favoured access to highly qualified students and to technology newly-developed in university laboratories.

An organisation dedicated to the fostering of precompetitive research is PRE-CARN Associates. This organisation is a non-profit corporation which has some thirty-five member companies and crown corporations. Each member organisation contributes an annual fee to the research programme. PRECARN promotes cooperation between industries, governments and universities on research projects in the broad area of artificial intelligence and robotics. Funding for projects comes, not only from members' fees, but also from the federal and some provincial governments (see Chapter 5 for detail). Individual projects are carried out by researchers from universities and participating companies.

Governments, at the national and provincial level, have been active in the development of various programmes and incentives intended to promote university-industry research collaboration. An interesting initiative was the creation, in 1984, of the Canadian Microelectronics Corporation. This is a non-profit organisation, largely funded by the federal government through NSERC, but with substantial in-kind support from Northern Telecom. The corporation provides specialised services to about twenty-five universities, enhancing their ability to conduct research in microelectronics. Special equipment and software is loaned to the universities and there is a sophisticated communication linkage between the universities and the corporation, which is located at Queen's University. University researchers may submit designs for integrated circuits to the corporation electronically. After checking for errors, the designs are sent to a Northern Telecom fabrication centre, chips are manufactured and then returned to the universities for use in research projects.

Since the early 1980s, NSERC has actively promoted a set of programmes now known as Research Partnerships. These programmes are specifically intended to bring together university researchers and industry, to facilitate the transfer of technology from the universities to Canadian companies and to contribute towards the development of a high quality industrial research base. This is done by means of three groups of programmes: cooperative research and development activities; technology diffusion activities; industrial research chairs. The latter activity is especially noteworthy. It involves a company (or several companies), together with NSERC and a university, in the funding of a research initiative which includes the salary of a university research chair holder, as well as financial support for infrastructure costs, equipment and general research expenses. The intent of the industrial research chair activity is to assist universities in building on existing strengths to achieve the critical mass required for a major research endeavour in science and engineering of interest to industry. Over 100 such chairs have been established since 1983. Industrial organisations have contributed about fifty per cent of the costs of these chairs.

Another NSERC activity which promotes university-industry interaction is it's programme of Industrial Research Fellowships. Through this scheme, Canadian companies are encouraged to employ recent doctoral graduates in their research and development departments. Fellows are expected to receive competitive salaries and to work on challenging projects. Their salaries are subsidised by NSERC for two years. The programme has two main objectives: to induce young scientists and engineers to consider a career in industrial r&d; and to encourage more companies to hire highly qualified persons, thereby increasing the innovative capability of the Canadian industrial sector. About sixty-two per cent of the Fellows have remained with their host company after the two-year period, about seventy-seven per cent have remained in the industrial sector. The federal government has recently provided additional funding to NSERC in order to double the size of this programme to 100 new awards each year.

Among provincial governments which have also actively promoted university-industry collaboration are Quebec, British Columbia and Ontario. The government of Quebec, for example, has made major contributions towards the funding of several new centres which help to strengthen university-industry links. One of these, le Centre de recherche informatique de Montréal (CRIM), is a consortium of universities and companies which undertakes research in various aspects of computer and information science. The province of British Columbia has established a non-profit Discovery Foundation which links the business, educational and government sectors. An example of its activities is the establishment of 'Discovery Parks' located adjacent to each of the three universities in British Columbia. These parks are the home to innovative companies which, in many cases, are making use of technology transferred from the universities. Ontario has created a University Research Incentive Fund to encourage universities and the private sector to enter into cooperative research ventures with each other.

Several Centres of Excellence programmes have been established in Canada during the past five years. Two will be mentioned here because an important feature of these programmes is collaboration between universities, industry and government. In 1988, the province of Ontario launched a Centres of Excellence programme through the Ontario Technology Fund. Through a competitive, peer-reviewed process, seven proposals for Centres were approved. These Centres all involve the participation by universities and companies. Frequently, several universities are collaborating with each other as well as with industry in these Centres. The fields of research of all seven Centres are in science or engineering.

The federal government followed the lead of the Ontario government by establishing it's own similar Centres programme, Networks of Centres of Excellence (NCE), in 1990. A competitive and internationally peer-reviewed selection process resulted in the creation of the fifteen NCEs listed below:

Canadian Network for Space Research
Institute for Robotics and Intelligent Systems
Canadian Institute for Telecommunications Research
Micronet – Microelectronic Devices, Circuits and Systems for Ultra Large Scale Integration
Neural Regeneration and Recovery Network
Respiratory Health Network of Centres of Excellence
Canadian Bacterial Diseases Network
Canadian Genetic Diseases Network
Canadian Aging Research Network
Mechanical and Chemimechanical Pulps Network
Ocean Production Enhancement Network
Protein Engineering Network of Centres of Excellence
Insect Biotech Canada-Biotechnology for Insect Pest Management
Network of Centres of Excellence on High-Performance Concrete
Centres of Excellence in Molecular and Interfacial Dynamics

Each of these NCEs typically involves participating researchers in many universities, companies, or government research laboratories distributed throughout Canada. Their aim is to generate synergism in research, and to accelerate the diffusion of knowledge and the transfer of new technology to the private sector. Responsibility for the programme of NCEs is shared by the three federal granting councils. The Centres will be evaluated over the course of the coming year to review their impact on advances in knowledge, effects in networking and contributions to the economy. In December 1992 the federal government announced an extension of this programme.

Numerous organisations promote university-industry linkages in Canada. Among the more active are the Canadian Association of University Research Administrators and the Corporate-Higher Education Forum. This latter organisation was established in the mid-1980s and is composed of the presidents and chief executive officers of some of Canada's more research-intensive universities and corporations. It's reports and studies have provided catalytic actions in the area of university-industry interaction.

Collaboration between university researchers

Research teams made up of cooperating university faculty members have not been a noticeable feature of Canadian universities. (There are, of course, some exceptions). For many decades, the programme of individual operating grants for research of the National Research Council and, after 1978, the Natural Sciences and Engineering Research Council, has been a much admired way of fostering research in the universities. If anything, however, this program has tended to discourage active collaboration between researchers in favour of an environment of individuality in which the size of one's personal grant has become an indication of standing in comparison with peers.

In recent years, there has been a trend towards promotion of cooperation between individual researchers. Initiatives such as the Ontario and federal Centres of Excellence schemes foster such collaboration, but also involve researchers in other sectors. Some other activities have been focused upon the encouragement of university researchers, in one or in several universities to work together. The province of Quebec has been very forward-looking in this respect. Since 1987, the Quebec research funding organisation, FCAR, has encouraged the formation of research centres involving one or more universities. The 'critical mass' characteristic of these centres favours the creativity and productivity of the participating researchers and their students. Some twenty Quebec centres exist in the various disciplines in the natural sciences and engineering. On a smaller scale, FCAR also supports teams of researchers at an individual university. In the natural sciences and engineering, nearly 300 teams involving almost 1000 faculty researchers receive research funding from FCAR.

An unusual organisation which has had some impact upon research collaboration is the Canadian Institute for Advanced Research (CIAR). A private, nonprofit corporation, CIAR was established in order to focus both intellectual and financial resources on research programmes in areas considered to be of critical importance. Funding for the Institute has come from industry, individuals, foundations and governments. The principal role of CIAR is to create networks of excellent researchers at universities in Canada and, occasionally, abroad in order to create a critical mass of talent in a limited number of programme areas. In effect, CIAR identifies groups of researchers who already receive high levels of research funding from other sources, such as NSERC, and provides full salary support to these individuals for a term of five years in order that they can devote their full effort to research. Funding is also provided to enable researchers to interact closely with each other. Programmes which have been established include artificial intelligence and robotics, cosmology, evolutionary biology, and superconductivity.

NSERC has introduced a new programme of collaborative project grants with the specific objectives of promoting collaborative research both within and between disciplines, and to train students in a collaborative research environment. Grants will be awarded for specific research projects with defined objectives. The anticipated budget for this new programme is relatively small but, even so, it is likely that there will be a significant impact upon the academic research community by fostering the idea that collaboration in research is a meritorious activity.

Interdisciplinary research

For many academic researchers in Canada, interdisciplinary activities often tend to be associated with the latest intellectual fashion, and interdisciplinary research is thought to lack rigour. Despite such views, scientific and engineering research increasingly draws upon ideas and inspiration from a variety of disciplines. Research on artificial intelligence is a case in point.

Although interdisciplinary research does take place in Canadian universities, there have been very few mechanisms designed to encourage such activities. The research grants programme of NSERC, if anything, tends to discourage interdisciplinary research because, given the structure of the peer review process, proposals for such research are usually adjudicated by a discipline committee. However, the new Collaborative Project Grants of NSERC, previously mentioned, will be open to interdisciplinary proposals, including those which may involve non-science disciplines.

A recent research funding initiative of the federal government is unusual in that it will require cross-disciplinary proposals. This initiative is the Eco-Research programme which, as the name suggests, supports advanced research and training in environmental studies. The programme is an element of Canada's so-called Green Plan, announced by the federal government in December 1990. The Eco-Research programme consists of three main components: research grants; university faculty chairs; and doctoral student fellowships. The research grants component is intended to support studies of Canada eco-systems affected by local, regional or global change. Research proposals must integrate the perspectives of the human sciences, the natural sciences and engineering, and/or the health sciences. Early indications are that this programme will generate a considerable interest in the universities, with the most common cross-disciplinary linkages being between the natural and the social sciences.

International research collaboration

Canadian university researchers tend to have close relationships with colleagues in other countries. Several factors favour such links. In the first place, a substantial fraction of the science and engineering faculty at the universities have come to Canada as immigrants. NSERC estimated, in 1985, that about one half of those faculty who held NSERC research grants had received their bachelor's degree abroad. Secondly, of those faculty who received their bachelor's degree in Canada, about twenty per cent received their PhD. from a foreign university. Overall, therefore, a big majority of the science and engineering faculty at Canadian universities have an international educational background.

International research collaboration tends to take place predominantly with colleagues in the USA, UK, France and Germany. However, increasing numbers of collaborations are occurring with colleagues in Japan, China, Australia and in eastern European countries.

The linkages which have been fostered by researchers with personal roots in other countries are diversified as a consequence of Canada's significant role in educating engineers and scientists from other countries. The substantial number of foreign doctoral candidates in science and engineering departments has already been mentioned. The fact that most of these students eventually return to their home countries means that the seeds are sown for subsequent research collaboration between foreign researchers and their former Canadian supervisors. This process encourages research links with Third World and developing countries in particular.

Further potential research collaborations arise from the fact that a significant number of young Canadian researchers spend two years or more as postdoctoral fellows in foreign laboratories. For example, NSERC encourages winners of its postdoctoral fellowships to hold their awards at leading research centres in other countries. In the year 1990–91, 128 new fellowships (out of a total of 206) were held outside Canada. Of these, seventy per cent were held in the USA.

In recognition of the long-term benefits to be gained by establishing linkages between researchers through such mechanisms as postdoctoral fellowships, Canada has recently introduced a new programme of international fellowships. These NSERC awards provide foreign doctoral graduates, at an early stage in their career, an opportunity to spend one or two years at a Canadian university engaged in research in collaboration with a Canadian host. Each year, seventy new fellowships are awarded. Over the years, the cumulative effect of these awards is expected to lead to an increase in joint research projects involving Canadian and foreign scientists or engineers.

The federal government has determined that research links between Canada and Japan should be strengthened, given the importance of Japan to Canada as a trading partner. Accordingly, the government has established a Japan Science and Technology Fund. The purpose of this Fund is to enhance Canada's scientific and technological base by expanding mutually beneficial collaboration with Japan. The Fund has three components which provide support to the industrial sector, to the university sector, and to federal or provincial government organisations. The university component provides up to seventy-five per cent of the costs of short or longer term visits by Canadian researchers to Japanese universities, government laboratories, or industrial research facilities, the Canadian component of bilateral research projects and Japanese language training for Canadian researchers involved in visits or collaborative projects. (This Fund along with examples of other international science and technology initiatives is detailed in Chapter Seven).

Canadian university researchers participate in numerous sizeable international research projects, and there is an increasing trend to this type of activity. Some recent and current examples of such collaborations are the Canada-France-Hawaii telescope, the World Ocean Circulation Experiment, the Ocean Drilling Program, and the Sudbury Neutrino Observatory.

Conclusion

This chapter has not elaborated on the evolving policy environment affecting university education and research, and the value placed by Canadians on higher education in general. This is discussed further in Chapter Thirteen.

Canadians can take some pride in the university system for the training of scientists and engineers which has evolved, particularly during the past 40 years. The system produces individuals who compare favourably in their abilities with their counterparts in the other Western industrialised countries. In some respects, one can say that this achievement has taken place in spite of a system of school education which lacks any formal national standards. Canada is unusual in being a federal state without a federal department of education. Even so, there is considerable public interest in educational matters. National educational standards are being debated by the Council of Ministers of Education and by their clients – parents and students. The Learning Initiative of the national Prosperity exercise has highlighted the need for greater coherence in the education system and has proposed a Canadian Forum on Learning to bring this about. The effects of such initiatives upon the role of the universities in training scientists and engineers could be to make their task simpler. The universities may be called upon less frequently to compensate in their teaching for the shortcomings of the schools. Students may receive a better education in the basic scientific and mathematical subjects in the schools and be better motivated for the rigour of the universities. The emergence of a greater and clearer role for the technical institutes and vocational colleges which may come about will also help the universities by ensuring that students attend the type of institution which best suits their abilities and talents.

The balance between the teaching and research roles of the universities will continue to be debated and will probably lead to improvements in the way in which undergraduate students are taught. There is likely to be little impact upon the training of researchers and the carrying out of research by the universities. In this connection, the most important factor will continue to be the availability of funding. It seems unlikely, given the present fiscal outlook for Canada, that the universities will increasingly have to look for funds needed to ensure their international competitiveness in science and engineering research.

Notes

1 For a thorough overview of the Canadian university system, see the article by Professor A D Gregor (ACU 1991)
2 In 1992, the federal government surprised the academic community by announcing it's intention to merge the Social Sciences and Humanitites Research Council, the Canada Council, and an academic relations division from the Department of External Affairs to create an as yet un-named new organisation.

References

ACU (Association of Commonwealth Universities) 1991 *Commonwealth Universities Yearbook 1991 Vol 2 The Universities of Canada*1013–1027
AUCC (Association of Universities and Colleges of Canada) 1991a *Directory of Canadian Universities 1991*

AUCC 1991b *Trends: The Canadian University in Profile* 1991 edition

Brychcy T Wilson S C 1991 *NSERC Strategy for Increasing the Supply of Highly Qualified Personnel in the Natural Sciences and Engineering - Emphasis on Women* Natural Sciences and Engineering Research Council CAE (The Canadian Academy of Engineering) 1991 *Engineering Research in Canadian Universities*

CEMB (Canadian Engineering Manpower Board) 1989 *The Canadian Engineering Manpower Inventory Databook*

ECC (Economic Council of Canada) 1992 *A Lot to Learn – Education and Training in Canada*

FCAR (Fonds pour la formation de chercheurs et l'aide à la recherche) *Rapport annuel 1990–91*

ISTC (Industry Science and Technology Canada) 1991 *Women in Science and Engineering. Volume I: Universities*

Kavanagh R J 1991 *New Scientists and Engineers from Canadian Universities* Natural Sciences and Engineering Research Council

King M C 1989 *E.W.R. Steacie and Science in Canada*University of Toronto Press

MacAulay J B Dufour P 1984 *The Machine in the Garden, The Advent of Industrial Research Infrastructure in the Academic Milieu* Science Council of Canada

NSERC (Natural Sciences and Engineering Research Council) 1991 *Natural Sciences and Engineering Research Council Act: Office Consolidation July 1991*

OCGS (Ontario Council on Graduate Studies) 1991 *Doctoral Graduation Rates in Ontario Universities: A Discussion Paper*

OECD *1991 Main Science and Technology Indicators*

Science and Technology Canada, Ministry of State 1987 *InnovAction - The Canadian Strategy for Science and Technology*

Smith S L 1991 *Report, Commission of Inquiry on Canadian University Education* AUCC

Thompson W P 1963 *Graduate education in the sciences in Canadian universities* University of Toronto Press

Tory H M (ed) 1939 *A History of Science in Canada* Ryerson Press

Chapter 7

International s&t collaboration

Michel Leclerc and Paul Dufour[1]

The role of s&t will continue to expand. It is probable that, not more than ten or twenty years from now, the locus of decision-making in many sectors will have to shift from the national to the international level. Sacrifices of national sovereignty may become necessary. Thus, the need to develop a capability for properly assessing costs and benefits, in both the political and scientific spheres, will be increasingly felt.

Canada, Science and International Affairs, 1970.

The above citation – whether prescient or just common sense – has proven itself in ways the authors could not have imagined. Indeed, the world's economic, social and political structures have been significantly altered by advances in scientific knowledge and progress in technology. Domestic policy has become internationalised to the extent that it is now often referred to as 'intermestic'. Corporate activity has been globalised such that several observers have difficulty distinguishing between national and international firms; according to some, the question of foreign ownership itself is now irrelevant, with the key factor revolving around corporate behaviour. Science, technology and innovation have become strategic elements in foreign relations, defence and trade policy, and are now seen as critical components to resolving global challenges such as environmental pollution, climate change and natural disaster reduction.

As a result of these situations, it has become facile to argue that national borders are meaningless. The capacity of nation-states to regulate their respective economies and innovation systems is overwhelmed by the enormous and rapid expansion of global trade and investment and the growth of financial and transportation highways and networks that straddle nations.

Indeed, some have argued that the role of the state will no longer be to advance so-called national interests, but rather to reconcile local aspirations with global limitations. Nevertheless, with respect to national innovation systems, political units do indeed still count.

The process of internationalisation can be described as three distinct phenomena. The traditional view rests on the movement of tangible and intangible goods, services, capital, ideas, and people across national borders. A second, and more interesting phenomenon refers to the increasing importance of non-national institutions such as global corporations and supranational political or scientific entities. The third alludes to problems, events, or situations that inherently involve more than one country, and that cannot be addressed effectively by domestic actions alone. Each form of internationalisation has a profound effect on the way that nation-states and their governments can influence, manage, and control domestic economic and social processes.

The international dimensions of s&t have been recognised widely in the past, but today, and in the future, these forms of knowledge have tremendous potential to liberalise global social and economic interests. This phase of internationalisation can be characterised by a number of trends, most of which are not mutually exclusive given the symbiotic relationship between scientific and technological knowledge. The new dynamics of trans-national science, technology and innovation activity include:

- *scientific advances and the creation of large technological systems*: for example, technology-driven communication, as witnessed by the rapid increase in the number of world-wide computerised communications and financial networks and transportation systems. The extension of scientific frontiers and increases in the scale and scope of scientific endeavour in some areas has seen the launching of major international scientific projects involving substantial public and private sector participation.
- *economic factors*: a world-wide movement toward corporate networking is taking place, measured by the surge in international technical cooperation agreements, strategic alliances, intra-industry and inter-industry research and development (r&d) consortia, and decentralisation of corporate r&d activities. There is also a trend toward regional economic integration in Europe, Asia and North America, with implications for s&t that have yet to manifest themselves fully.
- *geopolitical factors*: the widespread turn to market liberalism as an anticipated solution to problems of growth and accumulation; the end of the Cold War, and the start of an international regime of 'cooperative competition'. With the international debt problem, development issues in the Third World and Eastern Europe are becoming urgent. Environmental questions are also paramount, as the industrialised and Third World grapple with the linkages between economic development and environmental sustainability. International diplomacy, trade policy and domestic strategies are all becoming intertwined as s&t are seen as partial but critical elements of solutions.

These and other factors demonstrate the changed nature of international relations and the evolving role of s&t in this new landscape. Governments are increasingly under pressure to grapple with the plethora of changes; so too are corporations now engaged in large-scale global activities.

International technology systems

As national systems of innovation are increasingly dependent upon the capacity of world markets to innovate, their effectiveness can no longer be dissociated from the external elements that condition their evolution. In fact, by intensifying the free flow of scientific exchanges, technoglobalism has made possible the 'delocalisation' of scientific production instruments, that is, the value-added activities held or controlled by companies outside of their national borders. [Dunning, 1988].

However, some data attest to a persistence, if not a predominance, of the so-called 'national' system as a relevant scientific development space. According to these data (Table 7.1), a weak level of internationalisation characterises most major industrialised countries, at least on the basis of patent statistics; small countries call more naturally upon external sources of technological development. As one commentator explains it;

'Small countries, because of the absolute limit of the resources they can devote to scientific and technological effort and the faster internationalisation trend of their own firms, have probably found out sooner than large countries, the natural limits of national technology support policies. In these countries, the step towards international technology support programmes was quickly made.'
Soete 1991.

Patent statistics, for example, only provide a limited overview of the national technological capacity. For its part, the Business Enterprise r&d(BERD)/ technologies payment ratio allows the relative portion of national and foreign sources in technological progress to be defined. This ratio reveals that in general, the payments made for foreign technologies represent only a modest portion of national r&d. Whereas in Japan and Sweden the amounts spent on importing technologies come to less than five per cent of the BERD and attest to a strongly introverted technological development, this ratio is close to fifteen per cent in France, while it reaches twenty per cent in Canada and Italy and over twenty-three per cent in Australia [Rodriguez-Romero, 1992]. In these latter countries, the import of technology is relatively large in relation to the national r&d.

Moreover, statistics on r&d activities tend to corroborate this 'internationalisation' of the research effort in Canada, the result mainly of a defensive industrial strategy centred on the imitation of an American industrial and technological model [Niosi, 1991]. With the exception of Italy, it is in Canada that the foreign financing of r&d among OECD countries was the most rapid during the 1980s. Moreover, according to 1988 data (Table 7.2), it is once again in Canada that the share of foreign expenditure, related to the expenditure of national companies, was the highest at more than twenty-six per cent.

In comparison, foreign expenditure only represented about point one per cent of the r&d expenses of Japanese companies. A faster rate of increase of foreign expenditure explains such an internationalisation of expenditure in Canada: between 1980 and 1989, domestic sources of BERD increased by one hundred and

Table 7.1
Foreign controlled domestic technology compared to nationally controlled foreign technology (based on U.S. patenting, 1981–86

Home Country	U.S. patenting from inside country by foreign firms (as a per cent of country's total U.S. patenting)	U.S. patenting by national firms outside home country (as per cent of country's total U.S. patenting
Belgium	45.7	16.5
France	11.8	3.8
FR Germany	11.5	8.5
Italy	11.2	3.0
Netherlands	9.5	73.4
Sweden	5.4	16.7
Switzerland	12.5	27.8
UK	22.3	24.5
W Europe	7.4	9.3
Canada	28.1	12.5
Japan	1.2	0.5
USA	4.2	4.4

Source: Pavitt, 1991

Table 7.2
r&d carried out by firms and financed by foreign sources in some countries (1978–1979=1.0)

	1978–79	1981–82	1984–85	1987–88	Foreign expenditure in per cent of national expenditure (1988)
Italy	1.0	4.1	10.1	18.0	9.6
Canada	1.0	2.3	5.2	8.5	26.3
Norway	1.0	2.8	4.3	5.2	2.7
Sweden	1.0	1.9	3.0	4.7	2.1
UK	1.0	-	3.1	4.4	16.0
Denmark	1.0	1.9	2.8	3.9	2.8
France	1.0	1.3	2.4	3.2	10.3
Japan	1.0	2.1	2.6	2.7	0.1
Germany*	1.0	0.7	1.0	1.2	1.5

* 1979, 1981–83, 1985, 1987

Source: OECD, 1991a

fifty-seven per cent in current dollars, while the expenditure of foreign sources grew by more than seven hundred per cent. The numbers shown in Table 7.3 provide, in this respect, additional information.

Between 1980 and 1989, the share of revenues from foreign sources in Canada moved gradually from three point two per cent to nine point three per cent of the entire Gross Expenditures in r&d(GERD). We also see a similar progression in all sectors of operation, with the exception of the higher education sector, where the portion of foreign expenditure dropped progressively between 1980 and 1989, moving from one point one per cent to only point five per cent. This trend confirms the findings of an international survey which argues that 'the internationalisation of university-corporate relations seem to be, currently, more an objective than a reality' [OECD, 1990a]. In fact, according to an investigation completed in the USA, the revenues from foreign companies represented only point four per cent of the total university r&d financing in 1986. (We discuss the university role below.)

The internationalisation of scientific activities is, in part, another form of the globalisation of trade dominated by a redefinition of the industrial boundaries that mark the passage of the model of the 'multi-domestic' company [Porter, 1986], to that of the 'global internalised' company [Michalet, 1991]. For companies, the main architects of the globalisation process, internationalisation means less an erosion of their autonomy than a restructuring of their markets and an increase in the intra-corporate trade dynamics.

The flow of investments, midway through the 1980s, was characterised by an increase in direct investment into the United States by European countries and Japanese multinational corporations [Hagedoorn J and Schakenraad J, 1991], as Table 7.4 demonstrates.

The redistribution of foreign direct investment (FDI) was accomplished mainly to the benefit of the USA between 1981 and 1989, while the share of direct investments to Canada equalled, in 1988–1989, two point seven per cent of that of the OECD, as opposed to one point three per cent in 1985–1987 and three point three per cent in 1971–1980. Canada appears therefore relatively untouched by this flow of investments and, consequently, poorly integrated in the international economy.

Table 7.3

BERD financing from foreign sources in Canada, by sector and in % of the total sectoral revenues, 1980–1989

	1980		1981		1982		1983		1984		1985		1986		1987		1988		1989	
	$	%	$	%	$	%	$	%	$	%	$	%	$	%	$	%	$	%	$	%
Firms	91	5.8	158	7.4	266	10.7	431	16.7	515	17.2	518	14.3	547	13.8	730	17.3	836	18.4	789	17.0
Universities	8	1.1	9	1.0	10	1.0	11	1.0	11	1.0	8	0.5	11	0.6	11	0.6	13	0.6	12	0.5
Non-Profit	1	3.0	1	2.6	2	4.4	1	1.8	2	3.1	3	4.0	4	4.7	6	6.2	8	6.7	8	6.3
PROs	1	2.4	1	1.9	1	1.7	1	1.7	2	3.0	2	2.6	2	3.0	2	2.6	3	3.7	5	5.9
Total	101	3.2	169	4.3	279	6.0	449	9.0	530	9.5	531	7.8	564	7.7	749	9.8	860	10.4	814	9.3

Source: Statistics Canada, 1991.

Table 7.4
Portion of large OECD countries in direct investment to and from overseas,
1971 to 1989

Percentage of total OECD flux

	1971–1980	1981–1984	1985–1987	1988–1989
	Investment overseas			
USA	46.4	20.1	25.3	16.9
Canada	3.9	8.0	5.4	3.5
Japan	6.2	13.2	16.2	27.6
E.C.	41.8	56.0	50.1	47.6
Belgium-Luxembourg	1.1	0.4	1.8	-1.0
France	4.8	8.1	6.5	10.8
Germany	8.6	9.7	9.4	8.7
Italy	1.2	4.6	2.7	2.6
Netherlands	6.4	7.8	5.6	4.9
Spain	0.4	0.9	0.6	0.9
UK	19.2	24.6	23.4	20.6
Sweden	1.6	2.7	3.0	4.4
	Investment from overseas			
USA	33.8	62.8	59.3	51.4
Canada	3.3	-2.4	1.3	2.7
Japan	0.9	0.9	1.2	-0.6
E.C.	61.5	38.3	37.3	45.8
Belgium-Luxembourg	5.5	3.6	2.3	4.6
France	10.1	6.4	5.7	6.6
Germany	8.4	2.9	2.2	2.9
Italy	3.4	3.5	3.0	3.7
Netherlands	5.1	3.0	3.2	3.9
Spain	4.2	5.7	5.9	6.1
UK	24.7	13.2	15.0	18.0
Sweden	0.5	0.5	0.9	0.7

Source: OECD, 1992

The issue of developing a sound, integrated technology policy in the face of liberalised trade and technology 'leakage' has received considerable play in Canadian policy circles. There is some question, for example, whether a 'North American' innovation system exists given the 'hub and spokes' arrangement between Canada, Mexico and the USA Indeed, recent work from the OECD tends to demonstrate that successful interlinking of local and regional networks with global networks of innovation and production is likely to be a key determinant of local and regional competitiveness in the 1990s. Differences between regions will become increasingly important as the globalisation process proceeds. As a result, much emphasis in Canada has been placed on promoting national networks of excellence, strategic technological partnerships and 'technology clustering' [Niosi,1992].

Changing international scientific dynamics

While technology and corporate behaviour is undergoing significant mutation at a global scale, a parallel re-orientation is taking place in the world scientific community. The scientific community itself is being transformed, as scientific production has increased in the liberal democracies. In the less developed countries, small increases in scientific activity can be detected, while the fraction of the world's scientific activity in the communist countries has declined. With a widening gap between the liberal democratic nations and the LDCs, it is likely that the periphery of science will become further removed from the centre. Nevertheless, scientific activity nowadays is performed within a scientific community which has members in more societies, spread over larger areas of the earth's surface, than ever before.(Schott, 1991)

Canada is dependent on access to scientific knowledge and technological progress. It is dependent as a result of the country's limited contribution to world science (approximately four per cent) and global technology (approximately two per cent). Canada needs skilled personnel to maintain its industrial and scientific base. In order to continue to have access to global research programmes and to new technological opportunities, however, Canada must continue to 'pay its way' with continued strong support of it's own domestic research base. Therefore, the management of its interests and appropriate support of it's scientific and technical communities in an internationalised world is critical. Understanding the roots of the country's s&t policies is a key link in how the country is now moving to deal with international s&t collaborative activity. (For more on Canada's historical tradition in science, see Chapter Two).

Canada's international s&t lineage

Because of it's colonial heritage, Canada has benefited from scientific and technical ties to the UK and France. It's geographic proximity and economic linkages with the USA has also resulted in significant growth and maturity of it's s&t infrastructure. The traditional characterisation of Canada as a small colonial country on the periphery of world scientific centres has considerable truth to it – at least until World War I.

Before World War I, Canadian scientific participation in the international forum was limited mainly to informal personal ties with British, French and American scientists. So-called 'inventory sciences' of geology, terrestrial magnetism, astronomy, and botany were rapidly developed as the nation called Canada came into being. Engineering and applied science became a useful glue to bind the nascent nation as it struggled in the later part of the 19th century to emerge onto the world stage.

Scientific societies and their journals were established along the lines of the European models and Canadian naturalists, geologists, and engineers took considerable pride in their establishment. Apart from memberships in the Royal Society of Canada (est. 1882), the scientific elite were also active in American and British professional societies, including the American Association for the Advancement of Science (which held an early meeting in Montreal in 1854) and the British Association for the Advancement of Science (which met in Canada in 1882 and 1886).

With the establishment of the National Research Council of Canada in 1916, a strong pattern for Canada's maturity and independence in scientific affairs emerged. The NRC was responsible for virtually all of Canada's formal non-governmental scientific exchanges in the early part of the 20th Century. Representation at various international scientific congresses and expositions resided with the NRC. The role was consolidated between the wars and especially during World War II, when the NRC became the principal scientific liaison of Canada with the Allies. Prior to World War II, many viewed Canada as a minor overseas branch of British research, and a convenient point of access for American technology [Phillipson, 1989]. The NRC's scientific liaison functions in Washington and London began the process of an independent scientific policy that was to reach fruition with Canada's acceptance into the exclusive economic summit club known as the G-7. Canada's physics community, for example, had played a pivotal role in work on atomic physics at an Allied lab at the Université de Montréal in the early 1940s; work that was later to translate into a uniquely-developed Canadian nuclear technology known as the CANDU reactor.

Many of the post-World War II international governmental arrangements in science were cultural in nature and followed the traditional functionalist school of contributing to the world pool of knowledge and to global peace. In the context of East-West tensions, Canada signed an early agreement on scientific and industrial cooperation with the USSR (1971); an agreement that has since been re-assessed given the current turmoil in the former Soviet Union and increased attention by the G-7 to assist in upgrading that region's science and technology infrastructure.

Belgium and France were also the targets of formal cultural and scientific agreements in 1971 and 1965 respectively. At the non-governmental level, the NRC maintained strong Canadian representation in the International Council of Scientific Unions, and it's affiliated unions and committees.(The NRC is today the adhering member to 28 ICSU organisations). From 1950 to 1970, Canada found itself in the top ten of countries participating in governmental and non-governmental organisations, with Canada hosting an average of two to three per cent of international scientific congresses held during that period.

In the late 1960s and 1970s, the burgeoning field of international relations in science and technology began to take on a complexion that required greater sophistication and structure to track and implement. Before then, little central coordination within the national government took place. Individual government departments and agencies entered into bilateral arrangements on their own, with little assistance other than token briefings on foreign policy dimensions. The NRC, which had a network of technical offices abroad responsible for scientific liaison, chaired a Standing Committee on External Relations, but it remained an informal arrangement. The birth of Canadian scientific diplomacy coincided with the tabling of the *White Paper on Foreign Policy* in 1968. In it, the Canadian government described it's participation in international scientific and technical activities;

> *'Canada's most effective contribution to international affairs in future will derive from the judicious application abroad of talents and skills, knowledge and experience, in fields where Canadians excel or wish to excel (agriculture, atomic energy, commerce, communications, development assistance, geological survey, hydro-electricity, light-aircraft manufacture, peacekeeping, pollution control, for example). This reflects the Government's determination that Canada's available resources...will be deployed and used to the best advantage, so that Canada's impact on international relations and on world affairs generally will be commensurate with the distinctive contributions Canadians wish to make in the world.'*
> Maynard Ghent, 1979.

It was only in 1970, when the Department of External Affairs established the Scientific Relations and Environmental Problems Divisions (re-organised into a Bureau of Economic and Scientific Affairs in 1971), that international scientific relations became more formally structured (at least, the governmental linkages – the individual and corporate relations were quite another matter).

At the beginning of the 1970s, Canada was already a member of more than 200 organisations of cultural, scientific or technical nature, Canadian scientific diplomacy was pursuing objectives of its own but which also served to support foreign policy; the diplomacy found in the latter an effective way to widen its field of activity (Schroëder-Gudehus, 1989).

With the creation of the Ministry of State for Science and Technology (MOSST) in 1971, a policy and advisory function (in international scientific activities) was included in it's mandate. This would eventually lead to numerous bureaucratic squabbles with other central agencies such as the Department of External Affairs (DEA) over jurisdiction. Compounding the issue was the fact that provincial governments also began to see the value of developing their own international contacts in the area of technological information. MOSST established an International Branch which assumed responsibility for the federal government's bilateral s&t agreements, major s&t visits to and from other nations, and Canada's involvement in multilateral organisations such as the OECD's Committee on Scientific and Technological Policy, the Commonwealth Science Council and the International Institute for Applied Systems Analysis.

In 1975, the Interdepartmental Committee on International Science and Technology Relations (ICISTR) was established to help coordinate Canadian federal responses to and participation in international s&t activities. This Committee, chaired by External Affairs, consisted of a number of federal departments and agencies that had well-developed outreach activities with foreign partners. It's mandate is now under review in the face of changing roles for government in international s&t relations.

In 1976, in order to clarify responsibilities, a Memorandum of Understanding was prepared between MOSST and the Department of External Affairs. The main discussion centred on whether s&t was an extension of domestic interests and an element in Canada's overall foreign economic policy. The MOU served to smooth out the international relations picture until 1983, when the International Division of MOSST was transferred to DEA.

During this period, the international s&t affairs of Canada came under scrutiny by the Science Council of Canada in a series of reports examining the role and nature of these activities. The 1973 report *Canada, Science and International Affairs* examined the nature of Canada's activities, it's approach and suggested a stronger training and expertise in international relations related to s&t. The 1979 report *Canadian Government Participation in International Scientific and Technological Activity* focused on the value of a series of bilateral agreements with other countries, namely Belgium, France, Germany. In 1984, a statement from the Science Council made strong recommendations concerning the role of the Science Counsellor system and the need for it's expansion. Canada's Scientific Counsellor system consisted of seven officers posted in embassies in Washington D.C., Tokyo, Bonn, Brussels (Belgium and the EC), London, Paris, and the OECD (the number hasn't changed for several decades, though the system has been expanded to include locally-engaged Technology Development Officers whose function is to seek out technological prospects and opportunities for Canadian firms).

In a government decision of 1982, the international s&t relationship and role of government received more attention. Many of its principals are still in operation today. The strategic framework focused on three elements;

1. the selection of major international bilateral and multilateral target activities

chosen for their importance to Canadian domestic s&t programs, foreign policy or economic goals
2. the adoption of a more narrowly focused approach to the development of future bilateral collaborative activities
3. the establishment of a simple mechanism to facilitate the planning and initiation of these activities

Within the ambit of the second element, all future umbrella government-to-government bilateral activities were to be developed on a step-by-step basis, with each nation initially identifying projects or well-defined sub-areas considered to offer the potential for mutual economic and scientific benefit. It was in this context that national economic interests came to the fore as the guiding principle for any future umbrella agreements. As a result, only one formal government-to-government s&t agreement has been signed since the government decision. In 1986, a Canada-Japan Science and Technology Agreement was signed where many of the 1982 guidelines applied. In 1989, a Complementary Study of mutual s&t interests was initiated by the two Prime Ministers of Canada and Japan, and undertaken on the Canadian side by the Science Council of Canada. The ensuing report highlighted the need for a senior-level bi-national foundation to coordinate international s&t activities, as well as the necessity for a catalytic fund to support joint projects. The fund was created in 1989 as part of the federal government's Going Global strategy; the proposed Foundation did not materialise.

Other issues discussed in the government decision of 1982 included attention to the newly industrialising nations of Mexico, Korea, Brazil and Venezuela, where Canadian economic interests could be fulfilled as well as contributing to development objectives. This development objective was already well-recognised with the establishment in 1970 of the International Development Research Centre, a unique organisation designed to initiate, support, encourage and conduct research into the problems of the developing regions of the world and into the means for applying scientific, technical and other knowledge to the economic and social advancement of those regions. Today, IDRC's annual grant from the Parliament of Canada amounts to $114 million. It has a network of six regional offices in various parts of the globe and it's support for scientific research extends to some 5000 projects in more than 100 countries. A sister agency, the International Centre for Ocean Development, was eliminated by a 1992 federal government budget decision.

Also recognised in the 1982 decision was the need for some catalytic seed fund that would provide support for new international s&t opportunities. Using the existing ICISTR, the fund was to be managed by several departments, with a view to supporting collaborative s&t activities undertaken primarily to meet Canadian foreign or economic policy needs. The fund was transformed in 1986 into the Technology Inflow Program to support technology prospecting by Canadian organisations abroad. (See below)

With this 1982 strategy, the Canadian government set a course that would see international s&t relations from a more limited perspective of national economic interests and that would eventually transform the government's intelligence network of science counsellors (later Science and Technology Counsellors and Technology Development Officers) into an extension of commercial interests. Foreign aid objectives and cultural matters developed along parallel tracks.

The early function of the scientific attachés evolved from permanent scientific liaison representatives sent by the NRC during World War II to address numerous scientific and technical matters directly related to the war effort and also to the more political and diplomatic challenges of science policy, trade and investment matters. Some of the reporting that went on was conditioned by existing s&t agreements; in others, visits of scientifically oriented teams and exchanges of information were predominant. In 1974, when the NRC's last liaison office closed in London and DEA assumed all responsibility for the Science Counsellors (following a 1967 Cabinet Decision that had transferred all NRC responsibility for international s&t relations to DEA), the extent of coverage by the appointed officers was large, and included agriculture, arctic research, electronic data processing, energy research, environment, urban development, industrial technology, marine s&t, medicine and public health, science policy, space technology and transportation systems.

The evolution in approaches between the NRC Scientific Liaison function and the DEA Scientific Counsellor role of 1974 can be illustrated by the increased attention to policy issues:

Duties for Counsellors (Scientific) - 1974

- advice to Canadian ambassadors on non-military science or technology that affected science policy
- reporting science policy information to Canada
- representation of Canada on inter-governmental bodies
- facilitating the international exchange of scientific and technical information
- assisting visiting Canadian scientists

This has changed considerably in 1992 as the network itself has been supplemented by other agents whose function is to procure information on and prospect for industrial and technological intelligence for business interests. Indeed, the network will increasingly come under re-examination as the traditional role of s&t counsellors is squeezed on the one hand by fiscal and personnel restraints, and on the other by increased demand for information and analysis from a more sophisticated clientele.

Sub-national linkages in international s&t

Clearly, one of the complicating features of the international s&t liaison function by the central government has been the rise of provincial or state activities in this area.

Science and technology are essentially conducted in institutional concentrations and tend to be highly concentrated and geographically specific. Geographic proximity is a determining factor in s&t linkages and it is no surprise to discover that a great deal of effort goes into developing the strategic knowledge assets of local communities. At least, that is the trend today. In Canada, because of the emergence of technology engines designed to spur local-regional economic development, much activity is devoted by provinces and municipalities to developing their technological infrastructures. Over the past decade, every provincial government has acquired a s&t strategy or policy.

More recently, the provinces have been experimenting with active s&t policies that involve international linkages. The provinces spend more than $100 million abroad, where they maintain fifty-seven independent offices in twenty-five countries. Chief among these are Alberta, Quebec, Ontario, and British Columbia. Ontario, with seventeen offices abroad costing $27 million has been promoting technological and strategic alliance linkages with the so-called Four Motors of Europe. Alberta has six offices costing $9.7 million, and over twenty agree-ments in energy and oil sands technology with international partners. Quebec in particular has published a White Paper on International Relations that examines various facets of it's global position, including s&t. A good deal of it's $53 million budget for international offices abroad is being targeted to promote Quebec's technology-based industrial clusters that were announced in December 1991 (e.g. aerospace, transportation, pharmaceuticals). Quebec's largest city, Mon-treal, also has released a major development plan that places considerable em-phasis on international linkages and investment. In addition, in June 1992, the Quebec granting council for scientific research announced a new three year pro-gram to assist in the financing of international scientific cooperation between Quebec universities, researchers and students and foreign colleagues.

The scale and scope of provincial interest in international s&t relations is not well-documented. As Ghent has argued;

> 'The development of provincial interest in international s&t was largely a product of the emergence of new world priorities in fields of traditional provincial concern. These once triv-ial jurisdictions gradually assumed enormous significance, not only as key factors in the do-mestic shift of power, but also in the concomitant development of a provincial international competence.'
> Ghent, 1980.

Canadian foreign policy has, as a result, become a national, not solely federal, matter.

Nowhere is this activity as pronounced as it is with transborder linkages with American states. It would be no exaggeration to suggest that there is more trans-border science and technology collaboration between Canadian provinces and U.S. states than there is among the Canadian provinces. This is merely a reflec-tion of the extensive inter-provincial trade barriers, where historically little for-mal collaboration existed in s&t between the provinces.

A 1974 study, for example, revealed about 766 state-provincial arrangements of which about a third were scientific and technological in nature. These covered a wide range of areas such as oil pollution prevention, transborder wildlife mi-gration, fisheries, crop degradation, energy and transportation; many of these are in areas that one would expect in dealing with common transborder issues [Swanson, 1974]. Also, as one would expect, the most active state-province ar-rangements are those adjacent to one another. A number of these arrangements were carried out under the aegis of federal agreements (e.g., Great Lakes water quality). [Halliwell et al; 1992].

Very little systematic work exists to document the scale and scope, and more importantly, the new trends or directions in state-province relationships. It is al-most certain that the scope has not diminished, but that with the mutual devel-opment of state s&t programmes and provincial s&t policies, it is clear that the linkages are much stronger in technological development and transfer dimen-sions. For example, Quebec has assisted several of the province's firms with ac-cess to MIT research through the latter's Industrial Liaison Programme. The Quebec programme, called 'Technological Bridges', is a good example of the new orientation to target access to developments in research and technology from abroad. With the creation of numerous metropolitan technology councils and the like in both Canada and the USA, one can expect that increased linkages of this sort will be the norm rather than the exception. In addition, as regional groupings of provinces and states review common concerns, technology is in-creasingly on the agenda (e.g., New England Governors and Quebec and Mari-time Premiers). Finally, some provinces see advantage in linking their expertise to that at the federal level and use federal resources and intelligence networks to strengthen their promotion of international investment or alliances in techno-logy (*e.g.*, Investment Canada, External Affairs and Ontario have collaborated to put together biotechnology investment promotion seminars in several states, in-cluding New Jersey and Texas; Ontario and BC are promoting their firms' capa-bilities to bid on contracts for the Superconducting Super Collider).

With this proliferation of arrangements at the provincial-regional level, an in-creased attention to coherence of national action is emerging. The jurisdictional questions that marked the early 1970s and 1980s between MOSST (now ISTC) and DEA (now EAITC) continue, though there are indications that these will be-come consolidated as the federal government works more closely with the pro-vincial governments, and augments its national technology acquisition and diffusion system.

Having established some discussion of the administrative and policy appar-atus dealing with the early evolution of international s&t relations in Canada, we turn to some of the scientific issues and technological activities increasingly affected by this international environment.

Formal cooperation

Bilateral agreements

Over the last few years, the Canadian government has stepped up it's interna-tional scientific cooperation activities. Between 1985 and 1990, for example, Canada signed close to 250 bilateral arrangements on scientific and technical co-operation with more than 50 countries, including many of the member countries of the European Community. (This list does not include defence-related or de-velopment assistance projects).

Table 7.5, an inventory of the Canadian bilateral arrangements, shows that formal exchanges between Canada and foreign countries take place mainly with Western Europe (thirty-eight point six per cent), Asia (nineteen per cent) and North America (sixteen point two per cent). The Middle East, Oceania and Africa total less than ten per cent of foreign exchanges.

Arrangements of cooperation or technical and industrial assistance represent seventy per cent of all types of agreements concluded with foreign partners, not-ably with the European and Asian countries. More than forty per cent of the ar-rangements incorporating patents or licensing have been signed with Western European countries. While there has been little systematic effort to review and assess the benefits of these bilateral governmental arrangements (unlike the U.S. system, for example, which requires an annual inventory and progress report on scientific diplomacy), periodic assessments have taken place, including the Canada-Germany (est. 1971) and Canada-Japan s&t agreements (est. 1986). By way of illustration, we highlight some of these bilateral s&t activities.

Canada's s&t relations with Europe

In 1992, the Government of Canada and the provinces had at least 370 formal in-ternational s&t 'arrangements' worldwide [EAITC '92]. Of these, 113 were with members of either the E.C. or the European Free Trade Association (EFTA), 26 with states of Central and Eastern Europe, and 31 with the republics of the for-mer Soviet Union. Other arrangements were with countries as diverse as Algeria

Table 7.5

Inventory of Canada's scientific and bilateral technological cooperation, 1985–1990

	Western Europe	Eastern Europe	North America	Latin America	Asia	Middle East	Oceania	Africa	Total
Cooperation or technical & industrial assistance agreements	62	20	35	10	35	5	1	5	173
Patent & licensing agreements	15	3	1	3	4	3	2	4	35
Exchanges of information	8	1	2	6	4	—	—	—	21
Exchanges of personnel & training	4	1	—	1	2	1	—	—	9
Technological transfer	—	—	—	1	2	—	—	—	3
Standardisation agreements on	2	—	2	1	—	—	1	—	6
Total	91	25	40	22	47	9	4	9	247

Source: External Affairs and International Trade Canada, 1991

and Yemen. These arrangements are classified into those having the force of law, such as treaties and agreements, and those such as letters or memoranda of understanding that imply moral rather than legal obligations. The 370 do not include those between universities, firms, and individual researchers. Of growing importance are agreements between Canadian and foreign municipalities having an s&t component. These are also excluded.

According to a review of Canada's s&t relations with Europe, undertaken in 1991, there were approximately 116 arrangements, many of these at the subnational levels. [Dufour and McInnes, 1992].

Under the terms of the 1976 Canada-E.C. Framework Agreement for Commercial and Economic Cooperation, two subcommittees were established to promote industrial, science and technology collaboration between Canada and the E.C. In keeping with the focused approach that the E.C. takes to r&d funding, Canadian collaboration has centred on scientific fields such as information technologies and telecommunications, as well as remote sensing, environmental research, medical research and marine sciences.

EUREKA, the European research program originally conceived by President Mitterand in response to President Reagan's Strategic Defence Initiative (or Star Wars programme) and officially launched in 1988 by seventeen European countries, has assembled over 300 projects to date with a total budget of $8 billion and to which are connected more than 1,600 companies and research institutes. Over the years, the initial network has been joined by extra-European countries: Indonesia, Argentina, India, Yugoslavia and, finally, Canada.

Few Canadian companies have succeeded at developing a true partnership within the EUREKA framework, even if the $20 million Technology Opportunities in Europe Programme (TOEP) has sometimes eased their task. Through the government, financial assistance has been provided to close to fifty Canadian firms that wanted to participate in this programme. In fact, only Calmos Systems Inc., Lumonics Inc. of Kanata, Zenon Environmental Inc. of Burlington and Unisearch of Toronto have collaborated in making projects financed by EUREKA. More than ninety per cent of the $12.5 million spent to this end was allocated to these four Ontario firms. In 1990, Quebec-based Gentec Inc., followed later by DMR Inc., have taken part in EUREKA. From it's beginnings as a mere subcontractor, Gentec has risen to the rank of full partner of European firms within the framework of the Euro-laser project. Thanks to a $1.8 million financial contribution, this firm has associated itself with a European consortium composed of Haas-Laser and Festkorperlaser of Germany, Quantel of France and IEQ of Italy (Asselin, 1990). TOEP has since been phased out as a result of an evaluation of potential users which demonstrated that these success stories were likely to remain exceptional.

The role of E.C.-funded research programmes is particularly influential on overall European research efforts. The Framework Programmes for Community Research and Technological Development were initiated by the E.C. in the early 1980s to promote European industrial competitiveness. The Framework Programmes encourage and focus collaborative pre-competitive and pre-normative r&d involving industries, universities and governments of member states in key technologies.

The Third Framework Programme, which spans the period 1990–1994, comprises fifteen major programmes and has a total budget of ECU 5.7 billion. While this is only about four per cent of European s&t expenditures, it is generally believed that the impact upon long-term r&d planning by European research entities is significantly multiplied. The programme focuses on a limited number of strategic r&d areas grouped around three axes: enabling technologies; management of natural resources; and management of intellectual capital. This approach represents a consolidation of thirty-seven programmes run under the Second Framework (1987–91).

The Commission of the European Communities is recommending to the European Council of Ministers that the Fourth Framework Programme (1994–1998) encompass all Community r&d programme activities, including those international cooperation activities now outside the scope of the current programme. With the signing of an agreement with the EFTA countries, they too now have the right to participate in E.C. research programmes under the conditions applying to E.C. member states.

The E.C. Council of Ministers has yet to approve the negotiation of an operational science and technology agreement with a non-E.C./EFTA country. However, it is now reviewing an application from Australia and discussions are underway to prepare the conditions for negotiating a new s&t agreement between the E.C. and Canada. This would give Canadian research entities access to E.C. funded research programmes.

Canada-Germany agreement on scientific and technical cooperation
The Canada-Germany Agreement was signed in 1971, and was preceded in 1957 by an agreement on the peaceful uses of nuclear energy. It's purpose was to strengthen ties between the two nations, to broaden the scope of scientific cooperation, and to improve the quality of life and economic well-being of Canadians and Germans [20 Years Cooperation in s&t, '91]. Since 1971, Canadian and German scientists have collaborated on some 300 cooperative ventures, ranging across twelve different fields of research. While the bulk of these projects involved basic science performed by governmental, university and private research institutes; more recently, particularly in the fields of advanced technology such as telecommunications and space research, industry has become increasingly involved.

Given the importance of Germany scientifically, it's restructuring of the best of the institutes from the eastern lander, it's trade and investment relationship in Central and Eastern Europe, and the leadership it is exercising in s&t relations with the republics of the former Soviet Union, Canada will likely continue to place a great deal of importance on it's s&t relations with Germany. The challenge for this relationship in the next few years may lie in Canada's adjustment in the face of Germany's attention to the east, and it's increasing role within the E.C.

Canada-France
Canada and France exchanged letters on s&t cooperation in 1973. In addition, there are some twenty-six other arrangements signed by provinces, and federal research agencies with French counterparts. R&D collaboration has ranged over a wide area. This is understandable given that France is, with the UK and Germany, one of the most dominant research nations within Europe. Major fields of cooperation include biotechnology, environment, marine sciences, space, medical research, cold regions technology, industrial materials, transportation, and telecommunications. France's role in the European Space Agency, and Canada's contribution through the Canadian Space Agency, is a particularly strong node of cooperation.

Canada-Norway framework for enhanced bilateral s&t cooperation
Canada and Norway negotiated an exchange of letters in 1986. Designed to facilitate and enhance s&t cooperation, the letters were extended for a further five

years from 1992 to 1997. The s&t collaboration initially took place across a broad front in just about every area of priority for Norway. Given Norway's industrial and resource-based economy, it's s&t interests match those of Canada's strengths. The areas covered include fisheries and ocean science, hydrography, energy and geoscience, environment, polar research and forest pathology. In addition to the umbrella arrangement, Fisheries and Oceans Canada has three s&t arrangements, and the Government of Newfoundland and Labrador is an active player in offshore energy research with Norway.

Obviously, this section cannot do justice to the other arrangements Canadian government institutions have negotiated with other European and non-European countries. Of note on the multilateral front is Canada's active participation in Agenda 21 research activities (following the 1992 UNCED meeting); it's participation in the Organisation of American States' Directorate of Scientific and Technical Affairs, where a range of bilateral mechanisms have been developed to collaborate with Western Hemispheric nations; its membership in the science and technology Task Force of the Pacific Economic Cooperation Conference; and, its technical partnerships in the Commonwealth Science Council, la Francophonie through such agencies as the Canadian International Development Agency and the International Development Research Centre.

Programmes

A variety of programmes are in place to ensure the continuity, if not the durability, of bilateral cooperation. In the following paragraphs we will briefly describe the main programs of international scientific cooperation set in motion by the Canadian government.

External trade being the cornerstone of the Canadian economy, it is therefore somewhat natural that Canada seeks to weave scientific and technical links with it's major commercial partners, so as to favour collaboration in that area and that of technology transfer. The commercial promotion strategy 'Going Global' is the multi-disciplinary response of the Government of Canada to the new challenges of multilateral competitiveness. This strategy, unlike previous policies, attempts to approach international business, science, technology and investment from an integrated perspective.

The five-year 'Going Global' programme [est. 1989], endowed with a budget of $93 million, is comprised of three main elements

1. a USA Opportunities strategy to help Canadians take advantage of the Free Trade Agreement signed in 1989
2. a Pacific 2000 strategy to help Canadians participate fully in the emerging markets of the region
3. a Europe 1992 strategy to prepare Canadians for opportunities and challenges inherent in the integration of the European Community

We discuss elements related to the first two of these planks.

Opportunities in the United States and the impact of free trade on science and technology

As noted earlier, despite the strong historically-based scientific and technical ties with the UK and the USA, no formal umbrella agreement exists between these countries, though there are a host of specific sector and scientific arrangements between scientific institutions and research organisations.

The USA-Canada collaboration has been marked with strong historical ties. The volume and scope of these linkages is far greater than that with other countries. In the area of research and logistics associated with the Arctic, for example, Canada and the USA share over 67 agreements covering security, wildlife and wilderness management, fisheries, energy and environmental pollution. In space sciences and research, Canada's Alouette I satellite, launched in 1962 by a NASA rocket, provided both countries with over ten years of scientific information about the behaviour of the upper atmosphere in the northern latitudes. Today, Canada is devoting thirty-three per cent of it's total space programme envelope to the Space Station Freedom, through the contribution of the Mobile Servicing System, a robotic system. In addition, Canada will provide three per cent of common systems operation costs scheduled in mid 1997. In return, Canada expects to have access to three per cent of the Space Station research capacity and of the polar platforms and is entitled to provide three per cent of the Space Station crew. Canada's RADARSAT program (due to be launched by NASA in 1994), the Earth Observation and Astronaut Programmes, and the WINDII experiment launched on board NASA's UARS are other example of bilateral Canada-U.S. collaboration in s&t (see Chapter Nine). The Great Lakes Water Quality Agreement, signed in 1972 by both countries, has led to over $8 billion on research and engineering aimed at improving the water quality of these lakes shared by 40 million Canadians and Americans.

On the multilateral side, Canada and the USA share several strong partnerships in such international programmes as the Human Frontier Science Programme, the proposed International Thermonuclear Experimental Reactor Project, the Intelligent Manufacturing Systems Programme, and is considering possible participation in the American $8 billion Superconducting Super Collider. The extensive sub-national linkages have been noted above.

The historic bilateral Canada-U.S. Free Trade Agreement that came into force in 1989 appears to be the forerunner of a larger North American regional economic grouping. With the initialing of a trilateral agreement in August 1992 between Canada, the USA, and Mexico, one of the largest and richest regional markets in the world is being constituted. Even without a formal North American trade agreement, the region has already undergone significant economic integration. Canada and the USA are each other's largest trading partner, and the USA is Mexico's largest trading partner.

Science and technology *per se* were not directly addressed in the Canada-U.S. Free Trade Agreement, nor in the North American agreement, but several issues related to investments, technical standards, subsidies, public procurement, and intellectual property have important implications for Canada's technological development. [Davis, 1992].

Some key issues raised have been the importance of reaching an understanding with the USA on the range of public practices in support of science, technology, and innovation that would be considered as fair and therefore would not be liable to countervail. Unfortunately, the issue of subsidies was intractable during the Canada-USA negotiations and could not be resolved. Each party is currently collecting information about the subsidy practices of the other with a view to completing the agreement within five to seven years. The way the subsidies question is resolved (likely to be overtaken with the resolution of the GATT Uruguay Round of negotiations) will have important implications for innovation policy as well as for more general society-state relations in North America.

Most Canadian firms test their international wings south of the border, where differences in language, culture, customer preference, standards, distribution, and general business practices are not great (though often important). In fact, the strong attraction of the U.S. market explains in part why Canadian governments must institute various measures to encourage trade and technology exchange elsewhere. More important than cultural proximity may be the roles played by many Canadian advanced-technology firms as suppliers of components or sub-assemblies to large American firms. This role limits the ability of Canadian firms to take the lead in innovating entire systems and, consequently, hinders their entry into new markets. Furthermore, there are some indications that take-up of government assistance programs for technology acquisition is smaller for U.S. markets than for other regions.

However, trade liberalisation modifies the parameters of firms' strategic choices in ways that are not yet obvious. There is evidence of correlation between the degree of technology activity in a firm and management attitudes towards trade liberalisation, suggesting that trade liberalisation should be accompanied by policies favouring the diffusion and adoption of new technologies in manufacturing firms. Furthermore, Canadian firms appear to prefer approaching internationalisation via strategic alliances. Canada has experienced a boom in the formation of technology-related alliances. A 1989 survey of 822 Canadian advanced technology companies found that forty-one per cent were involved in strategic alliances. Firms with alliances had established an average two point seven alliances in the previous three years. A large majority (seventy-one per cent) of the alliances were with foreign partners. The rate and density of alliance formation is highest in biotechnology. Changing market structures, increasing complexity of products, accelerating costs of producing and absorbing technological innovations, and unexpected entry of competitors into previously secure product-market segments are frequently cited reasons for domestic firms' involvement in strategic alliances. The most frequent form of strategic alliance was the r&d consortium and the most often cited reason for entering an alliance was market growth. In electronics, which is over forty per cent foreign-controlled in Canada, a study of patterns of strategic alliances shows that Canadian firms prefer European partners, suggesting positioning to enter the European market. The study confirms that technical alliances are viewed as being ways of securing complementary assets by opening windows on new technologies and new markets. [Niosi, 1992].

Trade liberalisation and the formation of regional free-trade areas does not necessarily imply that firms will increase r&d investments, as many believe. A 1989/90 survey of over 800 advanced technology firms in Canada reported that 'the attitudes of respondents toward the Free Trade Agreement with the USA were both contradictory and discouraging" [Ernst & Young 1990]. Seventy per cent of firms were generally optimistic about the effects of the Agreement, but only forty-two per cent of firms thought that they themselves would benefit from it. When asked what actions they were planning to undertake in response to free trade opportunities, twenty-eight per cent indicated that no actions were planned, and thirty-four per cent said they didn't know. Thus, sixty-two per cent of the surveyed firms had no particular strategy with respect to trade liberalisation with the USA.

Strengthening the Canada-U.S. bilateral relationship in s&t requires a heightened sense of partnership. Calls are currently being made to move the Canada-U.S. relationship in this direction. William Winegard, Canada's Minister for Science, argues that 'techno-shock,' brought about by globalisation and the challenge of Japan, make a partnership in s&t between the United States and Canada necessary [Winegard 1991]. The two countries are seen to share an innovation 'system' which is now in competition with others around the world. A properly developed Canadian-U.S. approach to scientific and technological development is an 'unexplored agenda'. This agenda, according to some, might involve joint planning to identify ways to maintain a world-class s&t base in North America. This could include increasing collaborative research in complementary areas, identification of strategic industrial sectors common to both Canada and the U.S., promotion of technological diffusion, collaboration in education and training, and also establishment of cross-national linkages between business and industry, joint development of strategic plans between Canadian and American industry associations, common approaches to multilateral research projects, and collaboration between Canadian and U.S. granting agencies. Also proposed are joint meetings between Canada's National Advisory Board on Science and Technology (NABST) and the President's Council of Advisers on Science and Technology.

Despite the increased attention to 'formalise' Canada's s&t linkages with it's American counterparts, there remains some concern over the benefits that can accrue to Canada in the absence of any well-defined national technology strategy.

Pacific 2000
Eleven of the economies experiencing the fastest growth in the world are Asian. Japan alone represents a larger market for Canada than the UK, France and Germany combined. Canadian trade with Japan could reach more than $40 billion annually in the early 2000s. In 2000, the 'four small dragons' – that is, Taiwan, Hong Kong, Singapore and Korea – will probably catch up with Europe as far as industrialisation is concerned, and will enjoy faster economic growth rates than all of the OECD countries. Canada's trade with the 'four dragons' could rise to $10 billion at the turn of the century.

Canada is already an important part of the commercial business scene in Asia and the Pacific. Three of the five largest export markets outside the USA are in Asia: *i.e.*, Japan, Korea and Taiwan. According to forecasts, Canada will continue to strongly attract the attention of business people and investors/entrepreneurs from Asia and the Pacific.

Canada-Asian cooperation ins&t therefore appears essential to Canada's economic development, even more so given that Canada is lagging behind in this regard. In fact, while European and U.S. companies have already created thirty-four r&d centres in Japan, Canadian companies have one (Northern Telecom opened a twenty researcher facility in 1991).

The Pacific 2000 program is comprised of four phases;

(a) a *Pacific commercial strategy,* based on the promotion of trade in the Asian and Pacific region; the penetration of the market via key sectors; a broader commercial representation of Canada in the region, and the grouping of bilateral or sector-based companies
(b) a *training in Asian languages and culture fund*
(b) a *Pacific 2000 projects fund,* created to strengthen the awareness and knowledge of Canada in Asia so as to guarantee the efficient distribution of Canadian products in that region
(d) *the Japan Science and Technology Fund (JSTF)*

The JSTF represents, without a doubt, the most ambitious element of the 'Going Global' strategy. The $25 million JSTF is jointly managed by External Affairs and International Trade Canada (EAITC), and Industry, Science and Technology Canada (ISTC), in collaboration with the Natural Sciences and Engineering Research Council of Canada (NSERC).

Created to promote the technological development of each of the partners and to favour the creation of strategic associations in a number of priority sectors in Canada, the Fund has a number of objectives designed to;

- increase the participation of Canadian scientists and engineers in relevant world-class Japanese research and technology development programmes and projects
- train highly qualified personnel in state-of-the-art Japanese research facilities and thereby fill identified gaps in Canada's scientific and technological capabilities
- facilitate access to Japanese technologies and industrial laboratories considered important to the competitiveness of Canadian industry; and
- collaborate in research, standards setting and similar initiatives to facilitate exports to Japan

The Canada-Japan Science and Technology Agreement (signed in 1986) provides the necessary umbrella for on-going s&t collaboration between the two countries. There are over 100 co-operative projects in place, covering such areas as agriculture, ocean sciences, advanced manufacturing technologies and space science.

The JSTF aims to widen the access of Canadian researchers to s&t and research institutes in Japan. It supports various activities such as short or long-term visits and exchanges, bilateral collaborative projects, linguistic training activities and joint workshops. The Fund is open to researchers and Canadian research institutes from industry, universities and the government, and supports a vast range of original activities, from basic to pre-competitive research.

International Assistance Mechanisms

Formal cooperation with foreign entities cannot be efficiently conducted without the permanent support of a solid network of professionals assigned overseas. To this end, the Canadian embassies and consulates abroad house officials in charge of s&t dossiers and who are responsible for both concluding and implementing agreements. In this regard, Canada has a network of scientific and technological assistance comprised mainly of working personnel based in foreign countries; scientific and technological counsellors and technology development officers working in some twenty embassies and consulates, for instance in the USA, Japan, UK, France, Germany, and the E.E.C., as well as with international agencies such as the OECD.

In a number of these missions abroad, Canada has recruited local Technology Development Officers, responsible for assisting Canadian small and medium sized-businesses in acquiring foreign technologies. Their primary duties consist of ensuring the following services;

- facilitate technology acquisition and technology transfer
- search out contacts, make introductions, and locate foreign firms interested in arrangements such as joint ventures and r&d collaborative projects
- gather information on specific s&t areas and direct specialists to more detailed information
- familiarise Canadian firms with the business practices and operations of science and technology organisations in host countries

Locally-engaged strategic venture officers have also been appointed in certain of the embassies to provide market and investment intelligence for Canadian companies. In addition, the Trade Commissioners play a similar role in over 30 other missions. These agents are responsible for answering specific requests concerning new or appropriate technologies. They also are an integral part of the Technology Inflow Programme (TIP). Through this demand-oriented programme, which, in 1991, had a budget of $5.6 million Canadian companies can obtain foreign technology. The network of Technology Development Officers, in conjunction with the over 200 industrial technology advisers of the National Research Council, assists clients in evaluating their needs and using the services and support offered by the TIP network. TIP has also undergone a recent evaluation and it's future financing and delivery are being reviewed, though the NRC has assumed the responsibility for financial support of the program.

Multilateral cooperation

In an era of large private and public research consortia, Canada cannot afford to be isolated from the supranational research programmes supported by considerable multi-annual funding. Canada has therefore associated itself with several big science programmes requiring the construction of major research facilities, such as the Space Station, astronomical observatories and high energy physics projects, which are organised around large-scale equipment, instrumentation and research teams. Another type of big science programme requires vast distributed multi-disciplinary collaboration based on long-range objectives. The Human Genome Project and the Ocean Drilling Program are examples of the latter.

Financed mainly by Japan (approximately $26 million), based in Strasbourg and chaired by the Japanese, the Human Frontier Science Programme (HFSP) is open to several countries. The assistance offered by this programme is available in the form of student grants and research and laboratory support. In 1990, eighty $40,000 grants were announced for young neurobiology or molecular biology researchers in a foreign laboratory. That year, eight Canadian researchers benefited from financial assistance. In 1991, for example, an international team directed by Dr André-Roch Lecours received a $280,000 grant from HFSP, with the aim of supporting a multinational team of specialists in reading and writing problems. The team will attempt to establish a link between certain brain injuries and the partial or total loss of the ability to read and write. Canada participated in the first experimental phase of HFSP. The budget for the 1989–1990 exercise was around $20 million U.S., and has since been increased with greater funding from the US and the EC.

A short list of some of the more important big science projects involving Canadian participation is shown. As Chart 1 illustrates, the scope of scientific involvement from the Canadian perspective is quite large and several key policy questions are raised as a result, including how to maintain funding for on-going, essential 'little science', and how to develop mechanisms to better consult on prospective funding decisions with both the domestic and international communities. Fortunately, the OECD has created a Forum on Megascience issues as a result of the March 1992 Ministerial Meeting of s&t Ministers. This forum will serve as a useful vehicle to help promote better coordination amongst OECD members and to provide a useful catalyst for better exchange of information prior to national governments' decisions that will affect the international scientific community. As initial case studies. the OECD is examining astronomy and deep drilling with global change research to be examined at a latter time.

Chart 1: Selected list of big science projects with Canadian participation
- European Organisation for Nuclear Research(CERN)
- Canada/France/Hawaii Telescope (CFHT)
- Human Genome Project
- European Space Agency (ESA)
- Global Climate Observing System
- World Ocean Circulation Experiment (WOCE)
- Joint Global Ocean Flux Study (JGOFS)*
- WINDII
- Ocean Drilling Program (ODP)
- Sudbury Neutrino Observatory (SNO)
- International Geosphere/Biosphere Program (IGBP)
- ZEUS/HERA
- KAON*
- Superconducting Super Collider (SSC)*
- Gemini 8m. Telescopes*
- International Thermonuclear Experimental Reactor (ITER)*
- Space Station Freedom*
- BOREAS*

  under discussion or in planning phase

Canada is rarely the leader in megascience projects, simply because of it's limited s&t resources in this area. While Canada has a legacy of initiating a few such projects, it is more often than not a participant in such large-scale research. A recent exception is KAON. The $1.2 billion KAON proposal entails an expansion of the existing sub-atomic physics facility in Vancouver, the Tri-University Meson Facility (TRIUMF). The federal government has offered to provide $236M towards the estimated cost of construction of the KAON Factory with the province of British Columbia supporting a significant amount as well. The balance would be funded by international contributions.

A number of countries have indicated an initial interest in participating, *e.g.*, Germany, the UK, USA, France, Italy, Korea. It is understood that the project is seeking at least $200M from foreign contributors. The announcement by foreign contributors of their actual level of participation will probably only follow the announcement of any potential arrangement between Ottawa and the provincial government. Canada must also examine other potential investments in high energy physics projects that potentially compete for scarce resources. Other such projects include continued involvement in CERN, the proposed Superconducting Super Collider, and the International Thermonuclear Experimental Reactor. In each case, Canadian scientists have considerable expertise and informal involvement.

Another megascience project involving distributed research efforts is the Human Genome Project. The USA, Japan, UK, Italy, France and Germany are the major participants in the human genome initiative. It is an initiative not without scientific or ethical controversy. Annual U.S. expenditures are expected to be $200 million (U.S.). The UK will spend $110 million(U.S.) over the next three years. The European Community has launched a $35 million per year initiative aimed primarily at model systems. Current annual expenditures by Japan approach $14 million. An international scientific coordination body, the Human Genome Organisation,(HUGO) has been established. It has an elected membership of 250 distinguished scientists from twenty-three countries, including eleven from Canada, two of whom are on the governing council. Efforts are currently being mounted to focus on data management - formats, consistency of representation, organisation, exchange of information, access, *etc.*

Canada established a formal programme of support for the human genome project in June 1992. The National Cancer Institute, the federal government and the Medical Research Council are contributing a total of $22 million over the five year period 1992–1997. Canada has already developed important basic strengths that can both contribute to mapping and sequencing and draw from the results of others.

The academic community and international science

Along with federal, provincial and corporate s&t investments in international collaborative projects, the academic community in Canada has been grappling with how to better deal with, and manage, the effects of internationalisation.

Non-proprietary scientific research, which is usually, but not always, located in a higher education environment, is the stuff of trans-national scientific relations. The Canadian university community, which has historically been the international window to the Canadian scientific community, is now scrambling to adjust to the rapid internationalisation of the Canadian s&t system. It is very difficult to provide comprehensive information about the wide range of measures to promote international exchanges and movement between universities. However, some examples will illustrate the scale of this international activity.

A survey of 89 members of the Association of Universities and Colleges of Canada [AUCC, 1991] found that nearly two-thirds emphasise an international role in the institution's mission statement. One-third maintain an inventory of international visitors on campus, and fifty-nine per cent note that internationalisation has resulted in or encouraged more interdisciplinary collaboration within their institution. However, while there are about six hundred formal exchange agreements between Canadian and European universities, for example, many are dead letters because the universities do not have resources to implement them. With the developing world, the situation is similar. Canadian universities have about 450 international development projects, over 100 with Latin America.

Nevertheless, many cases of active collaboration exist. With the USA, Canadian universities have maintained long-standing and extensive linkages. In the sciences and engineering, examples include collaborative research arrangements between the Universities of Calgary and Northwestern in computer sciences, between Dalhousie and Kansas State in chemistry, between the University of Saskatchewan and South Florida in mechanical engineering, and between University of Toronto and Wisconsin in environmental sciences. With Mexico, similar collaborative research and teaching arrangements exist between the University of Montreal and Mexico's National University in mathematics, in forest engineering between the University of Quebec at Trois Rivières and Guadalajara, and in water resources between Laval University and the State University of Mexico. Various projects sponsored mainly by Canada's development agencies (the Canadian International Development Agency, and the International Development Research Centre) also create linkages between Canadian and Mexican institutions.

At the level of researchers and students, the links between Canada and the USA are very strong. A register of Canadian students studying abroad in the natural sciences and engineering shows that over seventy per cent are in the U.S. The Register is maintained by the Natural Sciences and Engineering Research

Council (NSERC) to assist Canadian institutions in locating qualified individuals for employment. NSERC also has an International Fellowships Programme for postdoctural students from abroad. It is designed to strengthen the research environment in Canadian universities and create or reinforce links between Canadian and foreign research establishments. In addition, NSERC sponsors international collaborative research grants, international scientific exchange awards, a research associateship programme with CIDA, NATO science fellowships and administers a programme of visiting fellowships in Canadian government laboratories. (see Chapter Six)

Informal cooperation

The study of internationally co-signed articles is another means of evaluating the degree of internationalisation of a national research system. This intensification in personal supranational relations is not unrelated to the growth of international commercial exchanges [Schott, 1991]. The proportion of articles signed by researchers from two or more countries is today considered an indicator of the breadth of international research [OECD, 1992].

Canada's contribution to this trans-national cooperation effort remains largely unknown and is just beginning to be the subject of more systematic studies [Leclerc M, 1992].

The present analysis is based on the Science Citation Index (SCI) database from the Institute for Scientific Information (ISI), as processed by Computer Horizons Inc. (CHI). This data base compiles the articles published in about 3,700 international journals endowed with a reading committee and representing ninety per cent of the major networks of international scientific communication [Garfield, 1983].

The articles are classified in eight fields: mathematics (MAT), physics (PHY), chemistry (CHM), engineering and technology (ENT), earth and air sciences (EAS), basic biology (BIO), biomedical research (BIM) and clinical medicine (CLI).

During the 1981–1986 period, as shown in Table 7.6, Canadian scientists published 21,016 articles in cooperation with foreign researchers, *i.e.* more than twenty per cent of all Canadian publications. This internationalisation index is slightly lower than the average recorded in most European countries, except for the UK, where the index equals seventeen point eight per cent.

Among the major industrialised countries (Table 7.7), Canada was, between 1981 and 1986, the country whose internationalisation index recorded the lowest annual increase: equal to three point two per cent, this increase evolved at a rate two point three times slower than the average world increase (seven point six per cent). Between 1981 and 1986, this index gradually rose from eighteen point eight per cent to twenty-two per cent.

Canada, whose volume of publications during this period totalled four point three per cent of world production, (Table 7.6), nonetheless contributed to five point six per cent of the internationally co-signed articles.

Canada's cooperation index, for the entire 1981–1986 period, therefore equals a ratio of one point three, which means that Canada's participation in international scientific exchanges, as measured by the bibliometric indicators, was thirty per cent higher than it's relative expected contribution, taking into account it's real weight within the global system of scientific production. On the

Table 7.6
Ranking of Canada and the major countries according to number of publications and trans-national publications, 1981–1986

	1 Publications	2 Cooperative publications	3 Internationalization Index (2/1)	4 % total publications	5 % total cooperations	(5)(4) Cooperation Index
USA	844,472	83,266	9.8%	34.8	22.3	0.64
UK	203,588	36,351	17.8%	8.4	9.7	1.15
U.S.S.R.	182,722	6,578	3.6%	7.5	1.8	0.24
Japan	169,404	12,451	7.3%	7.0	3.3	0.47
Germany	151,364	32,038	21.1%	6.2	8.6	1.38
France	121,693	27,009	22.2%	5.0	7.2	1.44
Canada	103,456	21,016	20.3%	4.3	5.6	1.30
India	65,228	4,953	7.6%	2.7	1.3	0.48
Italy	58,294	14,584	25.0%	2.4	3.9	1.62
Australia	52,338	8,596	16.4%	2.1	2.3	1.09
Netherlands	43,421	10,806	24.8%	1.8	2.9	1.61
Sweden	43,048	11,359	26.4%	1.7	3.0	1.76

Source: Science Citation Index and Computer Horizon Inc.

Figure 7.1: *Canada's Articles Internationally Coauthored, 1989*

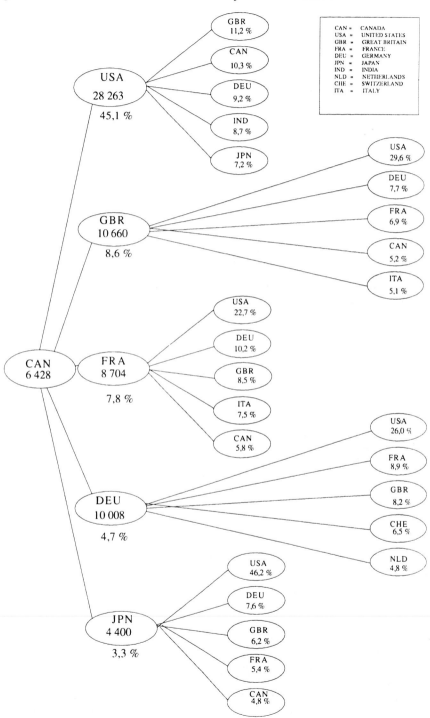

Source: Banque des indicateurs de la science (B&S), Ministère de l'Enseignement Superiéur et de la Science (Québec)

Table 7.7
Evolution of the internationalisation index of the major OECD countries,
1981–1986

	1981	1982	1983	1984	1985	1986	Average annual increase (%)
U.S.A	7.5	7.9	8.6	9.3	9.7	10.2	6.3
Japan	5.2	5.7	6.0	6.7	7.1	7.5	7.2
Canada	18.8	19.5	19.1	20.5	21.5	22.0	3.2
UK	15.5	16.4	17.0	18.8	19.0	20.3	5.5
France	17.7	19.6	21.6	23.2	24.7	26.0	8.0
Germany	17.0	18.3	20.0	22.0	23.4	26.2	9.0
Netherlands	21.5	24.7	24.0	27.0	25.1	26.6	4.3
Europe	18.0	19.6	20.7	22.7	23.5	25.6	7.3
World	13.6	14.7	15.7	17.2	18.2	19.6	7.6

other hand, we note that countries such as the USA, the former Soviet Union, Japan and India play a modest role in international scientific exchanges, considering their real contribution to world science.

As Figure 7.1 shows, international scientific cooperation outlines the main pivots of geopolitical relations within the G-7 economic zone.

Following the example of the major industrialised countries of this zone, Canada cooperates extensively with the USA which accounts for, according to 1989 data from the BIS databank, forty-five point one per cent of it's transnational scientific exchanges. Among the large industrialised countries, only Japan devotes a higher fraction of it's international scientific exchanges with the USA and this share totalled forty six point two per cent in 1989. With it's ten point three per cent contribution to international scientific cooperation, Canada remains, just after the UK, the second most important partner of the USA.

Canada's second scientific partner, the UK, absorbs eight point six per cent of the 6,428 publications internationally co-signed by Canadian researchers, while these data represent five point two per cent of the cross-border exchanges of British researchers. As for France, five point eight per cent of it's 8,704 publications written with foreign cooperation were done with Canadian researchers, the latter for their part performing seven point eight per cent of their cross-border exchanges with France. Canada's fourth scientific partner, with four point seven per cent of it's international exchanges, is the former West Germany, which has done only three per cent of it's international scientific exchanges with Canada, and ranks tenth on Canada's list of main partner countries after Italy, Austria, Japan and Poland.

Finally, Japan appears as the fifth priority zone for Canada's foreign scientific exchanges: three point three per cent of these exchanges are done with Japan, for whom Canada is also the fifth partner, internationally co-signing four point eight per cent of all of its publications.

The specialisation by discipline of international scientific cooperation varies from one country to the next, as much for the expertise specific to each country as for the politico-scientific networks woven by each of them. As Table 7.8 shows, Canada's cooperation activities are mainly concentrated in the areas of clinical research (twenty-four point seven per cent), biomedicine (seventeen point eight per cent) and physics (sixteen point six per cent), because of the dominant role of the Western European and American partners in it's trans-national scientific exchanges [Leclerc et al., 1992].

In the period between 1981 and 1986, these three fields accounted for fifty-six per cent of Canadian exchanges. Moreover, five fields of research flagged during these years in proportion to global foreign exchanges, *i.e.*, mathematics, physics, chemistry, engineering and technologies and biology. On the other hand, the fields of clinical research, biomedicine and earth and air sciences have progressed by three per cent, two per cent and one per cent respectively.

There is also the persistence, throughout this period, of a certain discrepancy between the relative distribution of cooperation between each of the fields within the Canadian scientific system and in the same fields on the global scene (column 2). In this respect, the mathematics field is particularly striking. With four point seven per cent of Canadian scientific exchanges in 1986, mathematics are in the last rank of international scientific activities led by Canada; yet Canadian mathematical activities represent eight point three per cent of global scientific exchanges in this field and significantly outstrip the most active fields of cooperation in Canada (CLI, BIM, PHY). Inversely, clinical research, which represents one quarter of Canada's international cooperation, is equivalent to less than six per cent of world exchanges carried out in this field.

Regardless of the field considered, Canadian cooperation remains largely limited to a few major countries (Table 7.9)

Thus, between 1981 and 1986, more than fifty per cent of Canada's international scientific exchanges carried out with the USA fall into the fields of clinical research, biomedicine and basic biology (much of this supported by the National Institutes of Health). In fact, in each of the fields, relations with the USA represent at least one third of Canadian foreign exchanges. The relations between Canada and the UK mainly involve chemistry (fifteen point seven per cent), and earth and air sciences and physics (ten point nine per cent). Canada-France cooperation concentrates mainly on chemistry (ten point five per cent), biomedicine (nine point five per cent) and physics (nine point three per cent).

Cooperation with other countries therefore appears relatively diluted from one field to another. In half of the fields, for example, the former West Germany represents less than five per cent of Canadian exchanges.

There are nonetheless some rare exceptions. Thus, in the field of engineering sciences and technology, Canada carries out five point eight per cent of its trans-

Table 7.8

Canada's scientific cooperation worldwide according to field of discipline, in %, 1981–1986

	1981		1982		1983		1984		1985		1986	
	1	2	1	2	1	2	1	2	1	2	1	2
MAT	7.8	11.2	6.6	9.9	6.7	9.4	6.5	9.6	6.4	9.1	4.7	8.3
PHY	17.4	4.5	16.0	4.0	14.5	3.3	15.1	3.6	16.9	3.9	16.6	3.7
CHM	10.3	5.6	9.8	5.0	10.4	5.1	9.3	4.7	10.4	5.6	9.7	5.0
ENT	8.0	8.8	6.8	7.7	8.7	8.4	8.2	8.7	7.4	7.9	7.1	7.3
EAS	9.3	6.6	11.0	7.3	9.6	6.3	11.1	6.8	9.2	6.3	10.3	6.7
BIO	10.1	8.0	8.8	7.4	9.8	7.5	10.1	7.6	9.4	7.5	9.1	7.5
BIM	15.8	5.6	16.0	5.4	16.6	5.2	16.1	5.4	17.2	5.7	17.8	5.5
CLI	21.3	5.4	25.0	5.7	23.7	5.5	23.6	5.8	23.1	5.4	24.7	5.5
CAN/% WORLD	—	6.0	—	5.8	—	5.4	—	5.6	—	5.6	—	5.4

1 = In % of Canada's international scientific cooperation
2 = Canadian cooperation in % of world cooperation in a field

Table 7.9

Canada's international scientific cooperation by country and by field, in %, 1981–1986

	USA	GBR	FRA	GERMAN	JPN	AUS	INDIA	NETH	SWITZ	.ITA	Total
MAT	45.9	9.4	5.3	5.5	1.9	3.6	4.5	1.6	0.75	1.3	79.7
PHY	36.1	10.9	9.3	6.6	4.2	2.0	2.3	2.9	3.1	3.1	80.5
CHM	32.0	15.7	10.5	5.3	4.1	2.2	3.6	1.9	1.4	2.7	79.4
ENT	40.3	8.9	4.5	3.2	5.8	2.0	6.3	0.62	1.6	1.0	74.2
EAS	48.8	12.2	5.7	5.0	1.3	5.4	1.5	2.4	1.1	1.3	84.7
BIO	52.1	8.0	5.1	2.6	3.3	4.2	1.7	1.0	0.65	0.60	79.2
BIM	53.3	9.8	9.5	3.6	2.7	2.1	1.2	1.5	1.3	0.94	85.9
CLI	57.3	9.3	8.2	2.0	2.4	2.7	0.34	1.5	1.6	0.97	86.3

Source: Science Citation Index and Computer Horizon Inc.

154

national exchanges with Japan; in the earth and air sciences sector, Australia absorbs five point four per cent of it's exchanges; and finally, close to five per cent of Canadian exchanges in mathematics are done with Indian researchers.

These indicators should be interpreted as only a partial measure of the scale and scope of international academic exchanges in the sciences. Numerous other forms of collaboration also exist between researchers, students and institutions that are not as easily quantifiable.

Indeed, it is difficult to provide strong overall measures of the degree to which Canada's s&t system is internationalised. Unlike Spain or Germany, for example, where specific budgets can be itemised along international cooperation lines, Canada possesses no strong measure of funds being spent specifically on international science and technology activities.

Conclusion

Nobel Prize winner Milton Friedman has argued that the internationalised economy is 'nonsense'. To argue that the world today is more internationalised than the early 20th century is to ignore history. As he quotes from Keynes:

> 'Before August 1914, the inhabitant of London could secure forthwith, if he wished it, cheap and comfortable means of transit to any country or climate without passport or other formality, could then proceed abroad to foreign quarters without knowledge of their religion, language, or customs, bearing coined wealth upon his person and would consider himself greatly aggrieved and much surprised at the least interference.'

Friedman goes on to suggest that the closing of the borders of the USA to immigration after World War II did more to fragment the world and disunite it than all the technological improvements in computers and satellites have done to unite it.

Friedman may have been exaggerating, but not by much. While technology and science have made great strides to unify the globe – through increased market transactions and big bangs, and communications and transportation have brought people, products and services much closer to everyone – this unifying force has been more than offset by government intervention and managed trade that is fragmenting the globe into isolated markets where the terms of exchange are by no means uniform. To be sure, we have seen the rise of truly global corporations that straddle nations and their citizens; we have witnessed the spread of international institutions that owe their allegiances to no single nation; and we have been the benefactors of global organisations that have assisted us in disasters, environmental contamination and other transborder phenomena.

But ultimately political units still count, though they are increasingly finding it difficult to master and manage the impact. While international trade theorists continue to argue for openness and market access, and highlight the pitfalls of neo-mercantilist behaviour, nation states continue to cling to the vestiges of traditional geopolitical power and manage trade, technology and investment to the best of their waning capabilities. The prototypical arrangement is the Global Partnership action plan signed by the USA and Japan in January 1992; two former enemies apparently on divergent tracks.

If bilateral arrangements don't work, then the trick is to operate through regional units of like geopolitical and cultural fabrics – thus, Europe 1993, the North American Free Trade Agreement and a proposed Asian Free Trade area. Sovereignty can still be respected in such arrangements since, in theory, one is only managing non-national economic behaviour. But what of other footloose activities that affect economic growth and national prosperity?

One such force that nations and their sub-entities have had to grapple with is science, technology and education. Investments in brain power are the only true remaining tools for comparative advantage left to governments.

The Japanese, as they go global, have probably been the most successful in managing their brain talent for national wealth by exporting their citizens to the far corners of the industrialised globe and bringing the new found riches back home. The investment, in some cases, has been so extensive that it has prompted an intensive examination by recipient countries to review the implications. Thus, American universities such as MIT have developed coherent guidelines to deal with international activities, and recent studies have explored the extent to which American investment in education has subsidised Japanese progress in s&t. Indeed, several commentators have argued that just as the centre of science shifted from Germany to the USA in the early 1920s, we may be seeing another shift of the centre (especially industrial research) to Japan, assuming, of course, that Japan is able to attract many more foreign researchers to stay in that country. [Nakayama, 1991].

While attention is being paid to the critical element of human talent and it's mobility, nation states are also adjusting their institutions of research so that they can be better positioned to reap the benefits of the advance of knowledge. This is not new; international scientific linkages have been used extensively in the past as instruments of power and prestige. Thus, today, the attention to large-scale facilities, including space programmes, are on the agendas of most industrialised nations. The question today, however, is how best to manage this international linkage in the face of massive costs and increased geo-political implications. Thus, the OECD, as noted, has discussed how to develop an international forum of governments to better plan and install early warning systems on proposed megascience projects requiring international contributions. International science is no longer the exclusive playing ground of scientists; now, governments are looking to engineer the maximum economic returns. The sheer cost of new instrumentation and keeping science equipped requires considerable attention to priority setting.

Then there is the question of controlling the ill effects of economic and environmental perturbations on national wealth and citizens. Governments therefore engage actively in international trade regimes like GATT and FTA and in multilateral forums to help mitigate environmental damage and reduce disasters (*e.g.* global change, planning for disaster).

On the technology front, managed trade is the name of the game. The capability to contain the leakage of subsidised technology is seen as a major challenge to industrialised nations. If science cannot be kept within national boundaries, then surely technology is easier to contain. Despite the immense literature on the subject demonstrating the futility of such strategies, nation states continue to aggressively pursue neo-mercantilistic policies. Thus, new international technology programmes that come along (*e.g.*, IMS, Advanced Telecommunications Research, EUREKA, *etc.*) are viewed with considerable suspicion by participating countries. How can home-based corporations reap the most benefit; how can a nation obtain fair share commensurate to its contributions; what are the motives of the initiating country; how can one gain access to an exclusive club, *etc.*?). These are the questions driving national interests.

There are numerous motives for international presence by corporations: local market access, technology access, capital (reduction of risk), cost issues, technological convergence, unusual market structures (customer preferences, standards and harmonisation), different trade regimes, shrinkage of product cycles. Along with this comes a host of policy issues centred on the so-called level playing field (*e.g.* symmetrical relationships, protection of IP, access to foreign s&t information).

All of this poses serious questions for the planning elite of nation states who have to develop sophisticated mechanisms for tracking, monitoring and pursuing global developments. Thus, in s&t, one sees the emergence of science adviser meetings among the G-7; an increased attention by the OECD in developing more effective consultative mechanisms, and more focus on technology prospecting abroad. Defence r&d conversion and dual use technologies are other issues of concern for some nations.

Until recently, Canada's domestic s&t policy system has been reactive (as it relates to international events). So too has Canadian foreign relations and trade policy. To the extent that Canada has an international policy guiding s&t activity, it is driven more by international events outside our borders than by carefully articulated domestic policy. This, of course, should not be taken to mean that domestic policies are unimportant; indeed, what one does unilaterally at home increasingly finds its entrails strewn over the international arena. Further, with the blurring of borders, there is a strong case to be made that domestic policy itself should be 'internationalised'.

For a middle-sized power with limited international clout (but still a member of the G-7, a regional trading bloc, the Arctic club, APEC, OAS, the Commonwealth and la Francophonie), maybe Canadians shouldn't expect too much. After all, do other countries with as many international memberships have these privileges? Doubtful. Compound this foreign relations dimension with that of international s&t linkages, and one can see the complexity. Canada's international science contribution is extensive (some would argue more than one might expect from our size); Canadian international technology transfer activity is heavy as a result of our memberships (particularly in the Third World) and intra-firm trade; and the country's corporate structure (foreign and domestic) is completely intertwined within global networks of investment and trade.

The new international challenges to s&t policy

Countless examinations over the past five years have highlighted the changing dynamics of foreign relations, technology, science, trade and investment. It is not that s&t has changed *per se* (although clearly this is happening with constant discovery and the 'fusion' of new technologies), but that the speed with which s&t are being developed, absorbed and adapted is such that corporations, research institutions, nations, their policies, and their existing institutions are having a hard time keeping pace with this dynamic.

Canada's foreign policy efforts have become more focused in the attempt to deal with the impact of globalisation. The 1991 'Foreign Policy Themes and Priorities' from External Affairs and International Trade provides a useful guide to the major trends in a dynamic, interdependent world. Technology plays a key role in this vision. As the framework argues, 'the leading role of technology in economic growth and prosperity in OECD countries will increase the importance of advanced infrastructures and of sophisticated and skilled human capital'. The International Trade Business Plan of the same department also attempts to provide some focus and direction to international commercial activities.

Among the objectives guiding Canadian foreign policy are broadening the efforts to facilitate the linkages between trade, technology and investment and increasing trade and investment in the knowledge-based and service industries. Environmental issues are also front and centre since environmental policy is a classic example of how domestic policy can spill over into the international arena and affect foreign policy, trade, investment and development assistance. However, like much of other federal government policy these days, non-market objectives are rarely cited and science per se is not explicitly recognised.

Given this ferment of global activity which increasingly places a premium on the currency of knowledge (in all forms), there is an opportunity for Canada to muster it's resources more effectively at home and influence the shape of things to come at the global scale. Some of the descriptions of the new domestic policy trends offered in the other chapters of this book suggest a direction that will ultimately require much more focused attention to Canada's international presence and interests in s&t.

Notes

1 Our thanks and appreciation to Victor Bradley of External Affairs and International Trade Canada, who kindly offered his comments and views on this chapter. Sections of this text are adapted from Halliwell (1992) and Dufour and McInres (1992).

References

Asselin P 1990 Gentec est devenue un partenaire à part entière du consortium européen qui pilote le projet Eureka *Le Soleil* February 23

Association of Universities and Colleges in Canada 1990 *Canada's Universities and the New Global Reality* Ottawa

CQVB(Centre québécois de valorisation de la biomass) 1988 *Euréka, mécanique du programme et possibilité d'y accéder pour une entreprise canadienne* Québec.

CSA 1991 *Background Paper on the Canadian Space Program* Montreal Canadian Space Agency

de la Mothe J Dufour P 1991 The New Geopolitics of Science and Technology *Technology in Society* 13 179–187

Dorozynski A 1990 Les Japonais ne copient plus: ils découvrent! *Science & Vie* 873 June 104–114 179

Dunning J H 1988 *Multinationals, Technology and Competitiveness* London Irwin Hymam

Dunning J H Cantwell J 1991 MNE's, Technology and Competitiveness of European Industries *Aussenwirtschaft* 1(46)

Epstein S 1992 *Buying the American Mind: Japan's Quest for U.S. Ideas in Science, Economic Policy and the Schools* Washington Center for Public Integrity

ESA(European Space Agency) 1991 *Partnership in Space - An Introduction to Canada's Special Relationship with the European Space Agency* September ESA BR-79

EAITC 1990b *Technology Prospecting Abroad: A Guide to Technology Opportunities in Selected Countries* Ottawa

EAITC 1990c *Extending the Network: The Science and Technology Mechanisms and Programs of External Affairs and International Trade Canada* Ottawa

EAITC 1990 *Canada: Going Global-Guide to Programs and Services* Ottawa

EAITC 1991 *The Japan Science and Technology Fund* Ottawa

EAITC 1991 *Foreign Policy Themes and Priorities, 1991–1992 Update* Ottawa

EAITC 1992 *Inventory of International Science and Technology Agreements* Ottawa

External Affairs and International Trade Canada and Bundesministerium für Forschung und Technologie Deutschland 1991 20 Years Cooperation in Science and Technology, 1971–1991 / 20 Jahre Zusammenarbeit in *Wissenchaft und Technologie* 1971–1991 September

Findlay G 1991 International Collaboration in *Science and Technology in the UK* Longman 253–283

Freeman C 1974 *The Economics of Industrial Innovation* Harmondsworth Penguin

Gagné J Leclerc M 1992 La coopération scientifique internationale du Québec *Nouvelles de la science et des technologies* 9(2):133–136

Garfield E 1983 Mapping Science in the Third World *Science and Public Policy* 10(3)

GAO(General Accounting Office) 1988 *r&d Funding Foreign Sponsorship of U.S. University Research* March, Washington GAO/RCED - 88–89 BR

Ghent J M 1980 Canadian Government Participation in International Science and Technology *J Canadian Studies* **X**(1):48–62

Ghent J M 1980 The Participation of Provincial Governments in International Science and Technology *American Rev Canadian Studies* 10(1);48–62

Hagedoorn J Schakenrad J 1991 The Internationalization of the economy, Global Strategies and Strategic Technology Alliances *Nouvelles de la science et des technologies* 9(2):29–41.

Halliwell J Davis C Dufour P 1992 *Scientific and Technological Collaboration in North America: In Search of New Paradigms?* paper presented to the annual meeting of the American Association for the Advancement of Science Chicago February

Heaton G 1990 *International r&d Cooperation: Lessons from the Intelligent Manufacturing System Proposal* The Manufacturing Forum Discussion Paper No 2 Washington

Howells J 1990 The Globalisation of Research and Development: A New Era of Change *Science and Public Policy* Oct :273–285

Howells J 1990 The Internationalisation of r&d and the Development of Global Research Networks *Regional Studies* 24(6):495–512

ISTC (Industry, Science and Technology Canada) 1991 *Canada's Science and Technology Relations with Europe* A paper prepared to facilitate discussion at the March 1991 focus groups organized by the APRO, ISTC and EAITC, Ottawa.

Latouche D 1991 *Science and Technology in Canada: An Essay on the New Strategic and International Context* prepared for the Science Council of Canada Draft May

Leclerc M et al 1991 La coopération scientifique France-Québec: une analyse des cosignatures *Interface* Sept.-Oc: 19–23

Leclerc M Frigoletto L Okubo Y Miquel J F 1992 The Scientific Co-operation between Canada and the European Community *Science and Public Policy* 19(1): 15–24

Leclerc M Okubo Y Frigoletto L Narvaez N Miquel J F 1991 La coopération scientifique internationale du Canada et du Québec: analyse des co-signatures *Interface* 12(5) :19–23

Leclerc M 1992 La coopération scientifique internationale vue du Canada *La Recherche* (forthcoming)

Logsdom J M 1991 International Cooperation in the Space Station Programme *Space Policy* 7(1):35–45

Macioti M 1991 *Science and Technology in the Nineties: The Challenge to Europe and its Relations with Canada* Presentation to the Canadian Conference on Electrical and Computer Engineering, Quebec, September 25

Magun S Rao S 1989 *The Competitive Position of Canada in High-Technology Trade* presented at the Canadian Economics Association meeting, Université Laval, June 2–4.

Maynard Ghent J M 1979 *Canadian Government Participation on International Science and Technology* Ottawa Science Council of Canada

McInres S and Dufour P 1992 *Setting the framework: The Changing Dimensions of Canada's International Science and Technology Arrangements* Canadian Political Science Association June 1992

Miquel J F 1990 La coopération entre le Canada et la France en science fondamentale in Leclerc M (ed) *Les enjeux économiques et politiques de l'innovation* Québec, Presses de l'Université du Québec, 201–222.

MIT Faculty 1991 *The International Relationships of MIT in a Technologically-Competitive World* Cambridge MIT

Michalet C A 1991 Global competition and Its Implications for Firms in OECD 1991 *Technology and Productivity* Paris; pp79–88

Myers L C 1989 *La station spatiale: la contribution du Canada* Ottawa Library of Parliament bulletin 87–5F

National Academy of Engineering 1991 *National Interests in an Age of Global Technology* Washington National Academy Press

Niosi J 1991 Canada's National System of Innovation *Science and Public Policy* April: 83–92

Niosi J Bergeron M 1991 *Technical Alliances in the Canadian Electronics Industry* Montreal CREDIT, UQAM.

Niosi J 1992 *Canadian-American Free Trade and Technological Development of Canadian Firms* report presented to the Science Council of Canada, April

OECD 1990a *Les relations université-entreprise dans les pays de l'OCDE* Paris, DSTI/ SPR/89.37

OECD 1990b *Technology and the Process of Internationalisation-Globalisation* Draft Background Report, Chapter 9, Technology/Economy Programme Sept. 17

OECD 1991b *Coopération internationale en matiére de science lourde* Paris, DSTI/STP(91)28

OECD 1991c *International Cooperation in Big Science* Paris DSTI/STP91(4) February

OECD 1992a *Science and Technology Policy Outlook and Review, 1991* Paris OECD

OECD 1992b *Technology and the Economy: The Key Relationships*Paris.

Ostry S 1990 *Governments and Corporations in a Shrinking World* New York Council on Foreign Relations

Pacific Economic Cooperation Conference, Science and Technology Task Force, 1990. *Pacific Cooperation in Science and Technology, Proceedings of the Seoul Symposium* Minden K (ed.)

Paquin G 1991 Les subtulités de la dyslexie distincte *Le Devoir* May 24

Pavitt K 1991 What Makes Basic Research Economically Useful *Research Policy* **20**: 109– 119.

Phillipson D C 1989 *International Scientific Liaison and the National Research Council of Canada: 1916–1974* Ottawa National Research Council of Canada

Porter M 1986 *Competition in Global Industries* Harvard Business School Press

Potter M W 1990 *International Science and Technology Cooperation: an Examination of Compatibility*Ottawa Carleton University

Reich R B 1990–1991 Does Corporate Nationality Matter? *Issues in Science and Technology* Winter 1990–91

Rodriguez-Romero L 1992 Interactions entre la R-D et les importations de technologies *STI Review* **9** April:47–74

Rycroft R W 1990 The Internationalization of U.S. Intergovernmental Relations in Science and Technology Policy *Technology in Society* 12: 217–233

Schott T 1991 The World Scientific Community Globality and Globalisation *Minerva* **XXIX**(4):440–462

Schroeder-Gudehus B 1977 Science, Technology and Foreign Policy in Spiegel-Rosing I de Solla Price D(eds.)*Science, Technology and Society: A Cross-Disciplinary Perspective* : 473–534 Sage Publications

Schroeder-Gudehus B 1989 Les relations culturelles, scientifiques et techniques in Painchaud, P (ed) *De MacKenzie King à Pierre Trudeau*: 581–607 Québec Presses de l'Université Laval

Science Council of Canada 1973 *Canada, Science and International Affairs* Report No. 20, Ottawa, April

Science Council of Canada 1984 *The Canadian Science Counsellors* a Council Statement Ottawa

Science Council of Canada 1989 *Canada-Japan Complementarity Study* Ottawa

Science Council of Canada 1992 *Reaching for Tomorrow: Science and Technology Policy in Canada 1991*Ottawa.

SVPC (Scinces de la vie spatiale au Canada) (1990)*Bulletin* (4)mai

Simeon R 1990 *Canadian Responses to Globalization, in Canada-Japan: Views on Globalization* K. Lorne Brownsey (ed.) Institute for Research on Public Policy

Skolnikoff E B 1992 *New International Trends Affecting Science and Technology* paper prepared for the Science Council of Canada May

Soete L 1991 *The Internationalization of Science and Technology (Policy): How Do Nations Cope?* MERIT, Maastricht, January

Statistics Canada 1991 *Estimates of Canadian Research and Development Expenditure (GERD), National 1963 to 1991, and by Province 1979 to 1989* Ottawa August

Steed G P F Kenney-Wallace G 1990 Present Conditions of Globalized Private Enterprise in Canada in *Canada-Japan: Views on Globalization* Lorne Brownsey K (ed.): 81–98 Ottawa Institute for Research on Public Policy

Swanson R F *State-Provincial Interaction: A Study of Relations Between U.S. States and Canadian provinces* Prepared for the U.S. Department of State, August 1974, cited in Ghent, op.cit

Winegard W C 1991 The Canada-United States Science and Technology Relationship in a Globalized Economy in W. Winegard et al, eds., 1–14

Winegard W C et al (eds) 1991 *Canada and the United States: an Emerging Partnership* Washington and Cambridge, Mass Brassey's and Institute for Foreign Policy Analysis

Chapter 8

Energy r&d policy in Canada[1]

Yves Gingras and Robert Dalpé

Introduction

Canada is a country richly endowed in natural resources. Energy has always been important to it's economy. Canada is a net exporter of energy resources with research and development undertaken quite early in order to localise and exploit natural resources such as coal, gas and oil. For instance, a Canadian public laboratory in mineral and energy was already performing r&d on coal and oil at the beginning of this century[2], and the largest and most innovative Canadian program in energy r&d, the development of a nuclear energy system, CANDU, can be traced back to 1942.[3] However, it was only after the 1973 oil crisis that the Canadian government recognised that energy resources were not unlimited and that a systematic energy r&d program could contribute to a more efficient exploitation of existing energy sources as well as to the development of alternative ones.

Before the oil crisis of October 1973, few countries had an energy r&d policy. Most did not even have an energy policy. Energy research and development was, of course, going on in industry as well as in governmental laboratories but there were little more than self-conscious attempts to take stock of and to coordinate overall activities. Governments usually think through institutions and the energy crisis prompted, around the world, the creation of organisations that could handle energy r&d policy matters. In 1974, the U.S. government created an Energy Research and Development Administration, with a mandate to define and coordinate the national efforts in this domain.[4] During the same year, Great Britain created new organisations such as the Energy Technology Support Unit and the Advisory Council on Energy Conservation.[5] In Canada, an Interdepartmental Panel on Energy r&d (PERD) was formed in January 1974 to coordinate activities at the federal level. At the end of 1974, the International Energy Agency (IEA) was created as an autonomous body within OECD to facilitate international collaboration among member countries in the management of perceived energy shortages. Among the many Committees created to monitor the energy situation was one specially devoted to energy r&d.[6] Other countries reacted more slowly to the energy crisis. In Australia, for example, a National Energy Research, Development and Demonstration Council was created only in 1978.[7]

When these institutions were created, there was little accurate information on the level of investment in energy r&d, which is an essential tool for the definition and implementation of any s&t policy. In 1975, IEA published a first compilation of energy r&d statistics designed to compare levels of activities in the member countries. This new concern about energy r&d was not matched by available data, which up until then had been collected and classified under different categories such as minings, commodities and science. There was a rush to identify and classify energy r&d. An early example in Canada, for federal government departments and agencies was produced in 1976.[8] Also, at that time the new Office of Energy Research and Development, within the Department of Energy, Mines and Resources produced its own inventory of energy r&d supported by the government, institution building and statistics gathering.[9] Before then, little comprehensive information existed on the state of energy r&d in Canada. Institution building and statistics gathering were thus the first steps toward a rational political intervention in energy r&d.

If the 1970s were the years of policy development, the beginning of the 1980s, following the second oil crisis were the 'golden years' of energy r&d policy in Canada. As we will see later, this period was characterised by a large increase of investments in energy r&d and, more broadly, a nationalist and interventionist energy policy, called the National Energy Program (NEP). Established in 1980, it had Canadianisation and self-sufficiency as important goals.

The interest for energy diminished in the mid 1980s with the decrease of oil prices. Energy was no longer a priority, and the NEP was abandoned in 1985 by the newly elected Conservative party. This deregulation process occurred in many other sectors. Government investment in energy r&d suffered from severe cutbacks. For example, the NRC's Energy Division, which was opened in 1975 with the mandate to develop programmes in renewable energy, was closed in 1985.[10]

In the late 1980s, the Canadian energy situation was characterised by a low increase of energy demand caused by stagnation of economic activity. Oil prices were relatively low, though unsteady. Coupled with a growing concern about the environmental consequences of petroleum usage, nuclear reactor accidents and large hydro-electric dams, these factors led both levels of government – federal and provincial – to revise downward their energy demand forecasts. In that respect, environmental protection has become a new and important variable for recent energy r&d policy.[11]

After briefly reviewing patterns of energy production and consumption in Canada, we will discuss in turn the role played in energy r&d by federal and provincial governments and by industry. (In 1989, the federal government accounted for forty-six per cent of the total investment in energy r&d, while provincial governments and industry each contributed twenty-seven per cent).

Energy production and consumption

The pattern of energy production and consumption has changed greatly during the past few decades in Canada. On the production side, the average annual growth rate from 1970 to 1990 was close to four per cent (see Table 8.1).

The increase of production was higher for natural gas, electricity – particularly for nuclear since 1968 – and coal, than for crude oil. On the consumption side, the most important trend is the low increase during the 1980s. Whereas the annual rate of increase was close to five per cent for the 1970s – when increase in demand was higher than increase in production – it was only one per cent during the 1980s. Furthermore, 1981, 1982 and 1983 saw a decrease in energy consumption

Table 8.1
Production, exportation, importation and availability of energy in Canada (petajoules)

		prod	exp	imp	avail
1970					
Coal		319	108	472	637
Crude oil		2,777	1,346	1,147	2,573
Natural gas		1,986	739	10	1,232
Gas plant		128	82	—	44
Primary elect		510	18	10	502
	Total	5,721	2,293	1,640	4,989
1975					
Coal		570	321	427	592
Crude oil		3,175	1,410	1,655	3,424
Natural gas		2,562	898	9	1,620
Gas plant		221	133	—	66
Primary elect		694	37	13	669
	Total	7,220	2,799	2,103	6,370
1980					
Coal		891	448	467	928
Crude oil		3,444	460	1,241	4,196
Natural gas		2,600	840	—	1,784
Gas plant		316	211	—	86
Primary elect		1,032	109	11	933
Steam		20	—	—	20
	Total	8,303	2,068	1,720	7,950
1985					
Coal		1,487	802	137	1,432
Crude oil		2,516	1,090	577	3,033
Natural gas		3,297	992	302	2,181
Gas plant		317	161	—	172
Primary elect		1,290	156	44	1,122
Steam		24	—	—	24
	Total	9,931	3,202	1,060	7,940
1990					
Coal		1,669	943	409	1,077
Crude oil		3,735	1,462	1,198	1,077
Natural gas		4,262	1,537	24	2,676
Gas plant		405	180	4	217
Primary elect		1,306	66	64	1,304
Steam		16	—	—	16
	Total	11,392	4,188	1,699	8,754

Source: Statistics Canada, Cat. 57–003 and 57–207.

due to energy conservation measures and the economic recession. The other important trend was a decrease in crude oil consumption in the 1980s, leading to a rising share for natural gas and electricity.

The Canadian energy industry remained a net exporter of energy during the last two decades. However, during the 1970s, when demand growth was higher than production growth, it's foreign sales decreased. For instance, in 1980 Canada was a net importer of crude oil. Exports increased during the 1980s. One-third of natural gas production was exported. Electricity exports increased up to 1987, but remained at a relatively low level, and declined after that year.

Table 8.2 presents data on energy production in Canada in 1990. This table reveals the high level of production in natural gas and crude oil. Also important is the regional distribution of energy production. Natural gas and crude oil are available only in Western provinces, where Alberta is the largest producer. Ontario and Québec produce almost exclusively electricity, from nuclear power plants and hydro-electricity respectively. As far as production is concerned, regional interest for energy r&d differs accordingly.

Table 8.3 presents provincial consumption of energy in 1990. Here also regional patterns are evident.

Eastern provinces rely almost exclusively on petroleum products and Québec shows greater consumption of electricity. Ontario and British Columbia use petroleum products, natural gas and electricity, while the Prairie provinces rely on petroleum products and natural gas. Energy r&d needs will then also differ according to consumption patterns.

The federal role in energy r&d

The Federal Panel on Energy r&d

In the mid 1970s, the federal government created institutions with a mandate to coordinate the federal energy r&d policy. In January 1974, a Task Force on Energy r&d was established, chaired by the Deputy Minister of Energy Mines and

Table 8.2
Production of Primary Energy, 1990 (petajoules)

	Coal	Crude oil	Natural gas	Gas plant	Primary electricity	Steam	Total
New Foundland	—	—	—	—	1254.0	—	124.0
P.E.I.	—	—	—	—	—	—	—
Nova Scotia	97.3	—	—	—	4.3	—	101.6
New Brunswick	14.8	—	—	—	31.9	—	46.7
Québec	—	—	—	—	482.4	—	482.4
Ontario	—	9.5	17.0	—	359.7	16.0	402.2
Manitoba	—	28.5	—	—	71.4	—	100.1
Saskatchewan	141.1	479.9	262.4	2.6	15.2	—	901.1
Alberta	667.23	057.83	546.9	390.5	7.4	—	7,669.7
British Columbia	748.9	83.8	427.9	11.7	206.3	—	1,478.7
Yukon, N.-W.T.	—	75.4	7.4	—	2.4	—	85.2
Canada	1,669.33	734.8	4,261.6	404.9	1,306.0	16.0	11,392.6

Source: Statistics Canada, Cat. 57–003.

Table 8.3
End Use of Major Fuel Types, 1990 (petajoules)

	Coal	Oil products and LPG's*	Natural Gas	Electricity	Steam	Total
Newfoundland	4.2	121.1	—	37.9	—	163.2
		(3%)	(74%)		(23%)	
P.E.I.	0.2	19.9	—	2.7	—	22.8
		(1%)	(87%)		(12%)	
Nova Scotia	64.1	174.0	—	34.9	—	273.0
		(23%)	(64%)		(13%)	
New Brunswick	14.3	191.9	—	47.7	1.4	255.3
		(6%)	(75%)		(19%)	(1%)
Québec	26.8	722.1	211.7	567.6	0.1	1528.3
		(2%)	(47%)	(14%)	(37%)	
Ontario	579.4	1 104.3	817.0	515.9	19.1	3035.7
		(19%)	(36%)	(27%)	(17%)	(1%)
Manitoba	7.1	102.2	86.9	63.1	0.2	259.5
		(3%)	(39%)	(33%)	(24%)	
Saskatchewan	115.5	136.6	189.9	49.2	—	491.2
		(24%)	(28%)	(39%)	(10%)	
Alberta	397.5	541.8	1 075.3	152.5	0.2	2167.3
		(18%)	(25%)	(50%)	(7%)	
British Columbia	8.1	379.0	290.7	206.6	—	884.4
		(1%)	(43%)	(33%)	(23%)	
Yukon and N.W.T.	—	24.0	3.1	3.8	—	30.9
			(78%)	(10%)	(12%)	
Canada	1 217.3	3 516.6	2 676.5	1 681.8	21.0	9 113.2
		(13%)	(39%)	(29%)	(18%)	

*Liquid Petroleum Gas Product
Source: Statistics Canada, Cat. 57–003.

Resources, and composed of Deputy Ministers or senior officers from seventeen departments and governmental agencies having responsibilities or interest in energy matters. It's objectives were to review federal energy r&d activities, define a coordinated programme and advise the government on the allocation of funds.[12] Tabled a year later, the report of the Task Force led to the creation of a coordination structure composed of an Interdepartmental Panel on Energy Research and Development (hereafter referred to as the Panel), assisted by an Office of Energy Research and Development (OERD) which played the role of secretariat to the Panel. Bringing together senior representatives of s&t branches of all the federal government ministries and central agencies involved in energy r&d, the Panel acts as a central policy and planning committee responsible for coordinating the programme of federal energy r&d and for recommending allocation of resources within the different sectors of energy.[13] The coordination activity also includes collaboration with provinces and with foreign countries through the international programmes administered by the IEA.

The government provides the Panel with it's own annual budget, which is distributed according to priorities set up in relation to the energy policy defined by the federal government. The ministries and agencies involved are in charge of implementing those aspects of energy r&d which relate to their domain. Though the most important federal institutions in matters of energy r&d are Energy, Mines and Resources and Atomic Energy of Canada Ltd. The Panel also includes other departments affected by energy resources such as Transport, Public

Works, Agriculture, Environment, Fisheries & Oceans, Indian & Northern Affairs, Health & Welfare and National Defence.

Because of the many departments involved in energy, Canada rejected the option of a central agency like the American ERDA, and adopted a cooperative structure designed to ensure coordination of efforts between the activities of the departments. This seemed to be an appropriate choice because, in addition to being divided between many departments, energy r&d is also divided between different provinces, which have control of their natural resources. An interdepartmental panel could define joint projects with provincial authorities and avoid unnecessary duplication.

In contrast to the Australian National Energy Research, Development and Demonstration Council which regroups members of government, universities and industry, the Panel is strictly a governmental structure. The lack of representation from industry at the level of definition and shaping of policy was noted by the IEA in it's 1978 Report, and it was suggested that industry should be associated in some way.[14] Though not officially present on the Panel, industries as well as provincial governments, are nonetheless consulted through 'formal and informal routes by scientists, r&d managers, Panel members and officers of OERD'.[15]

Though the report of the Task Force served as a starting point for the definition of a national programme of energy r&d, there was another important document produced at the time on the subject by the Science Council of Canada. Created in 1966 as an advisory body on science policy, the Council produced many documents on sectorial aspects of s&t policy. In the context of one of it's policy projects, the Science Council established a committee on national resources in September 1971 to study those aspects of science policy connected with the production, distribution, conservation and end use of energy resources. The committee, composed of five members from governments departments and agencies, five from industry and two from universities, issued its report on *Canada's Energy Opportunities* in March 1975, suggesting an expansion of energy r&d activities in the sectors of conservation, conversion and more efficient use of energy.[16] Four years later, the Committee published another report recommending eleven demonstration programmes, ranging from oil and gas production in ice-congested water to nuclear, bioenergy and solar energy.[17]

Though energy r&d had obvious links with science policy and thus with the federal government's Ministry of State for Science and Technology (MOSST), the official responsibility for developing an energy r&d policy lay with Energy, Mines and Resources.

With rich and diversified energy sources, Canada has frequently taken stock of it's energy situation and produced at least nine studies related to energy policy between 1944 and 1985, but energy r&d was not an important preoccupation before 1973.[18] In the summer of that year, the federal government published *An Energy Policy for Canada: Phase 1* which, as one commentator put it, 'can be read as the last document of the sixties'.[19] The document still took for granted the necessity of high level energy consumption and thus recommended more efforts on the development of nuclear energy and research on synthetic oil.[20] However, the report became obsolete a few months later with the oil crisis, so that the definition of an energy r&d policy adequate to the new situation would only come with the report of the Task Force in 1975. According to this document, a composite goal for a national energy r&d program should be 'to develop the scientific and technical capability to achieve self-reliance in energy with minimal environmental, social or economic costs and maximum industrial or quality of life advantages'.[21] For the next decade, this objective of self-sufficiency was at the core of the National Energy Policy (NEP).[22]

The NEP was the most important energy policy statement of the 1980s. It's main objectives were self-sufficiency and Canadianisation of the oil industry. [23] Canadianisation of the petroleum industry was achieved through the Petroleum Incentive Programme, which offered grants to stimulate oil and gas exploration in relation to the degree of Canadian ownership, and through Petro-Canada's strategy to acquire foreign-owned competitors.[24] These actions meshed well with the Foreign Investment Review Agency (FIRA), an institution created in 1974 by the same Liberal government charged to oversee foreign investments in Canada. Other NEP priorities were frontier and oil sands development, conservation and oil substitution. An important federal objective was to establish a national price for petroleum products, despite the fact that energy was a provincial jurisdiction. This particularly affected Alberta, the main oil producer, which objected to this aspect of the NEP, but was well received by the central provinces, who were consumers and thus paid less than the international price for their oil. Energy r&d funding increased dramatically under NEP, as the programme nevertheless established alternative fuels, energy efficiency and new energies as priorities for r&d.

The election of the Conservative Party in November 1984 marked a shift in federal r&d policy. Canadianisation was no more a priority, the NEP was abandoned, and competitiveness and environment were the new goals.[25] Concerning energy r&d, *Energy Options*, published in 1988, stated that a 'commitment to research, development and management of technology is critical to enhancing Canada's energy choices and environmental quality into the 21st century'.[26] The mandate of the PERD was adapted to reflect this new trend and it's objective is now 'to provide the s&t for a diversified, economically and environmentally sustainable energy economy'.[27]

Federal Energy r&d Expenditures

In its survey of the state of energy r&d at the federal level, the 1975 report of the Task Force showed that research on nuclear energy was by far the main activity in energy r&d and that it amounted for more than three quarters of the total expenditures over the years 1972–1975.

The expenditures on nuclear energy were dedicated to the development of the state-of-the-art CANDU nuclear power plant. The Atomic Energy Commission of Canada Ltd., a Crown Corporation was responsible for the development of the CANDU nuclear power plant. The OPEC oil embargo of 1974 led the Canadian Government to a broadening of energy r&d expenditures. This was precipitated by a recognition that r&d could help lead to increased security of supply. The Program of Energy Research and Development was established in 1975 and mandated to focus on energy conservation and on the identification of alternative energy supplies that would increase security of supply. The provinces also established programs aimed at the increased security of supply.

With the advent of these federal and provincial intiatives, energy r&d expenditures were applied to a variety of energy sources (e.g. conservation, coal, renewable energy and fossil fuels). By 1987, r&d expenditures devoted to nuclear (fission and fusion) energy represented 43% of total federal and provincial energy r&d expenditures (Table 8.6).[28] It should be noted that AECL's budget greatly decreased in 1987 and 1988. AECL activities concern application of nuclear technology in energy and in other sectors, such as the biomedical industry. (see Chapter Four). In the energy sector, AECL has developed the CANDU technology and was also involved in smaller projects such as the Tokamak and the SLOWPOKE.[29]

Starting with the fiscal year 1975–1976, the Interdepartmental Panel on Energy r&d would distribute additional funds according to the objectives of diversification and self-sufficiency which were at the core of the federal energy policy. It should be noted that the Panel has the formal responsibility for coordinating and

Table 8.4
Energy r&d Expenditures in Canada (millions, current dollars)

	Federal (% administered by PERD*)	Provinces	Sub total federal provinces	Industry**	Total
1974	105,3 (0,0)	35,6	140,9	n.d.	—
1975	109,4 (1.0)	34,9	144,3	n.d.	—
1976	111,3 (9,0)	41,6	152,9	n.d.	—
1977	127,1 (16,4)	69,4	196,7	113	309,7
1978	150,4 (22,3)	92,2	242,6	161,1	403,7
1979	157,1 (23,7)	101,1	258,2	186,6	444,8
1980	204,6 (19,2)	103	307,6	259,7	567,3
1981	251,0 (31,0)	107	358,0	402,6	760,6
1982	345,1 (35,6)	67,1	412,2	404	816,2
1983	403,1 (40,3)	85,3	488,4	347	835.4
1984	407,5 (41,8)	110,5	518	363	881
1985	396,8 (28.8)	95,9	492,7	414	906,7
1986	352,4 (27,0)	115,1	467,5	411	878,5
1987	409,6 (21,7)	86,0	495,6	386,4	882,0
1988	404,8 (22,0)	90,3	495,1	437,9	933,0
1989	417,8 (21,6)	73,9	491,7	433,1	924,8

*PERD: Panel on Energy Research and Development
**To eliminate double counting, the amount for industry includes only self—funded activities. The large, provincially-owned electric utilities are included under industry.
Sources: EMR, Office of Energy Research and Development; Government of Canada, Statistics Canada, Science Technology and Capital Stock Division, *Industrial Research and Development Statistics* (Catalogue 88–202) , 1981 to 1988.

implementing the Program on Energy r&d which continues to be the Federal Government's cornerstone of investment in energy r&d in all areas except nuclear (CANDU) fission. The Panel establishes the strategic direction for the Program and allocates resources to energy r&d activities on the basis of established priorities and strategic directions.

Comparing Tables 8.5 and 8.6, we see that the major part of these funds go to research on nuclear fission coordinated by AECL which receives its budget directly from Parliament.

The most important federal energy r&d performers are AECL, the Panel and the Ministry of Energy, Mines and Resources, where the Mineral and Energy Technology Sector get slightly more than half of Panel funds. For university research, the main source of funds is the Natural Sciences and Engineering Research Council (NSERC).

Concerning Panel funds, on the basis of Table 8.5, we can distinguish three periods in the evolution of the federal energy r&d budget allocation. The first period covers the years 1975–1980, during which the Panel concentrated more than half of it's resources on renewable energy and conservation, followed by oil sands and heavy oil which received seventeen per cent of the $142 million distributed by the Panel over this six year period. These three domains translated into r&d measures the objectives of conservation and enhanced production of petroleum put forward by the government in *An Energy Strategy For Canada* published in 1976.[30] During this period, energy r&d accounted, on average, for fifteen point eight per cent of the total r&d budget of the federal government.

In addition to doubling the budget of the Panel in 1977–1978, the federal government also gave additional funds to NSERC with the stated requirement that these new resources be used to support university research in areas of national importance of which energy was one. Accordingly, NSERC created, in 1977–1978, a Strategic Grants programme focused on environmental toxicology, oceans and energy.[31] From $2.3 million, the budget gradually rose to $32.3 million in 1984–1985 and the number of eligible sectors rose to eight, energy always remaining the most important sector in terms of allocated funds (fifty-four per

Table 8.5
Allocation of panel on energy r&d resources for the 1975–1990 period
(millions, current dollars)

	Energy efficiency	Oil sands, heavy oil, coal	Fusion	Renewable energy	Alternative fuels	Conventional energy*	Coordination	Total
1975–76	0,114	0,410	0,0	0,0	0,429	0,0	0,020	0,973
1976–77	1,977	3,017	1,909	1,150	0,938	1,716	0,160	10,867
1977–78	4,957	4,017	1,090	4,915	2,179	2,680	1,025	20,863
1978–79	8,408	5,417	1,450	10,236	2,959	3,860	1,238	33,568
1979–80	7,902	5,417	0,310	15,427	3,054	3,830	1,388	37,328
1980–81	7,607	6,136	0,310	15,574	3,750	4,722	1,238	39,337
1981–82	15,290	8,479	2,884	21,355	14,936	10,008	4,996	77,948
1982–83	26,850	12,421	5,200	28,500	34,418	12,691	3,054	123,134
1983–84	32,290	16,747	10,329	36,141	35,526	29,302	2,234	162,569
1984–85	33,785	20,575	7,767	39,680	35,003	29,584	1,790	168,184
1985–86	18,100	20,132	9,492	20,067	22,102	22,612	1,751	114,256
1986–87	16,374	22,740	8,935	10,149	17,991	22,082	1,291	99,562
1987–88	15,316	21,404	8,374	8,281	15,136	19,195	1,108	88,814
1988–89	15,828	21,332	8,374	9,001	14,219	19,143	1,180	89,077
1989–90**	15,854	11,189	8,374	11,397	22,635	18,463	2,186	90,098

*Includes oil, natural gas and electricity.
**Changing definitions of categories moved oil sands and heavy oil in "alternative fuels.
Source: Energy, Mines and Resources Canada, Office of Energy Research and Development.

cent in 1979, when three sectors were eligible and twenty-nine per cent in 1983, when eight sectors were eligible). With this programme, the federal government's energy r&d policy was thus extended to cover basic research in order to develop scientific expertise in energy areas of potential importance in the long term. (Of course, one should recognise that considerable activity in energy-related research was already well-established in Canada's university research community). It also assured the continued training of scientists in a sector of national importance. In order to secure a certain relevance to industrial needs, however, fifty per cent of the members of the evaluation committees for strategic grants on energy were drawn from industry – the highest proportion of all the strategic grants committees.[32] Analysis of strategic grants awarded in bioenergy and solar energy revealed that it contributed to an intensification of research in these fields.

The second period runs from 1981 to 1984, and follows the implementation of the National Energy Program of 1980. While still concentrating on renewable energy and conservation, the Panel gave new priority to research on new liquid fuels for transportation – such as natural gas, alcohol, gasification and liquefaction of biomass and coal – which was the main consumer of petroleum.[33] During this period, the budget of the Panel grew rapidly from $39 million in 1980 to $78 million in 1981 and $170 million in 1984. This raised the part of energy r&d in the total r&d budget of the federal government to an average of twenty per cent for this period, with a peak at twenty-two per cent in 1983. All sectors were beefed up and nuclear fusion r&d came of age during this period, so to speak, with the construction of a Tokamak reactor at Varennes near Montreal, a joint project of the federal government and Hydro-Québec. This project originated from a group of researchers from Québec which, after years of negotiation, convinced the Panel and NSERC to define a national fusion programme and to fund a magnetic confinement apparatus to gain knowledge on fusion in anticipation of when commercial fusion reactors will appear.[34] Research on inertial confinement is also going on in NRC laboratories on a smaller scale. Later on, a fusion

Table 8.6
Energy r&d Expenditures by Federal and Provincial Governments
(millions, 1989 Canadian $)

	Fossil fuels	Fission/ Nuclear	Coal	Renewable energy	Supporting* technologies	Conservation	Fusion
1978	83.3	213.7	37.1	51.3	5.6	31.3	7.9
1979	82.0	192.8	29.2	57.1	13.4	31.2	7.3
1980	61.6	187.5	23.6	59.3	10.5	40.9	9.9
1981	71.7	173.8	30.1	92.1	9.8	59.9	10.3
1982	105.0	216.6	44.6	76.9	14.9	69.8	9.5
1983	160.6	213.4	49.2	84.4	24.3	87.8	15.3
1984	271.7	202.7	44.0	66.1	19.7	88.2	10.0
1985	158.8	219.3	36.1	39.0	10.5	92.3	12.9
1986	204.3	208.5	56.4	22.7	9.5	37.4	12.7
1987	156.9	173.9	30.7	19.4	8.1	37.7	19.8
1988	144.5	149.8	36.2	17.7	6.3	40.0	18.5
1989**	144.0	149.4	36.3	17.6	6.2	39.7	18.4

*Analysis of energy system and others.
**Estimated
Source: International Energy Agency 1991 *Energy Policies and Programmes of IEA Countries – 1990 Review* Paris OECD

fuels technology group was established jointly by Ontario Hydro, and the Ontaria and the federal governments. Both the Varennes group and the Toronto are also involved in international collaboration and participates in feasibility studies for the ITER project, an international project for a magnetic confinement fusion reactor.

The third period, from 1985 to 1990, began with the election of the Conservatives in November 1984 and was characterised by major cutbacks in energy r&d and the abandonment, in 1985, of the National Energy Policy followed by a stable distribution of energy r&d investments. A major objective of the new government was to diminish the budget deficit and reorient the energy r&d activities. The government's deficit reduction program resulted in the significant reduction to Program of Energy R&D's (PERD) budget. As well, the National Research Council's portion of PERD funds was eliminated and reallocated to other departments. This had a major impact on the r&d subject areas that were the responsibility of the NRC (e.g. conservation, hydrogen, renewable energy and fusion). These activities were subsequently sustained at minimum viable levels from the PERD Program budgets of EMR and AECL. Between 1985 and 1987, energy conservation and renewable energy budgets diminished by fifty per cent and seventy-five per cent respectively. The budget of the Panel was reduced by thirty-three per cent in 1985 and by a further sixteen per cent in 1986. It decreased slightly in 1987, where it was stabilised, in current dollars, which indicates, of course, a decrease in real money. The proportion of the federal energy r&d administered by the Panel decreased from forty-one per cent in 1984 to only twenty-seven per cent in 1986 and even lower, at twenty-two per cent, in 1987.

It should be noted that the activities in this area are supportive of the economic development, energy diversity and environmental policies of the government. A significant portion of the activities in this sector are designed to ensure that exploration, production and utilization these resources are carried out in ways that minimize the damage to the environment.

In 1989, some sectors were redefined and oil sands and heavy oil research was moved to the 'alternative fuels' rubric. The increase of that sector is thus more apparent than real and does not constitute a major shift in priorities. In the near future, environmental concerns will likely bring an increase of investment in alternative fuels through the Canadian government's commitment to decreasing carbon dioxide emissions. The Green Plan, an important $6 billion environmental programme, also provides funds for projects devoted to energy research and represents another significant source of funding .

If we compare the distribution of the Panel's funds with the distribution of the total amount of money invested in energy r&d by the federal and provincial governments (Tables 8.5 and 8.6), we can see what the real effect of the Panel has been on energy r&d policy. The Panel has been highly successful in directing r&d to those areas which are of vital importance to Canada (e.g., regulations, health and safety standards, longer term energy supply options, etc.). PERD has increased the size and scope of all energy r&d activities in Canada, the variety of possible approaches, the pool of expertise available to explore the research, the mechanisms by which r&d can be accomplished and the ability to undertake the project.

In order to stimulate r&d in industry and the diffusion of innovation, the federal government instituted, in 1972, a contracting-out policy limiting the amount of r&d conducted within government laboratories.[35] In the case of energy r&d, this policy stimulated contracting out thirty-six per cent of the budget of the Panel in 1976, seventy per cent in 1985 and sixty per cent in 1989. For certain programmes, such as those in renewable energies, projects which are not financed entirely by the programme are now preferred: financial participation of firms is a requisite. From this perspective, the Mineral and Energy Technology Sector and CANMET of EMR, which, in 1991–92, received more than half of Panel funds, continued to follow the principles of the government's r&d contracting out policy. They gave funds to industry, university and other public laboratories. Whereas CANMET has facilities in coal and petroleum research, the administration of conservation and renewable energy research programmes is done for the most part without intramural r&d facilities and research is contracted out or cost-shared through contributions.[36] This strategy had the advantage of offering great flexibility in energy priorities and the possibility of relatively rapid reorientation. It's weakness, however, was the lack of stability for organisations which had to adjust their r&d projects according to changing priorities. The creation of 'centres of expertise' dealt with this problem: for example, in solar energy, four research centres in universities and one at the Ontario Research Foundation received the mandate to specialise in a sector and to offer services to industry[37] CANMET, the federal public laboratory in oil and coal, spent close to half of it's budget in contracts with industry.[38] CANMET's mandate is to find safer, cleaner and more efficient methods to develop and use Canada's mineral and energy resources. Important projects concern the development of more efficient upgrading technologies for heavy oil. As with other large federal public laboratories, CANMET has an industry-led advisory board, with a mandate to recommend how it can more adequately serve the Canadian industry.[39] Federal r&d institutions have had therefore to adjust themselves to energy priorities, as well as to s&t policy.

Through IEA, Canada participates in international r&d projects. Over the years, it has been involved in nearly fifty projects coordinated by IEA.[40] In 1991–92, for instance, Canadian organisations were active in thirteen projects contributing nearly one million dollars. More than one third of these funds go to two coal research and combustion projects, the rest covered participation in heat pumps, fusion buildings, solar, wind and alcohol research. In addition to these IEA related projects, Canada also participates with Europe, Japan, Russia and the USA in an international project on nuclear fusion, ITER. Of course, though harder to estimate, there are also international activities through the initiatives of researchers that lead to scientific projects with foreign collaborators.

The provincial role in energy r&d

As is shown in Table 8.4, direct provincial spending in energy r&d is less than ten per cent. Seventy per cent of these provincial expenditures are spent by the government of Alberta. It should be noted that these statistics do not take into account provincially-owned electricity, which are important r&d spenders: forty per cent of industry energy r&d spending came from utilities and particularly these state-owned firms. At the provincial level, the distribution of energy r&d investment is even more skewed than at the federal level, for each province depends on a particular source of energy for it's development. Moreover, the fluctuations over the years are more important than at the federal level because the r&d of the provinces depends more critically on specific projects such as the James Bay project in Québec, or the tar sands project in Alberta.

Oil sands is a vast resource in Alberta, and the province created, in 1974, the Alberta Oil Sands Technology and Research Authority (AOSTRA) to ensure the full exploitation of this resource which accounts for ninety per cent of the total energy r&d budget of the province over the period 1976–1981. AOSTRA, as a crown corporation, now funds research in processes for the recovery and upgrading of oil sands and heavy oils, and also enhanced recovery methods for conventional oils. The most important projects are performed in collaboration with industrial partners and relate to *in situ* oil sands recovery. For instance, one project involves Amoco Canada Petroleum and Petro-Canada in an air-steam injection project at Gregoire Lake. AOSTRA also has programmes which offer funds for university researchers and develop projects in collaboration with the Alberta Research Council. AOSTRA is considered by the federal government as the principal source of funds for research on oil sands. The Alberta Research Council is also involved in energy research in coal and hydrocarbon processing. The Alberta Office of Coal Research and Technology assists the coal industry. Saskatchewan and Manitoba have also devoted r&d spending in similar domains.

By contrast, the province of Ontario depends heavily on nuclear energy with it's 17 reactors generating (in 1986) forty-five per cent of it's electricity – and eighty-six per cent of Canada's total nuclear electricity, which accounts for sixteen per cent of the total production of electricity in the country. This explains the concentration of Ontario's energy r&d on nuclear and supporting technologies (which includes transmission and distribution of electricity). In comparison with other provinces, however, Ontario also makes important efforts in the conservation and renewable energy sectors. Ontario Hydro is the most important provincial energy r&d spender for projects related to it's own electricity system. While it's mandate was, up to the 1980s, to support corporate needs, it now includes the support of 'provincial economic development, especially the electrical needs of Ontario industry'.[41] The role of the Research Division is the development of new technologies which can improve power production and utilisation efficiency. In that respect, it has important activities in the testing of products and services that use electricity. One third of it's r&d is devoted to the efficiency, reliability and safety of nuclear generating units. For instance, projects include methods and tests for the evaluation of equipment. Other important topics are the transmission and utilisation of electricity and fusion.

Québec concentrates it's efforts on hydro-electricity production and transportation. It's main research centre is the Hydro-Québec Research Institute (IREQ) which also studies fusion technology. Since the involvement of Hydro-Québec in hydro-electric megaprojects in the 1970s, the most important challenge has been the development of technologies, such as the 735kv transmission system, to permit the transportation of energy over long distances. More recently, IREQ and Hydro-Québec have had to deal with new technological challenges generated by environmental concerns, such as the installation at Grondines of an underwater line under the St. Lawrence river. IREQ also developed testing equipment used by industry. According to a recent policy statement, hydro-electricity is still the priority and r&d projects concern production and transportation of electricity, and utilisation, since its creation in 1987, in collaboration with the NRC, of the Laboratoires des technologies électrochimiques et des électrotechnologies (LTEE).[42]

In the Atlantic provinces r&d efforts are focused on coal, renewable and energy conservation projects. This great diversity of priorities among the provinces, and the fact that natural resources fall under provincial jurisdiction, calls for a constant collaboration between the federal and provincial governments which often takes the form of joint funding in projects. We have already mentioned the joint fusion projects in Ontario and Québec and there are other similar joint endeavours AOSTRA and the Canada Centre for Mineral and Energy Technology (CANMET) on the treatment of oil sands, *etc.* Some of these projects are part of international endeavours through the IEA.

During the past few years, most provincial governments have been greatly effected by growing environmental concerns: Québec, because of it's Great Whale hydro-electricity project; Ontario, because of it's nuclear projects; and Alberta, because of new norms concerning oil emissions. As a result, provincial governments have shifted their energy policy. For instance in 1990, British Columbia stated energy efficiency and environmental protection as its goals.[43]

This new context has had the effect of delaying megaprojects and upgrading conservation and renewable energy r&d projects. Thus, a new interest has emerged, for example, in small hydro plants. They have less environmental impact and their potential in Canada is evaluated at nearly 8,000 MW, concentrated mainly in Ontario (2,300), British Columbia (1500), Québec (1300) and Newfoundland (1200).[44] Utilities are encouraging private companies to develop this potential and sell them the electricity.

There are few solar energy installations in Canada.[45] In 1991, the total power generated was about 500kW. These installations are used by the Canadian Coast Guard for lighthouses, radio beacons and by railways and telecommunications companies to power remote radio repeater stations. Canadian r&d in this sector is focused on hybrid solar-diesel electric systems to reduce fuel consumption. Ontario Hydro has designed and operates (since 1986) the largest facility, a 10 kW solar diesel plant at Great Trout Lake. According to the Canadian Photovoltaics Industries Association, there are 47 suppliers of photovoltaic equipment and information. Actual projections to the year 2000 do not expect installed solar energy potential to rise substantially. In Canada, the harnessing of wind energy has been more important than solar energy, the National Research Council having been the main actor of r&d in this sector. EMR estimates the installed capacity in the country at about 7.5 MW. Research has concentrated on the vertical axis turbine, the most important of which is a 4MW engine installed at Cap Chat, some 400km north east of Québec city. Of the fifteen installations, three are in Prince Edward Island at the Atlantic Wind Test Site (for a total of 600 kW), six in Alberta (350 kW), two in Ontario (200 kW) and two in Québec (65 kW at Kuujjuaq), with the other two being in the Northwest Territories and on Bell Island.

Though Canada is not the ideal place for geothermal energy, r&d in this sector focuses on energy extraction systems and support for technology and engineering developments. Two projects using underwater energy sources for thermopumps are in operation: one at Carleton University in Ontario, and the other at Springhill, Nova-Scotia.[46] In a related sector, a tidal energy power plant of 20MW was constructed in 1984 at Annapolis in the Bay of Fundy. Overall, one can conclude that provincial and federal governments follow developments in all major alternative energy sources, though they are expected to play a minor and mainly complementary role in the energy matrix of Canada.

Of course, Canadian universities are active in all sectors of energy r&d. If we take their participation in NSERC's energy strategic grant programme as a measure of their activities, we see that among thirty-four participating institutions, only one third are responsible for two-thirds of the projects over the period 1978–1985. Among them, the University of Toronto is the most active and is present in all sectors of energy research, followed by the Universities of British Columbia, Waterloo, Alberta and McMaster. Like Toronto, these institutions cover a wide spectrum of sectors and conduct research in renewable, oil and gas, conservation, storage, coal and nuclear. Others are more concentrated in their specialities as with the University of Calgary which focused more than half of it's projects on petroleum related research, and Québec's Institut national de recherche scientifique-Energy which concentrates on nuclear fusion, while fission research is being concentrated in Ontario, the only province extensively deploying the CANDU reactor. Coal research is undertaken essentially at the Technical University of Nova Scotia, Queen's, and the Universities of Toronto and British Columbia. Research on electricity is found mainly at McMaster, Laval University and Ecole Polytechnique. Renewable energy research is going on in most institutions. Over all, one can say that, except for the largest institutions which cover all fields, expertise in universities tends to correlate with provincial needs in energy.

Energy r&d in industry

A fluctuating commitment to energy r&d is also visible in industry. Whereas the federal government can pursue long term objectives, industry is driven by the market. Through the federal r&d contracting out policy, industry carries on research for the federal government. In addition to these funds obtained for specific r&d projects, industry invests it's own money in energy r&d. Table 8.4 reveals that industry, including utilities, performs close to half of Canadian r&d in energy. As Table 8.7 shows, the major research commitment was on technology related to the production of petroleum.[47]

Though its contribution to energy consumption has fallen from sixty-one per cent in 1973 to thirty-six per cent in 1986, fossil fuels are still a strategic source of energy .The federal government and the province of Alberta invest important amounts of money in this sector – especially after the implementation of the National Energy Policy in 1980. In response to the growing cost of energy, industrial investment in conservation technologies grew steadily between 1977 and 1984 to augment the efficiency of the production processes and of transport vehicles. On the other hand, industry's research on new energy was rather limited

Table 8.7
Energy r&d expenditures by industry (millions, current dollars)

	Fossil fuels	Nuclear	Coal	Renewable energy	Supporting* technologies	Conservation	Total
1977	72.8	6.9	0.7	6.2	21.1	5.3	113
1978	85.8	13.3	4.9	5.5	30.9	20.7	161.1
1979	108.4	–	5.4	5.6	44.5	22.7	186.6
1980	134.7	20.1	3	6.3	51.3	44.3	259.7
1981	255.7	19.9	11.5	18.2	59.2	38.1	402.6
1982	239	33	7	17	61	47	404
1983	179	41	6	16	51	54	347
1984	181	48	9	18	65	50	362
1985	220	45	7	21	69	58	420
1986	159	58	12	51	77	65	422
1987	114	47	11	22	91	65	350
1988	121	33	11	20	175	78	438
1989	114	30	11	19	158	101	433

*Includes transportation and transmission, others.

Source: Statistics Canada. Science Technology and Capital Stock Division. *Industrial Research and Development Statistics*. 1977 to 1988.

and the government invested in this sector specifically to help this young industry finance and commercialise it's products.[48]

In the nuclear domain, industry concentrates it's investment on uranium exploration and production, but the main actor in this sector is the federal government which, through AECL, maintains the technical base for Canada's nuclear
reactors. This is probably the sector in which the government will have the most
difficult choices to make, as only Ontario is really dependent on this technology.
Coal research is another important sector that has been neglected by an industry
that does not possess sufficient r&d resources. However, coal is considered a potentially important source of diversification and it's use has been increasing over
the last ten years, contributing thirteen per cent to energy consumption in 1990,
compared to only five per cent in 1973. Accordingly, the federal and provincial
governments have invested in this sector to help in the modernisation of the
technology in order to ensure clean burning of coal.

Over the last few years, the pattern of investments in energy r&d by the industrial sector has changed. First, due to the oil price decreases, industrial research
on fossil fuels has slowed down drastically since 1985, contributing to a decline
in industrial commitment to energy r&d. Secondly, r&d investment is larger in
supporting technologies, because of the role of provincial electric utilities.
Thirdly, r&d in conservation is increasing.

In 1989, seventy per cent of energy r&d was performed in three industries:
electrical power, refined petroleum and crude petroleum (including oil sands/
heavy oils).[49] R&D spending in electrical power was performed by provincial
electrical utilities. In the petroleum industry, the two most important performers
are Petro-Canada, a state-owned firm, and Esso Petroleum.[50] Petro-Canada r&d
projects involve extraction of heavy hydrocarbons and their processing.[51] Esso
Research Centre is performing research on subjects related to Esso's business.

Conclusion

In addition to helping industrial sectors which cannot by themselves invest suf-
ficiently in energy r&d – like coal and new energy related industries – or to invest in sectors considered as particular to the Canadian situation in terms of
natural resources – like tar sands – or in terms of scientific and technical capability – like the vertical axis wind turbine developed by the National Research
Council – the role of the federal government energy r&d programme is conceived in terms of achieving longer terms goals, such as energy self-sufficiency
and diversification of it's energy source in order to become less dependent on
non-renewable energy sources. Recently added to these goals was the question
of environmental impact of energy production and consumption.

In 1985, a government document estimated that the distribution of energy r&d
investments in the public and private sectors showed a reasonable equilibrium
between short and long term objectives. Though this statement was probably
true for the private sector, it is doubtful that the government cutbacks in energy
r&d implemented that same year, left the public sector with a balanced
programme.[52]

Whereas the private sector's investments are legitimately targeted at short
term goals, the role of the government should be to provide for longer term options. It is therefore doubtful that further diversification and less dependency on
oil will be achieved by reducing budgets in new energy and conservation technologies. In fact, except for the nuclear energy sector, the distribution of the federal government's investments in energy r&d in 1986 had the same structure as

that of the private sector, and reflected the government's economic renewal policy with it's emphasis on short term economic benefits. Moreover, given that about sixty per cent of the budget of the Panel on Energy r&d was contracted out to industry, the reductions have had more effects on the private than on the public sector and contributed very little to 'economic renewal'.[53]

Though this general survey of the emergence and development of energy r&d policy in Canada is not intended as an evaluation of the policy itself, or of its benefits[54], one cannot fail to observe that the effective role of the Interdepartmental Panel on Energy r&d was less to drastically reorient the priorities of the mid-1970s in the face of a crisis situation than to open new avenues without disturbing the existing distribution of power among the departments and agencies active in the energy sector. AECL, for instance, was not really affected by the Panel's decisions. Moreover, being less entrenched in the governmental structure than the individual departments, the Panel was more susceptible to seeing it's budgets reduced. And the effect of these restrictions could only be to diminish the degree of coordination among the various projects and to weaken the sectors of energy r&d which were depending on the Panel's budget. In fact, in order to really strengthen the coordination of energy r&d at the federal level, the Panel should be responsible for the effective coordination of all the energy r&d related budget instead of the twenty-two per cent left in 1989.

The problem of energy r&d policy is part of the larger problem of the appropriate governmental organisation for horizontal activities which pass through the usual departmental and vertical structures. As the case of the Ministry of State for Science and Technology (MOSST) has shown [55], coordination faces the obstacle of the autonomy of each department which does not want to lose control of a part of it's mandate, be it of broad and horizontal interest as science, technology or energy. From this perspective, the solution adopted for energy r&d policy in Canada – an Interdepartmental Panel with it's own funds – is certainly a more appropriate structure than a MOSST without a portfolio for real coordination of s&t activities at the federal level. In matters of long term planning, however, Canada is ill-equipped since the dismantling, in 1992, of the Science Council, the only independent organisation that existed to think about long term scenarios for research and development. Without a proper institution to plan the future of energy r&d as part of a coherent s&t policy, or for that matter an industrial strategy, energy r&d will continue to be defined only through the energy policy.

Notes

1 Thanks to Jacques Rivard and Elaine Gauthier who assisted us in the different phases of this research. Thanks also to M. Gilles Mercier of the Office of Energy r&d in Ottawa and Jean-Marc Carpentier for providing us with information on energy r&d.

2 Ignatieff A 1981 *A Canadian Research Heritage. An Historical Account of 75 Years of Federal Research and Development in Minerals, Metals and Fuels at the Mines Branch* Ottawa. Supply and Services

3 Bothwell B 1988 *Nucleus. The History of Atomic Energy of Canada Limited* Toronto. Toronto University Press

4 Vietor R K 1984 *Energy Policy in America Since 1945* pp321–322 Cambridge U. Press

5 Surrey J Walker W 1975 Energy r&d A UK Perspective *Energy Policy* June:90–115

6 OECD 1978 *Energy Policies and Programmes of IEA Countries. 1977 Review* p2 Paris OECD. Only 19 of the 24 OECD member countries were part of the IEA.

7 Lowe J 1983 Who Benefits from Australian Energy Research *Science and Public Policy* December :pp.278–284.

8 OECD 1975 *Energy r&d* Paris OECD

9 The Senate of Canada. First Session. Thirteen Parliament 1974–1976. *Proceedings of the Special Committee of the Senate on Science Policy* Issue No. 18. Wednesday. August 11. 1976. p. 22

10 National Research Council. 1986 *Technologies des énergies de remplacement au Canada. Programme de R et D énergétiques du CNRC. 1975–1985*

11 In 1988. the Canadian Government applied norms concerning the decrease of carbon dioxide emissions. Ontario-Hydro stopped, in January 1992, the expansion of new nuclear reactors, invoking a drop in demand for electricity. In March 1992. Hydro-Québec lost a \$17 billion contract in New York State due to a decrease in demand for electricity and environmental concerns.

12 Drury M 1975 *Brief to the Senate Committee on Science Policy* November 1975. Appendix 8 to the Minutes of the Senate Committee on Science Policy. 3 December. p. 1:96.

13 *Summary Report of Activities . 1985–1986. Federal Panel on Energy Research and Development* Energy, Mines and Resources. Ottawa Canada. OERD 86–03

14 OECD 1979 *Energy r&d. 1978 Review* Paris.OECD

15 Energy Mines and Resources *Summary Report of Activities. 1985–1986. Federal Panel on Energy Research and Development* Ottawa.Energy Mines and Resources OERD 86–03. p.1

16 Science Council of Canada 1975 *Canada's Energy Opportunities* Report no. 23. March

17 Science Council of Canada 1979 *Roads to Energy Self-Reliance:the Necessary National Demonstrations* Report no. 30. June

18 For a review of the period 1944–1980 see Nelles H V 1980 Canadian Energy Policy. 1945–1980: A Federalist Perspective in Carty R K Ward W P(eds) 1980 *Entering the Eighties. Canada in Crisis* pp.91–117Toronto. University of Toronto Press. For the recent period see Doern G B and Toner G 1985 *The Politics of Energy. The Development and Implementation of the NEP.* Toronto Methuen

19 Nelles H V *op. cit.* p98

20 *Ibid* p99

21 Report of the Task Force on Energy Research and Development. 1975 *Science and Technology for Canada's Energy Needs* Ottawa. April p.8

22 *An Energy Strategy For Canada: Policies For Self-Reliance* EMR. Ottawa. 1976; *The National Energy Program* EMR. Ottawa. 1980. *The National Energy Program: Update 1982* EMR. Ottawa. 1982

23 Energy. Mines and Resources *op. cit.* 1980. Doern G B Toner G 1985 *The Politics of Energy* Toronto. Methuen

24 For the evolution of the NEP, see Bott B 1990 Ten Years After the NEP. *CPA Review* November and Niosi J. Duquette M 1987 La loi et les nombres: le Programme énergétique national et la canadianisation de l'industrie pétrolière *Revue canadienne de science politique* **20**(2) juin

25 Energy, Mines and Resources 1988 *Energy and Canadians into the 21st Century* (Energy Options).

26 *Ibid* p103.

27 EMR. Office of Energy Research and Development. *1989 Guide to the Activities of the Panel on Energy Research and Development of the Government of Canada*

28 OECD. 1989 *Politiques et programmes énergétiques des pays membres de l'AIE. Examen*

29 AECL *Annual Report 1987–88.*

30 EMR. 1976 *An Energy Strategy For Canada: Policies For Self-Reliance* Ottawa

31 Dalpé R Gingras Y 1990 Recherche universitaire et priorités nationales: l'effet du financement public sur la recherche en énergie solaire au Canada *La revue canadienne d'enseignement supérieur.* **25**(2):p 27–44

32 NSERC 1979 *Report of the President. 1978–1979* p16 Ottawa ; NSERC 1985 *Completing the Bridge to the 1990's. NSERC's Second Five Year Plan* p65 Ottawa.

33 The details of these projects are given in *Energy Research and Development by the Federal Government. Summary of Realizations. 1975–1985*EMR. OERD. 85–01

34 On the tokamak project see Gingras Y Trépanier M 1989 Le Tokamak de Varennes et le programme canadien de fusion nucléaire: anatomie d'une décision *Recherches sociographiques* **30**(3):pp421–446 and Constructing a Tokamak: Political, Economic and Technical Factors as Constraints and Resources *Social Studies of Science* **22**(4) November 1992

35 Treasury Board 1977 *Policy and Guidelines on Contracting-out the Government's requirements in Science and Technology*Ottawa

36 It should be noted that the merger of CANMET with the Energy Conservation and Renewable Energies Branch was done in 1988 in order to create the Mineral and Energy Technology Sector.

37 Dalpé R Gingras Y *op.cit*

38 Anderson F Dalpé R A Profile of Canadian Coal and Petroleum Research Communities *Scientometrics* forthcoming

39 EMR 1991 *CANMET Business Plan 1991–1994* See also Anderson F Dalpé R The Evaluation of Public Applied Research Laboratories. *Canadian J.Program Evaluation* **6**(2): p107–125

40 OECD 1990 *Energy r&d. 1989 Review*. Paris

41 Ontario Hydro. *Ontario-Hydro Research Division Annual Report 1990* p4

42 Energie et Ressources 1988 *L'énergie force motrice du développement économique*

43 BC 1990 *BC Energy Policy: New Directions for the 90's..*

44 EMR. 1991 *Small Hydro. Technology and Market Assesment*

45 For more details see Ontario Hydro 1991 *Alternative Energy Rev*. September

46 1992 *Annuaire du Canada* p. 407

47 The data for the years 1979 and 1980 are not directly comparable with the other data because the definition of energy r&d used by IEA is different from the one used by Statistics Canada

48 Whitman K 1985 *Programme fédéral de recherche et de développement énergétiques* exposé présenté au comité sénatorial permanent de l'énergie et des ressources naturelles. miméo. 4 mars p.5

49 Stat. Can.*Statistiques sur la recherche et le développement industriels. 1989* cat. 88–202. Table 32

50 Anderson F Dalpé R A profile of Canadian Coal and Petroleum Research Communities *op.cit*

51 In February 1992, the Canadian government privatised twenty per cent of Petro-Canada. Most of Petro-Canada's research centers were closed in the 1980s and research is now mostly done in the research center originally owned by Gulf and bought by Petro-Canada; Senate.*1990 Petro-Canada. Report of the Standing Senate Committee on Energy and Natural Resources*; see also Riopel P 1989 *Maîtrise technologique et transfert international: le cas de Pétro-Canada* PhD. thesis. Université du Québec à Montréal

52 EMR 1987 *La sécurité énergétique au Canada. Document de travail* juin p.87. Ottawa

53 Whitham K *op. cit*.:p17

54 Such an evaluation has already been published by the OERD after ten years of operation and at a time when the programme was seriously threatened by the government's budget restrictions; *An Assessment of the Economic and Energy Supply Benefits From the Energy r&d Program Administered by the Federal Interdepartmental Panel on Energy r&d* EMR. OERD. October 1985

55 Aucoin P French R 1974 *Savoir, Pouvoir et politique générale* Conseil des sciences du Canada. Etude no. 31. novembre ; see also, by the same authors 1974 The Ministry of State for Science and Technology *Canadian Public Administration* vol. 17, Fall 1974, no. 3, pp. 461–481.

Chapter 9

Canada in space

Ron Freedman and Jeffrey Crelinsten

Introduction

From the outset, Canada has had a stature in world space research that is far beyond the size of its technical infrastructure or its gross domestic product ... Canadians have ventured into space half out of curiosity ... and half from economic practicality.[1]

The story of Canada's involvement in and with space is about the interplay between curiosity and commerce. This is a recurring theme in Canada's proud space history and had its origins far earlier than most people would imagine.

We can trace Canadians' interest in space over 150 years to the early part of the 19th century. The first magnetic observatory in Canada was established in 1839, almost three decades prior to Confederation. Working from a site on the grounds of the University of Toronto, Sir Edward Sabine used data from stations in Toronto, Madras, Melbourne and St. Helena to determine that 'magnetic disturbances occurred world-wide and were related to the number of sunspots, which varied with an eleven year cycle[2].

A number of unique geographic features made Canada's interest in space almost inevitable. Canada is 'home' to the north magnetic pole, making the country a focal point for map making, for centuries. It is also a locale of the spectacular *aurora borealis* ('northern lights'), a scientific phenomenon which has intrigued humankind throughout the millennia. The aurora results from the interaction of charged particles, arriving from the sun, with the upper atmosphere, the ionosphere. Finally, the sheer vastness of the country imposed a reliance on space by early explorers, who depended on the sun and stars to help them position themselves and draw maps of the sprawling, often featureless land.

The 1882–83 International Polar Year was the first time that organised scientific observations were coordinated on an international basis. The IPY research programme included measurements of the meteorological, magnetic and auroral phenomena in northern Canada. A second IPY was established in 1932, and used new scientific equipment, including kites and balloons, to take measurements high above Earth's surface. For the first time, radio was used to communicate scientific measurements taken at different sites. Long hypothesised by scientists, the correlation between solar radiation and the ionosphere was demonstrated during the solar eclipse of 1932, in southern Canada, when radio techniques were used to measure the movement of the ionosphere[3].

As with much of technology and science, space and related research in Canada received a boost during and after World War II. Scientific interest in the ionosphere gained real world importance during the war. Short wave signals owe their ability to traverse vast distances to their being reflected back to Earth from the underside of the ionosphere. Incoming solar rays change the shape of the ionosphere and its distance from Earth's surface, so too does the transmission of short wave radio signals. Being able to monitor and predict ionospheric conditions would help allied technicians to plot the source of German radio emissions. In 1941, the National Research Council built and installed an 'ionosonde' at Chelsea, Québec, where F.T. Davies used it to predict the optimum operating frequencies for allied communications. By 1945, there were 7 such ionosondes in operation[4].

What we now call 'remote sensing' was also an early Canadian interest. The first known aerial photograph in Canada was taken by Captain H. Elsdale from a balloon over the Citadel in Halifax in 1883. War planners relied increasingly on aerial reconnaissance in World War I. After the war, it was adapted for aerial surveying. By the early 1920s, Canada had become a leader in remote sensing and had a large number of active survey companies. The vastness of the country and the density of vegetation made traditional ground-based mapping a near impossibility on a large scale. Even with the huge increase in capability that came from aerial mapping, it was to be many decades before cartographers could claim to have mapped the entire country. One of the problems involved the interpretation of aerial photographs. During the 1930s, NRC scientists developed a machine for plotting maps from highly oblique aerial photographs, as well as a stereographic plotter.

After World War II, the NRC initiated studies of cosmic rays, and an NRC group later developed rocket-borne instruments that measured the interaction of high-energy particles with the upper atmosphere. The federal government established the Defence Research Board (DRB) in 1947. DRB took an active interest in space s&t. Together with the National Research Council, DRB began to fund an expanded program of Canadian space research. Research capabilities were established or expanded at several universities, including Saskatchewan, Toronto (Institute of Aerospace Studies), Western Ontario, Laval and York.

DRB also inaugurated the Defence Research Northern Laboratory (DRNL) at Churchill in northern Manitoba. The location was ideally suited for studying the aurora. Auroral research was carried out there using rockets developed at another DRB research establishment at Valcartier, Québec. The Valcartier research was aimed at developing a solid rocket propellant for use at low temperature. That work eventually led to the development of the highly successful Black Brant family of sounding rockets, which are still manufactured for upper atmosphere research and sold by Bristol Aerospace of Winnipeg. In fact, though situated on Canadian soil, the Churchill range was established and maintained for twenty-five years, first by the U.S. Army and then by the USAF. In 1966, management of the range was turned over to Canada and was funded by the National Research Council and NASA. In 1970, NRC took over full responsibility for the range. At its peak, the range operated with a staff of 200 and at a cost of more than $4 million. In 1969, for instance, 40 rockets were launched at Churchill.

Churchill was finally shut down by NRC in November 1984, under the pressure of declining budgets. In 1992, private interests, led by CAE Electronics, are studying the feasibility of re-opening Churchill to meet an anticipated demand for small satellite ('Smallsat') launch services in the 1990s.

DRB also established the Defence Research and Telecommunications Establishment (DRTE) in 1947, which was later to become the home of the first Canadian satellites. Until the advent of telecommunications satellites in the 1970s, communications with most of northern Canada depended on unreliable HF radio transmission. It was simply too expensive to string telephone wires and repeating stations thousands of uncharted miles to serve small populations. DRTE scientists worked to improve the reliability of radio communications.

> *Thus the DTRE scientists, while working to improve radio communications, at the same time seized a unique opportunity to investigate the fundamental nature of high-latitude geophysical phenomena. Because the disturbances were recognised to be associated with features on the sun, the research was extended to include solar-terrestrial relations[5].*

When NASA was created by the U.S. government in 1958, a call went out to Canadian and other scientific groups to develop payloads that could be launched by the Americans. Scientists working at DTRE proposed that a satellite be built to sound the ionosphere from above, the direction from which solar rays 'pushed' on it. DTRE scientists eventually built the Alouette satellite, the world's first 'topside sounder'. It is with Alouette that Canada first entered space with its own satellite, confirming its position in the modern era of space.

The modern era of space

Until the 1960s, Canadians concentrated on gathering information **about** space, by using ground-based equipment and high altitude balloons and sounding rockets that carried instruments to the upper layers of Earth's atmosphere. In the 1960s, Canada was able to develop a satellite design and construction capability, and was soon conducting research **from** space. Beginning with the 1984 Space Shuttle flight of Marc Garneau, Canada's first astronaut, the country developed the ability to conduct research in space.

In 1958, as part of the International Geophysical Year, the USA supplied Nike-Cajun rockets to launch the first scientific payloads built entirely by Canadians from the Churchill DRNL test range. The instruments were designed to determine the electron density and temperature of the ionosphere for comparison with ground-based measurements. The National Research Council contributed cosmic-ray detectors whose purpose was to detect the intensity of charged particles reaching the upper atmosphere.

Only five years after the 1957 launch of the world's first satellite, Sputnik, Canada became the third country in the world to build its own satellite, Alouette I. Fifteen years later, in 1972, with the launch of Anik A1, Canada became the first country in the world to operate a commercial, domestic, geostationary communications satellite system. Today, Canadians routinely use space technology for telecommunications, weather forecasting, resource management, mapping and search and rescue operations, to name several of many applications.

By world standards, Canada is a middle player in the space arena. Why should a country of only twenty-seven million people have such a vital space sector? The reason for Canada's commitment to space lies in the ability of space to meet real Canadian needs. Canadians are **users** of space. Because Canada is so vast and its population so small and wide-spread, Canadians were quick to realise the potential of space s&t. For example, the cost of providing basic telephone service to communities in northern Canada would be extremely expensive by normal ground-based methods. But telecommunications satellites eliminate the need to construct land lines in inhospitable areas. With satellites, all communities cost much the same to service, irrespective of distance. Canadians use space for more than telephone communications. Today, for example, Canadians are using satellite images to help update topographic maps - a technique which is vastly less expensive than revising maps using aerial surveys.

People often marvel at the scope of Canada's activities in space. Canada's space sector employs about 3,000 people. In 1990, it had sales of about $350 million. About forty-five per cent of the sales are exported - seventy per cent to the USA – making Canada one of the few countries in the world where space industry revenues exceed expenditures by government. Furthermore, the industry is made-in-Canada, with over ninety per cent Canadian ownership.

In these ways, and many others, Canada has come to rely on space technology for economical solutions to real problems. In the process, Canadian companies have developed the expertise to compete with the world's best. In areas such as remote sensing data acquisition and analysis, communications satellite systems, and space-based robotics, Canadian industry has developed an international reputation for excellence.

The Canadian Space Plan

In 1986, the Canadian Government adopted a five year Long-Term Space Plan (LTSP), tailored to Canada's special needs. The plan expired in 1991, but it's successor was delayed by start-up problems with the new Canadian Space Agency. Following a large-scale consultation exercise with the space community, a new plan is scheduled for release in 1993. Under the current plan, Canadian expenditures on space would total $3.6 billion between 1988 and 2000 (Figure 9.1). The majority of those funds would be spent by the Canadian Space Agency. Expenditures are set to peak in 1992/93, as the result of the maturation of a number of large projects, in particular RADARSAT and Mobile Servicing System.

The 1986 long term plan focused on communications and remote sensing applications and thereby encouraged the growth of internationally-competitive firms in those areas.

The plan recognised the maturing of the space telecommunications industry, a sector which now generates enough income to finance it's growth and technology development. The largest component of government space expenditure (thirty-five per cent) has therefore been directed to Canada's contribution to the International Space Station Program, the Mobile Servicing System (MSS) and related science and industrial development activities (Figure 9.2). The MSS springs from the successful Canadarm, the robot arm which is installed on every U.S. Space Shuttle orbiter.

As a result of participation in the Space Station project, Canadian scientists and engineers will be granted access to the unique space environment for sustained periods of time. This will further research into two new fields of space research, life science and materials science. Investigators hope to learn more about the behaviour of living organisms and materials in space, so they can better understand their workings here on Earth. Canada's historic commitment to remote sensing will continue in the 1990s. The innovative RADARSAT remote sensing satellite will account for the next largest component of government space expenditure between 1988 and 1992/93, after the space station expenditures. Space science, communications and other government space activities account for the balance of expenditures under the plan. The new LTSP will likely focus future efforts on 6 areas (Table 9.1).

Almost all of Canada's major space initiatives have been undertaken in partnership with other countries. This is because Canadians recognised they do not have the resources to develop an independent capability in all of the areas where they have a need. International partnerships have opened doors for Canada to take part in a wider range of projects than it could do on its own. They have also

Figure 9.1: *Canadian Space Program Expenditures*

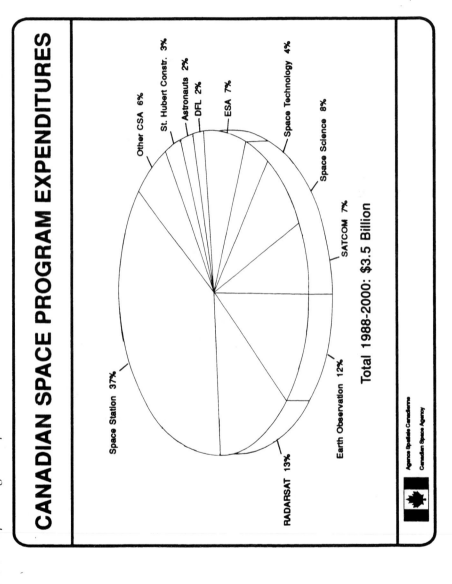

CANADIAN SPACE PROGRAM EXPENDITURES

Other CSA 6%
St. Hubert Constr. 3%
Astronauts 2%
DFL 2%
ESA 7%
Space Technology 4%
Space Science 8%
SATCOM 7%
Space Station 37%
RADARSAT 13%
Earth Observation 12%

Total 1988-2000: $3.5 Billion

Agence Spatiale Canadienne
Canadian Space Agency

Figure 9.2: *Canadian Space Program Expenditures
Cash Flow (1988–89 to 2000–01)*

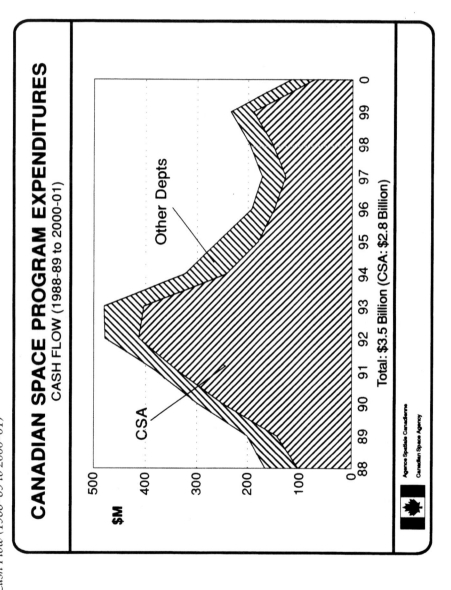

helped to lower costs and gain access to new ideas and technologies developed in other countries. Thus, international cooperation is a significant aspect of any future activities under the LTSP. The LTSP will also seek a better balance among small, medium and large programmes and among space science, technology and applications.

Throughout Canada's first quarter century in space, the USA has been a close partner. The European Space Agency (ESA) is also a key associate. Canada is the only non-European nation to have a close and formalised association with ESA. Cooperative agreements signed in 1978 and 1984, renewed in 1989, have allowed Canada to participate in major ESA programs in communications (*e.g.*, Olympus) and remote sensing (*e.g.*, ERS-1) and to have a voice in shaping these programmes to meet Canadian needs. The original agreement with ESA was renewed in 1989 for ten years. Canada pays an annual membership fee of $8 million for the right to participate in ESA optional programmes. Canada is accorded a number of contracts which is proportional to it's financial contribution. It also has the right to attend and vote at most Agency meetings.

Canada has worked closely with ESA on a number of collaborative projects. For example, Canada invested $102 million in the Olympus telecommunications satellite project. In return, Canadian firms were awarded contracts to develop solar array sub-systems and microwave amplifiers and components. Canada's

Table 9.1
New Focus Areas of the Long Term Space Plan

Space Infrastructure

Ongoing participation in Space
    Station
Space robotics and power systems
Serviceable carrier
Facilities upgrading
Materials and processing centre of
    excellence

Radarstat

Proposed RADARSAT II and III
    follow-ons
Antenna technology
Software
On-board and ground-based
    processing
Solar array research
Data user program
International partnerships

Earth Observation

Passive optical imager for vegetation
    and forestry applications
Canadian data network for "one stop
    shopping"
New instrument development (e.g.
    passive optical imager)
Applications development

Satellite Communications

Satellite Communications

KA-band satellite program
Small satellite applications
Earth terminals manufacturing
Navigation and positioning
Network integration

Space Science

Solar-terrestrial relationships
Astronomy and Astrophysics
Atmospheric pollution
Microgravity sciences
Microgravity facilities

Space Technology

Enabling technologies
Spacecraft systems (power,
    antennas)
Robotics technologies
Space qualification technology
Small satellites
Launch vehicle development
Technology transfer

Department of Communications will use Olympus to demonstrate digital tele-communications services in the 20/30 GHz frequency bands on a commercial basis, and experimental services at 10/20/30 GHz.

Canada invested six per cent of the total cost of the experimental ERS-1 radar remote sensing satellite, in order to gain knowledge that would be applied in the Canadian RADARSAT satellite. Canadians will receive data from the ERS-1 and follow-on ERS-2 satellites until the end of 1994, when RADARSAT is scheduled for launch.

Canada is also contributing to the cost of the Hermes space plane project, slated for it's first manned flight in the year 2000. Canadian industry is expected to receive up to approximately $25 million in Hermes contracts, depending upon the scope of the project finally adopted by ESA.

These and other partnerships - with Japan, the UK, France, Germany, the U.S.S.R. and Sweden - have confirmed Canada's international stature and en-sured its niche in the world's space age economy.(For more detail on Canada's international S&T relationships, see Chapter 7).

Satellites

Canada has the second largest land mass in the world, rich in natural resources. Compared with most countries, the Canadian population is small and wide-spread, and therefore costly to service with traditional telecommunications tech-niques. Canadians were therefore quick to appreciate the importance of satellite technology. With the launch in 1962 of the Alouette I research satellite, Canada became the third country in the world to design and build it's own satellite. In the years following Alouette, Canada developed a succession of satellites for communications, science and resource management. During that time, the Ca-nadian satellite industry matured and is now capable of acting as prime con-tractor for satellite design, manufacture, testing and operation. Along the way, Canadians pioneered in the application of satellite technology in fields as di-verse as tele-medicine, tele-education, remote sensing and satellite direct broad-casting (DBS).

The early years

Canada began it's satellite programme with a series of four research spacecraft that studied the ionosphere: the layer of the upper atmosphere which contains electrically-charged atomic particles that affect long-distance radio transmis-sion. The success of Canadian scientific instruments launched from Churchill in 1958 led to a proposal to build a complete satellite, which would be launched by an American rocket. Alouette I produced considerable data about the upper sec-tion of the ionosphere, including geographic details and information about the daily, seasonal and longer-term variations in that part of the atmosphere.

The success of Alouette I led to an agreement between the USA and Canada to build and launch a series of International Satellites for Ionospheric Studies (ISIS). Under this agreement, Alouette II, originally intended as a backup in case Alouette I failed, was modified to improve it's performance. Alouette II became the first satellite in the ISIS series.

ISIS I, launched in January 1969 and ISIS II, launched in April 1971, were the second and third satellites in the series. Both satellites exceeded their design life and produced useful scientific data for longer than originally planned.

The communications era

By 1969, the success of Canada's research satellites encouraged Parliament to pass an Act incorporating Telesat Canada, the world's first domestic communications satellite company. Until recently, Telesat Canada was jointly owned by the Government of Canada and thirteen of Canada's common carrier telephone companies. In February 1990, the Government announced it's intention to privatise Telesat completely. In 1992, it accepted a proposal to purchase it's shares from a consortium of companies headed by Bell Canada, and including Spar Aerospace Limited. The new company, named *Alouette Telecommunications* –harkening back to Canada's first satellite – will operate a commercial system of satellite communications to serve all Canada. It's system of Anik satellites, inherited from Telesat, carries voice, video, data and facsimile services. There have been four generations of Anik satellites and each has made a significant contribution to international technology development.

Anik A1 was launched in November 1972. It was the first domestic geostationary communications satellite in the world. Because it orbited above the equator at a speed matching the rotation of Earth, it was always visible from any point within its coverage area. Thus it could provide 24-hour-a-day telecommunications service. Anik's twelve microwave channels together could carry the equivalent of 11,520 one-way telephone circuits, or twelve colour television programmes. It operated on the 6/4 GHz frequency. Anik A2 was launched in April 1973 and was joined by Anik A3 in May 1975. All three satellites were identical and were designed to last three years.

By 1981 some channels had failed on Anik A2 and A3. In 1981, Telesat accomplished a world first. Anik A2 and A3 were co-located in the same orbital position. This permitted the still usable channels on each to be operated as if they were aboard a single spacecraft. Anik A1 and A2 were 'retired' in 1982 after almost ten years in space. Anik A3 ceased operation in late 1984.

Hermes

At the same time as the Anik series was being planned, work was under way on a new generation of spacecraft, the Communications Technology Satellite, later renamed Hermes (not to be confused with the ESA project of the same name). After being launched in 1976, Hermes revolutionised space communications by demonstrating that spacecraft could operate effectively at higher frequencies (14/12 GHz) than previous systems such as Anik. By using 14/12 GHz, Hermes eliminated interference with ground-based transmissions and could use earth stations with dish antennas measuring less than one metre wide – less than one-quarter the size of other dishes.

Hermes was a joint project of the USA and Canada. It was a high power satellite – for many years, the most powerful non-military communications satellite – which was energised by solar cells attached to two retractable 'sails'. The sails contained 27,000 solar cells and during launch, folded like an accordion beside the satellite. When in orbit, the solar sails unfurled. A sophisticated tracking mechanism kept the sails pointed at right angles to the sun's rays and at the same time held the satellite's antennas pointed toward Earth. During it's operation, Hermes was used to carry digital television signals, paving the way for a new generation of spacecraft. Many of the innovations pioneered by Hermes found their way into later satellite designs.

Satellite applications

The Hermes satellite ushered in a new era of satellite applications. Many experiments were carried out on new ways to use satellites; for example, in the field of 'telehealth' – health services and health education via satellite and ground-based telecommunications.

In one project between the University Hospital in London, Ontario and a small hospital in Moose Factory, Ontario, doctors operating on patients in the remote hospital could consult surgeons and other specialists in London. The specialists could remotely control a television camera in the operating room and zoom in on the patient. X-ray images, electroencephalograms, electrocardiograms, fluoroscopy and other medical data could be relayed for immediate 'teleconsulting'.

Memorial University of Newfoundland used Hermes to provide continuing education programmes for health professionals and health education programmes (nutrition, pre-natal care, diabetic diets, etc.) for local communities. The University also tested a system to provide medical support services (as well as direct dial telephone and data service) to off-shore oil drilling rigs.

Tele-education was the focus of several Hermes experiments. So successful were the experiments, that a number of provincial educational broadcasting networks now distribute their signals by satellite. Today, for example, broadcast signals from TV Ontario - the Ontario Educational Communications Authority - reach over ninety-five per cent of Ontario's population via the Anik C1 satellite.

One of the most significant capabilities which Hermes proved feasible was direct broadcasting by satellite (DBS). Because of Hermes' high-power, high-frequency transmissions it was now possible to use much smaller less expensive earth receiving stations to pick up its' signals. This made it possible to broadcast directly to the end user. End users need only have small (0.6 to 1.6 metre diameter) receiving dishes, which could cost as little as $400 in mass production. Dishes could be mounted in backyards, on roof-tops or even in windows. Hermes avoided the need – and cost – to route signals through large central receiving and switching stations. DBS satellites are now coming into use around the world, and have profited greatly from the Hermes experience.

The new Aniks

Canada's success with the Anik A satellites and with Hermes, led to the development of the Anik C and D and of the sophisticated Anik E satellites. Launched between 1982 and 1984, the Anik C series included Canada's first commercial spacecraft operating solely at 14/12 GHz. Anik C3, for example, was capable of carrying the equivalent of thirty-two colour television signals, twice the capacity of Anik A.

Two Anik D satellites were the first to be built by a Canadian prime contractor, Spar Aerospace Limited. They were designed as replacements for the Anik A series. They were the biggest Telesat had sent into orbit. Anik D1 - Canada's fifth commercial communications satellite - was launched in August 1982. It has 24 6/4 GHz channels, together capable of carrying the equivalent of twenty-four colour television signals.

The next generation of communications satellites is the powerful Anik E series. Anik E is of the dual-frequency design and operates on both 6/4 and 14/12 GHz. The two Anik E satellites are the largest and most powerful spacecraft built and launched to date for domestic satellite communications. Each of two identical satellites (Anik E1 and E2) carry the equivalent of fifty-six television channels. The two satellites are designed for a ten year life.

Anik E2 was launched on 4 April 1991 on an Ariane 44P launch vehicle. But on 12 April the satellite's main sending and receiving antenna would not deploy. Only $240 of the 300 million project costs were covered by insurance. On 15 April, the Ku-band antenna freed itself, but the C-band antenna was still useless. Technicians determined that a thermal blanket had shifted during takeoff and snagged the antenna. On 2 July, after dozens of procedures and the addition of 25 Telesat and 35 contractor personnel to the typical 60–70 person staff of the Flight Operations team, the emergency team increased the satellite's spin rate to dangerous levels in order to break the antenna free. The manoeuvre worked, and in September 1991 Anik E2 entered commercial service and began to deliver television signals across Canada.

Brasilsat

Canada's satellite expertise and willingness to share her knowledge, led in 1982, to the sale to Brazil of two Canadian-built telecommunications satellites and ground stations. Spar Aerospace Limited was the prime contractor for the Brasilsat satellites. The Brazilian Domestic Satellite System (SBTS), built by Spar, links together the far-flung reaches of Brazil. For the first time ever, by means of their own satellite, Brazilians were able to make telephone calls across the country as easily as across town in Rio de Janeiro. Television programmes could now be seen simultaneously throughout the country.

One of the hallmarks of the Brasilsat programme was that Spar undertook to work with Brazilian engineers and technicians in the construction of the satellite. A similar programme put Brazilian technicians in a position to operate the ground stations which were part of the contract. This substantial investment in training and technology transfer placed a special stamp of international cooperation on the SBTS programme.

Spar competed for follow-on work to build Brasilsat II in 1990, but lost the contract to Hughes Aircraft of the U.S. Brazil opted for the Hughes offer of older technology at a lower price, over Spar's more sophisticated, but more costly technology.

Mobile satellite communications

Most current types of communications satellite require the use of stationary receiving dishes. For the most part, telecommunications customers who frequently change their locations – such as truck operators, construction companies or hydro workers – are unable to use these satellite services. Canada's Department of Communications has pioneered the development of a new mobile satellite (MSAT) concept to meet the needs of these users.

MSAT will provide mobile radio, telephone, data, vehicle tracking and paging services to users in remote locations where there is no access to normal telephone services. The satellite will provide coverage of all Canada up to 200 miles offshore, for a nominal twelve year period. Industries expected to benefit from this new service include; minerals, oil and gas, construction, forestry, agriculture, transport, and, public services (*e.g.*, air ambulance). MSAT will complement existing cellular telephone systems.

Telesat Canada formed a subsidiary company, Telesat Mobile Incorporated (TMI), to operate the service in Canada. With it's partner company in the USA, AMSC, TMI will offer wide-area satellite services to land, marine and aeronautical mobile users beginning in 1994.

When fully operational, MSAT will consist of two satellites, one operated by Canada and the other by the USA Each satellite will act as back-up to the other. Up to 60,000 mobile units can be accommodated on the first-generation MSAT system. The terminals used with MSAT will be connected to small roof-mounted antennas. There will also be terminals light enough for a person to carry into remote locations.

David Florida Laboratory

Satellites are vastly complex and expensive pieces of equipment. Because the price of failure is so high, it is critical that every satellite, component and sub-component be thoroughly tested prior to launch.

The David Florida Laboratory (DFL) is Canada's national facility for spacecraft assembly, integration and testing. Owned and operated by the Canadian Government, the DFL is located in Ottawa. Its' facilities are available on a cost-recovery basis to the Canadian and foreign aerospace and communications communities.

The DFL has made a substantial impact on satellite communications in Canada and on the development of Canada's communications and space industries. The DFL is also in growing demand as a facility for world-wide environmental testing and integration of spacecraft. Built in the early 1970s to assist in the development of Hermes, the joint Canada-U.S. Communications Technology Satellite, the laboratory has since supported a number of major programmes and projects. The most extensive use of the laboratory facilities has been in support of Anik C2, D1 and D2; the Canadarm (Shuttle Remote Manipulator System); Brasilsat S1 and S2; and the European Space Agency's Olympus satellite.

The facilities offered by the DFL include three large 'clean' rooms suitably equipped for the assembly of satellites and other space hardware. It also contains a range of thermal vacuum chambers and an infrared testing system for verifying the thermal design and workmanship of spacecraft. Equipment at the DFL can test the ability of space structures to withstand the vibration experienced during launch. Other facilities examine the electrical and electronic properties of transmitting and receiving equipment.

DFL was used to perform the environmental testing of the Anik E1 and E2 satellites launched in 1991 and is currently preparing for the testing of the MSAT and RADARSAT satellites, as well as Canada's contribution to Space Station Freedom, the Mobile Servicing System.

SARSAT

When a plane crashes, or a ship is endangered at sea, the greatest challenge facing search and rescue crews is to locate the site of the accident. Canada is playing a key role in developing SARSAT (Search and Rescue Satellite), an international search and rescue satellite-aided tracking system. Canada, the USA and France have developed SARSAT in cooperation with the Soviet Union's COSPAS system. Compatibility between COSPAS and SARSAT allows the participating countries to use both systems for distress alerts.

When an accident such as a plane crash occurs, the craft's emergency locator transmitter (ELT) is set off automatically. Radio repeaters on board COSPAS-SARSAT satellites in low polar orbit detect and relay the ELT signals to a ground station. Through computer analysis of the signal shift of the ELT, operators can locate the position of the crash – usually to within 20 km, and often much better. In Canada, the ground station reports the accident location to the Canadian Forces Rescue Co-ordination centre, which dispatches search and rescue aircraft.

Each of the four COSPAS-SARSAT satellites circles the earth in polar orbit about every 100 minutes. This ensures that signals can be received frequently. In Canada, because of the congruence of satellites at the north pole, signals are likely to be heard within an hour.

Canadian companies are involved in all aspects of the COSPAS-SARSAT system. They have designed and manufactured a variety of instruments, including:

- Repeaters carried on U.S. satellites
- Ground stations for a number of countries, including the USA, France, UK and Brazil
- The Canadian SARSAT mission control centre
- Distress beacons
- Airborne direction finders.

One company, Canadian Astronautics Limited, is a world leader in the supply of receiving stations for COSPAS-SARSAT signals.

By December 1990, COSPAS-SARSAT satellites were responsible for saving over 1,800 lives around the world from 700 aviation, maritime and land distress signals.

Weather applications

Weather forecasting in Canada is the responsibility of the Atmospheric Environment Service (AES) of the Department of the Environment (DOE). AES has a long history of meteorological satellite data reception, beginning with the first direct-readout image received in Canada, in December 1963. AES now operates a network of ten ground-based reception and readout facilities across the country and cooperates with the Danish Meteorological Institute in the operation of another in Greenland. These facilities receive data from various geostationary and polar-orbiting meteorological satellites. Satellite meteorologists analyse and interpret imagery on a routine basis in preparing weather analyses and forecasts. In addition, satellite data are incorporated into numerical weather prediction (NWP) models along with data from conventional sources.

AES scientists are carrying out research into improving and extending the utility of satellite data for meteorological applications. Examples of such applications include the measurement of atmospheric temperature and moisture profiles and sea-surface temperatures from microwave and infrared sounders, detection of rainfall and rainfall rates from visible and infra-red imagery, and measurement of sea-surface winds from active microwave instruments (scatterometers). Studies are also under way to investigate the utility of data from passive microwave radiation sensors for application to short-range regional weather forecasting.

AES's mandate also includes the responsibility for operational monitoring and forecasting of ice and ice motion in Canadian waters. This information is important for the shipping and oil and gas industries which operate in Canada's far north. Satellite imagery from the USA and other polar-orbiting weather satellites is analysed by ice forecasters, in combination with data from other sources, to produce daily ice forecasts. The satellite data give forecasters a view of the 'big picture'. Close-up views of the ice situation in smaller regions of special interest are provided by airborne SAR (Synthetic Aperture Radar) – the same type of instrument that will be carried on board Canada's RADARSAT satellite. AES will study the operational utility of satellite-borne SAR measurements when these data become available, with a view towards ultimately incorporating them into its' operational systems. AES also carries out research into the use of passive microwave data to identify ice edge and extent, ice concentration and ice age.

Environmental applications

Many other applications studies using environmental or earth-observation data from space are carried out in DOE. These include research into the use of passive microwave radiation measurements to estimate snow depth and snow water-equivalent. This information is important for predicting spring water supply for hydroelectric generation and crop moisture conditions.

AES's Canadian Climate Centre has used U.S. NOAA satellite data since 1978 to map the surface water temperature of the Great Lakes and Atlantic seaboard on a regular basis for climatological analysis and fisheries applications, among other studies. They also collect satellite imagery of clouds on a long-term basis as part of an international cooperative project to study the role of clouds in global climate modelling.

For several years, AES has been conducting a programme of research into the measurement of atmospheric trace gases (*e.g.*, ozone, nitrogen oxides and aerosols) from space. This work is directed at the global problems of ozone depletion and atmospheric pollution. It also has applications in the study of the global warming phenomenon. In 1984, a 'sunphotometer' designed and built at AES was utilised aboard the U.S. Space Shuttle by Canadian astronaut Marc Garneau. The sunphotometer measures atmospheric haze and trace-gas concentrations. A follow-on Shuttle experiment with astronaut Steve MacLean took place in 1992.

DOE scientists are using satellite imagery to monitor habitat and land use changes in various regions of the country. For example, potential impacts of climate and other changes on wildlife habitats and protected areas are being studied. Satellite imagery is also used to monitor water resources in Canada. In addition, satellites are used to relay water conditions from a network of over 400 automated sensors and data-collection platforms on rivers and lakes across Canada to central locations for analysis and distribution. This information is used by a wide range of public and private organisations concerned with industrial and municipal planning, agriculture, marine transportation, flood forecasting, waste water treatment and wildlife conservation.

Space technology

Canadarm

As early as 1969, an invitation had been extended by NASA for Canada to participate in the Space Shuttle development program. The National Research Council headed a task force to explore options for involving Canadian industry. Interest eventually focused on a remote manipulator arm for the Shuttle cargo bay. In 1974, Canada agreed to develop the Canadarm (Shuttle Remote Manipulator System) for the U.S. program. As it's name suggests, it is essentially a robotic arm with all the motor functions of a human arm. The design specifications for Canadarm were rigid. The arm had to weigh less than 480 kg and had to reach to around fifteen metres in order to do useful work. It had to be able to move more than 30,000 kg – the mass of a fully loaded bus – and use less electricity than an electric kettle. It also had to work automatically or manually under the direction of astronauts.

The Canadarm was a joint undertaking of the National Research Council of Canada and Spar Aerospace Limited. Although it performs perfectly in space, the Canadarm cannot lift its own weight on earth! As a result, it could never be fully tested before it was launched. Engineers had to await its first flight before they were sure it would work.

The arm's software development alone took four person-years. After the software was developed, CAE Industries was able to build a training simulator for the astronauts. In all, there are 15,210 electrical and electronic parts that are vital to the success of the arm. Parts were supplied by 40 subcontractors throughout Canada. Each Canadarm is designed for up to 100 missions without major maintenance.

The Canadarm uses a snare mechanism to grab on to a special fixture mounted on satellites and spacecraft. Today, all satellites need to have this special fixture. In 1989, the arm was used to capture the LDEF (Long Duration Exposure Facility) satellite – the first satellite designed to be retrieved in space.

Canada contributed the cost of the first Canadarm on the understanding that all future arms would be purchased. Spar has now supplied arms for all the U.S. Shuttles and also has an ongoing contract to service them. Such was the success of the Canadarm that Canada offered to provide a greatly-improved version as it's contribution to the International Space Station, 'Freedom'. The Mobile Servicing System (MSS) is now under construction.

International space station

The International Space Station 'Freedom' is the largest international development project ever undertaken. Canada is a partner in Space Station, along with the USA, Europe (European Space Agency) and Japan. Space Station will provide a permanent base in low earth orbit for conducting space research in fields such as astronomy, life sciences and materials research. The advanced technologies being used to construct the station are expected to find application in many industries on earth.

Canada will be a user and operator of the orbiting complex throughout it's lifetime and Canadian astronauts will regularly conduct experiments aboard Space Station.

Mobile servicing system

The Mobile Servicing System (MSS) is Canada's contribution to the Space Station. The MSS will play the main role in Space Station assembly and maintenance, moving equipment and supplies around the Station, releasing and capturing satellites, supporting astronauts working in space and servicing instruments and other payloads attached to the Station. It will also be used for docking the Shuttle orbiter to the Space Station and then loading and unloading materials from its cargo bay.

The entire system will consist of equipment located both in space and on the ground. The MSS will be a 'roving space robot', riding aboard the Station on a travelling base provided by the United States. Installations for ground operations and logistics will be located in Canada. The MSS Manipulator will be about the same size as Canadarm but over three times as strong. It is designed to handle loads as big as the Shuttle orbiter as it approaches to re-supply the Space Station every three months.

Canada's programme for designing, developing, operating and maintaining the MSS encompasses a number of activities. These include developing both advanced technologies and the space systems incorporating them, integrating and testing equipment, training flight crews to use the MSS and providing operating and logistical support throughout the 30-year operational life of the Space Station.

The MSS programme consists of both a space segment and a ground segment. Three flight elements constitute the space segment, that part of the system which is attached to Space Station: the Mobile Servicing Centre (MSC), which includes a Space Station Remote Manipulator System (SSRMS); the MSS Maintenance Depot (MMD); the Special Purpose Dexterous Manipulator (SPDM); and, various sub-elements common to each system, such as control stations, and power and data management systems.

The MSS ground segment will include an MSS operations support facility which will be concerned with the operations, maintenance, logistic and training aspects of MSS use. In addition, a separate facility will accommodate Canadian users of the Space Station, providing access to experimental data and facilitating 'telescience' – science which is conducted remotely, using computer and telecommunications technologies.

A third aspect of the ground segment, the Manipulator Development and Simulation Facility (MDSF), will support the design and development of the space-based manipulator systems, on-going MSS operations and future engineering enhancements.

Unlike the Canadarm, MSS will be able to let go of it's movable base and crawl hand-over-hand to different faces of the Station's truss structure in order to service hard-to-reach objects. When operated by astronauts, it will act as an extension of their own senses. Astronauts will control MSS by voice commands. MSS will have it's own vision system and even a sense of 'touch'. Artificial intelligence will give MSS a limited ability to operate autonomously.

A separate smaller robot, called the Special Purpose Dextrous Manipulator (SPDM), will have two arms – each two metres long – for more delicate jobs such as cleaning surfaces, replacing faulty components and working on the Station's electrical circuits, fuel lines and cooling systems. SPDM can work either as a companion to the big arm (connected to it's end) or alone (attached to the Station's truss structure). Its' arms can work together or separately. The combination of the two robots gives the MSS the skill to do many kinds of work aboard Space Station, from handling massive satellites to small repair jobs ... from the heavy 'lifting' of a crane to the delicate positioning of a surgical clamp.

Estimates of the economic spin-offs from MSS are that it could generate industry sales of up to $5 billion, resulting in the creation of 80,000 person-years of employment. Initial estimates placed Canada's investment in the Space Station project at $1.2 billion to the year 2001. However, programme delays due in part to a number of re-designs mandated by the U.S. Congress will inevitably mean that programme costs will escalate beyond the original targets.

Strategic technologies

One of the chief tasks for the Mobile Servicing System is to reduce the amount of time humans spend working outside the pressurised modules of the Space Station. To do this, the space segment of MSS will need to incorporate many advanced tele-operation and robotics technologies. These include:

Vision Systems: The MSS vision system will be used for collision detection and avoidance, automatic payload capturing, detection of unexpected objects and inspection of payloads

Sensors: MSS will use a variety of sensors which will simulate the ability of humans to feel objects. Force and torque sensors will measure the forces which an arm applies on a surface. Tactile sensors will enable MSS to 'feel' surfaces and edges

Robot Programming: Due to the complexity of the MSS robotic systems, the simple programming schemes used on factory floors to teach robots to perform repetitive tasks are inadequate. For the MSS, a complex hierarchical robot programming system is required

Human-Machine Interfaces and Telepresence: In a complex system such as the MSS, the human-machine interfaces must make robotic devices as 'friendly' and easy to use as possible. Such devices may include systems that recognise verbal commands (voice recognition) and convert computer outputs into spoken words (speech synthesis)

It is expected that many of the technologies and capabilities developed for MSS will find their way into other industrial products and services. A Space Station Industrial Development Programme has been established to manage MSS intellectual property and to promote the future application of spin-off technologies in Canadian industry.

User development programme
By developing the MSS as it's contribution to the construction of Space Station, Canada will be entitled to have a Canadian astronaut on board for the equivalent of three months every two years. To ensure that the Canadian industry and research communities are in a position to exploit the Station's microgravity environment, the government is funding the User Development Program (UDP) as part of the Space Station Programme.

Although the production of space goods and services was not high on industry's list of priorities in the early 1980s, interest in this field is increasing rapidly. Interest is especially high in two new fields of microgravity science: life science and materials science. Approximately $73 million has been allocated to UDP from 1992–1997. UDP awarded $20 million worth of contracts from 1985 to 1992 and at the end of 1991 had 60 contracts under way.

Canadian astronaut programme

Despite tremendous advances in space robotic technology, it is acknowledged that there are certain space tasks for which people are needed. For this reason, all the leading space nations maintain an astronaut corps.

In 1983 the USA invited Canada to fly an astronaut aboard Space Shuttle mission 41-G, scheduled for launch in October 1984. This invitation led to the decision to establish a permanent astronaut corps. The purpose of the Canadian Astronaut Programme was to train and maintain a corps of astronauts who would coordinate and conduct Canadian experiments aboard the Space Shuttle. A nation-wide competition was held under the auspices of the National Research Council to select Canada's first astronauts. Twenty-one years after the successful launch of its first satellite, Canada was to have its own team of space workers.

In fact, the use of the term 'astronaut' is a slight misnomer. In NASA parlance, astronauts are Shuttle pilots or mission specialists who are extensively trained in the operation of the orbiter vehicle. Pilots[6] are in charge of the flight systems and overall mission parameters. Mission specialists are trained for Extra-Vehicular Activity (EVA) and are primarily responsible for the operation of orbiter systems (such as the Canadarm) and the deployment and retrieval of payloads in orbit. A third category of space worker, payload specialists, is expert in the operation of experiments being flown aboard the Shuttle. As such, they receive minimal training in the operation of other orbiter systems.

Applications to join the new Canadian astronaut team were issued in 1983 and were accepted until July of that year. The competition was opened to all Canadian citizens and was advertised extensively in newspapers across the country. Eventually, 4,300 applications were received for six positions. Between July and December all applicants were subjected to a careful screening process. The selection criteria included academic background (with an emphasis on science, engineering or medicine), professional experience, health and communications skills. In December 1983, the final selection was announced.

Six outstanding candidates were selected, five men and one woman. They included a neurologist, electrical engineer, laser physicist, physiologist, biomedical engineer and an engineering physicist. Most had worked in fields related to space science or technology. In addition to their outstanding professional accomplishments, they excelled in other areas as well. Their outside interests included hot air ballooning, yachting, gymnastics, badminton, scuba diving, and flying. Two had even represented Canada in the Olympic Games!

In response to the NASA flight invitation, Marc Garneau, a Commander in the Canadian Navy, became prime payload specialist for Canada's first manned space flight. Bob Thirsk was named as Garneau's backup for the mission. Garneau conducted a series of experiments to gather information about the motion sickness which afflicts many people in space, and the body's adaptation mechanisms to living and working in microgravity. Garneau also conducted experiments designed to understand 'orbiter glow' – a puzzling phenomenon which had been observed on previous Shuttle missions, in which a luminescent glow of light was observed to surround the orbiter. Other 41-G studies conducted by Garneau during his eight day mission investigated the constituents of Earth's upper atmosphere and the effects of the upper reaches of the atmosphere on the erosion of materials in low earth orbit.

The tragedy of the Challenger accident in 1986 severely delayed scheduled Canadian follow-on Shuttle missions. The next Canadian astronaut to fly on board the Shuttle was Roberta Bondar. With Ken Money as her backup, Bondar participated in the first International Microgravity Laboratory (IML-1), a Spacelab mission which flew in January 1992. Spacelab is a scientific laboratory which is carried in the cargo bay of the Shuttle. IML is designed as an international project involving participation from Canada, the USA, Europe and Japan. It is designed to test the effects of space on humans, as well as to carry out a number of experiments with crystals, plants and insects.

The third Canadian mission had astronaut Steve MacLean launched in October 1992. Bjarni Tryggvason trained as MacLean's backup. The primary experiment was a full evaluation of the Canadian-designed Space Vision System. This state-of-the-art machine-vision technology will assist astronauts with the precision handling of the Canadarm and future Space Station manipulators.

The astronaut programme will also be supporting Canada's billion-dollar participation in the International Space Station. When Space Station is in full operation in the late 1990s, there will be a Canadian on board for the equivalent of a three-month period every two years. Canadian astronauts will operate Canada's contribution to Space Station, the 'mission-critical' Mobile Servicing System (MSS) and perform scientific work and commercial experiments on behalf of Canadian researchers.

To date, Canadian astronauts have been trained as payload specialists. Plans call for Canadians to be trained as mission specialists for work aboard Space Station. In 1992 two Canadian astronauts, Marc Garneau and Chris Hadfield, will be assigned to begin four years of training with NASA to become mission specialists. At the end of their training they will be qualified to fly Shuttle missions in the same way as U.S. astronauts.

Canada's space industry

Canada's dynamic space industry is an example of how a growing world-wide demand for technology and technological improvements can be turned to national advantage. Increasingly, improvements to business practices and products are being made possible by new techniques and processes arising from space research and development. Canada's head-start in space-related activities and it's position as a world leader in space communications and space hardware manufacturing have provided many Canadian firms with the opportunity to compete aggressively in the demanding global marketplace.

The space industry operates from coast to coast in Canada and is ninety per cent Canadian owned. Fully seventy per cent of all sales are made to off-shore customers. The success of Canada's space programme has been mirrored by the growth in its' space industries, from a minor role in the late 1960s to it's international status today. The industry employs over three thousand people and sells in excess of $300 million in goods and services annually. It has an average annual rate of growth between ten and twenty per cent.

Between 1988 and 1993, the Canadian Space Programme will spend over $1.3 billion on its' various technology, science and applications projects. This investment is expected to increase the breadth and depth of Canada's industrial space capability.

A 1987 survey showed that five firms with more than 100 employees accounted for more than four-fifths of the total business activity in the space sector that year. However, it is significant that there was a large number of small space firms represented in the industry. This indicates that there is ample scope for new opportunity in the Canadian space business. Since the 1987 survey, other companies – a number of them from the U.S. and Europe – have established space businesses in Canada.

Space products, together with related consulting services, made up the bulk of all space-related sales in Canada in 1987. Almost two-thirds of all space products were produced for the space segment and one-third for the ground segment of space activities.

Total research and development spending of space companies was greater than one-quarter (twenty-five per cent) of their sales in 1987. R&D sponsored by the firms themselves averaged eleven per cent of sales. An important feature of Canadian space industry R and D activity is the high degree to which firms have participated in joint r&d undertakings with other firms and with Canadian universities.

Space station team

The Space Station programme is emphasising economic spin-offs and ensuring that these are distributed throughout the country. The federal government has set ambitious economic goals for the Space Station Programme, calling for the generation of $5 billion worth of spin-off business in space and non-space business in the first fifteen years of the programme.

The industrial team that is responsible for designing and constructing the Mobile Servicing System is led by Spar Aerospace Limited, the company which acted as prime contractor for the Canadarm. Spar has assigned major development roles in the programme to Macdonald, Dettwiler and Associates Ltd., SED Systems Inc., Canadian Astronautics, CAE Electronics and the IMP Group. Numerous other companies are expected to work with this group. In fact, a special programme called STEAR (Strategic Technologies for Automation & Robotics) has been put in place to ensure that firms across Canada have access to the benefits of MSS.

The STEAR programme stresses joint venturing and collaborative efforts. As such, some projects are jointly sponsored or contracted-out through other government agencies or non-profit organisations. STEAR is supporting advanced technology development in four areas which will bolster MSS. These are:

● Automation of Operations
● Automation of Power Management
● Autonomous Robotics
● Enhanced Space Vision Systems

By 1992, some 44 STEAR contracts had been awarded and 57 companies, research organisations and universities had benefited from STEAR-supported contracts. STEAR contracts had a total value of $15.2 million.[7]

Remote sensing

Canada has an important national stake in remote sensing, the process of extracting valuable information about natural resources from Earth images taken by sensors on aircraft or satellites. In remote sensing, instruments sensitive to visible light are augmented by sensors that can see in the ultraviolet, infrared and microwave regions. This vastly expands the amount of available information and thereby greatly increases what can be learned about the region of Earth surveyed.

With it's large unpopulated areas, wealth of natural resources and long history of map-making, Canada is virtually tailor-made for the application of remote sensing.

A pioneer in remote sensing

Canadian expertise in remote sensing goes back many years, beginning with early airborne photographic surveys. These surveys helped create visual images of remote parts of the country which were otherwise difficult or expensive to survey on foot. These images were used to create maps and to help open remote areas for resource development. Following World War II, Canadian companies were pioneers in the application of aerial prospecting techniques – a form of remote sensing – which used aircraft sensors to find hidden mineral deposits.

The Canada Centre for Remote Sensing (CCRS) was established in 1972 to coordinate federal government policy in remote sensing through a national committee network linking federal and provincial government departments and agencies and private sector companies. CCRS now develops and demonstrates new applications for remote sensing data. It receives, processes and distributes satellite data and operates specially-equipped aircraft.

Today, CCRS works in partnership with governments, universities and Canada's rapidly-growing remote sensing industry to provide products and services which have been used by over 60 countries. Many developing countries have shown special interest in Canadian expertise and in remote sensing education and training programmes provided by Canada.

In 1972, CCRS was involved in the reception of the first data from an Earth observation satellite (the U.S. LANDSAT-1 series). The LANDSAT satellites circle Earth at a height of 700 km in a polar orbit. As Earth turns below the satellite's fixed orbit, each successive orbital track is shifted westward up to 3000 km. This arrangement means that all of Earth's surface can be imaged in a few weeks, while more northerly locations are imaged more often. The outer-space perspective allows groups such as ranchers, foresters and other resource managers to examine detailed images of vast areas of the surface of the earth.

Some provincial governments are using satellite imagery for forest-fire management. The images are more up-to-date than maps and show the current state of the land, pinpointing new access routes and forest clear-cut zones crucial for optimum fire fighting. Space images are being used to estimate acreage of forests cut on Crown land and to help ranchers manage their range better by aiding in identifying regions where cattle grazing could be increased or decreased. They are also used to trace the path of pollutants in the air or water, to direct attention to mineral deposits, and to estimate crop acreage. Satellite data can be used in coastal engineering, to estimate factors such as sedimentation rates in rivers.

For projects that require high-resolution study of individual trees, buildings, lots and roads, remote sensing by aircraft is utilised, sometimes in conjunction with satellite data. In cooperation with industry, CCRS operates a fleet of three aircraft fitted with a range of optical, radar and laser sensors.

But for the big picture, satellites have revolutionised the way we look at ourselves from on-high. Satellite capabilities increased dramatically in the late 1980s, when France launched a new generation of remote sensing satellite, SPOT (Systéme probatoire d'observation de la Terre), which yields images at ten metre resolution. A special capability also allows SPOT to measure the heights of objects for mapping and mineral exploration purposes.

Remote sensing industry

Canada's remote sensing companies are world leaders in the collection, processing and interpretation of remote sensing data. Remote sensing is concerned with information about natural resources, which is collected with the aid of sensors mounted on aircraft or spacecraft. There are approximately 115 companies in Canada which are actively involved in remote sensing. What is more, almost half of them do business outside Canada.

Canadian firms are world leaders in a number of fields of remote sensing technology and value-added services. One group of companies is a world leader in certain aspects of sensor technology, including sensors which use lasers and radar to gather their data.

Other firms excel in the manufacturing of complex image production systems.

These sophisticated computer-based systems convert raw or semi-processed remote sensing data so that it can be used to create photographs. A third group of companies specialises in building equipment that converts the computer data into formats which are useful for analysts in industries such as forestry, agriculture, and mineral exploration.

Lastly, in excess of thirty Canadian companies have expertise in the interpretation of remote sensing imagery to meet the varied needs of end users in industries as diverse as ocean engineering and urban planning. Many of these companies have worked in other countries.

The Canadian remote sensing industry has annual sales exceeding $200 million, about half of which is sales to customers in other countries. Canadian firms have captured about fifty per cent of the world-wide market for ground receiving stations and twenty per cent of the world market for image analysis systems, and over ninety per cent of the market for airborne synthetic aperture radar services. It is expected that the launch of Canada's RADARSAT radar remote sensing satellite in the early 1990s will spur the development of the remote sensing industry in Canada.

Radarsat

On September 13, 1989 final approval was granted for the construction of RADARSAT, thereby launching a new age in civilian remote sensing. The satellite will carry a powerful active microwave SAR (Synthetic Aperture Radar) instrument. Unlike most remote sensing satellites that capture sunlight reflected from Earth, RADARSAT can transmit and receive radar signals and thus 'see' through clouds and darkness, obtaining detailed images of the earth. RADARSAT is planned for launch in 1994.

From almost 800 km above Earth, the satellite will circle the globe from pole to pole, scanning the entire surface in 500 km swaths and producing high-resolution images of Earth's lands and oceans. RADARSAT will cross the same point at the equator every 24 days and cover the Arctic daily. It is planned to operate in space for a minimum of five years. Processed data will be available only a few hours after RADARSAT passes over an area.

The total cost of building and operating RADARSAT is placed at $495.8 million. The Canadian government has committed $378 million and provincial governments an additional $52.9 million to construct RADARSAT. Private sector funding through Radarsat International (RSI) will total about $58.3 million.[8] RADARSAT is a Canadian-led project involving the USA, which will launch the satellite.

RADARSAT will be built in Canada by Spar Aerospace Limited. Ball Aerospace of the U.S. is building the spacecraft bus. A number of Canadian firms are building the spacecraft instruments and ground operations centre. A new company, Radarsat International (RSI) has been formed by a group of private investors to collect, process and distribute Radarsat data to international users. RSI has also assumed responsibility from CCRS for collecting and distributing international remote sensing satellite data, such as LANDSAT and SPOT data.

Space science and international programmes

Space science

Space science is a cornerstone of Canada's space programme. Space science began as the study of the universe, the solar system and the region of space near

Earth. In recent years, it has expanded to cover the field of microgravity science and its' two new sub-disciplines, materials research and space life sciences.

Space science had its early beginnings in Canada in the 19th century, when an astronomical observatory was constructed in Québec City. In the late 1880s, astronomical stations were used to help plot the route of the trans-Canada railway. In 1918, work was completed on the Dominion Astrophysical Observatory near Victoria, British Columbia. As a result of research conducted there, Canada was for many years recognised as the leader in studies of stellar radial velocities and of spectroscopic binary stars.

During the 1930s, scientific teams began to use ground-based instruments to study the *Aurora Borealis* (or 'Northern Lights'). It was these glowing lights in the sky which first enticed Canadian researchers into space. After centuries of speculation over the origin of this mysterious illumination, scientists began using balloons (and later rockets) to carry instruments into the upper atmosphere for a first-hand look.

Though incomplete, the picture formed to date indicates a complex interplay of magnetic and electrical fields, energetic particles arriving from the sun, and atmospheric atoms all in a maelstrom of activity. Canadian scientists are cooperating with researchers around the world adding detail to our image of the earth-space border zone. They have determined that the planet is surrounded by a teardrop-shaped magnetic field, rounded on the sun's side and stretching far out into space on the night side. They have traced incoming streams of particles from the sun encountering this field and being diverted to the earth's polar regions to create auroral lights. Instruments have detected intense electrical currents associated with the magnetic field and have attempted to 'circuit-trace' their paths in space.

In depicting near-earth space, university and government scientists, in close liaison with industry, have pioneered novel research methods and improved the quality of older ones. Passive measurements are being replaced by more active techniques. Recently, a team from the National Research Council launched a rocket bearing an explosive charge and chemicals and temporarily 'switched off' a segment of the Northern Lights. A 50-km diameter 'waterhole' formed in the aurora while changes in the particle stream and the magnetic field were recorded. Reversing the procedure, another rocket with a different composition of chemicals produced the first man-made Northern Lights.

Canadian science and industry teamed in developing a new generation of imaging devices for upper atmospheric research. Instead of laboriously collecting and collating data, these instruments can 'read' signals generated in the ionosphere producing direct images of events of electro-magnetic structures. Scanning devices of this kind are particularly useful in coarse surveys by indicating interesting aspects to be studied in detail during later missions. A highly sophisticated version of these devices has been designed and built in Canada to be flown aboard the Space Shuttle in upcoming missions.

Canadian scientists have worked extensively with colleagues in many countries on projects such as these. The Swedish VIKING auroral investigating spacecraft, carried a Canadian-built ultraviolet imager, and the European Space Agency will use another instrument when the International Solar Polar Mission explores the sun's polar zones. Agreements have also been reached with Japan for that country's satellites to carry Canadian instruments.

The Space Science Programme of the Canadian Space Agency supports scientists in university, industry and government laboratories. Approximately 50 companies share in $19 million of annual funding to conduct research in physics, astronomy, life sciences and materials processing.[9]

Microgravity science

In the 1970s and 1980s, as human beings began to spend more time in the microgravity conditions of space, opportunities began to emerge in two new fields of space science, Materials Processing in Space (MPS) and Space Life Sciences (SLS). Canada's participation in the International Space Station, 'Freedom', will provide Canadian researchers with long-duration access to the conditions of space, especially 'microgravity'.

Spacecraft in orbit around the earth are actually in 'free fall'. Gravity is still acting on the craft (and any cargo), but because everything on board is falling at the same rate, it's usual effects disappear. Because there are minor perturbations in the orbit caused by the operation of thrusters and equipment and by the movement of objects on board, spacecraft do not achieve true 'gravity-free' conditions. Rather, small gravitational effects (micro-gravity) still act on the craft.

Nevertheless, the near-absence of gravity suggests exciting new possibilities for space-based research and eventually, manufacturing. By eliminating the effect of gravity in space, researchers can investigate the effect of gravity in everyday earth-bound activities, such as manufacturing metal alloys or forming protein crystals. Canadian scientists and engineers are in the forefront of this type of science.

For a number of years, Canada's Space Station User Development Programme (UDP) has sponsored microgravity research in universities and industry. UDP supports SLS research in areas such as:

● Assessment of back pain in astronauts
● Ocular counter-rolling
● Transcranial doppler for measurements of blood flow velocities
● Opto-kinetic stimulation towards understanding space adaptation syndrome
● Otolith/spinal stimulation
● Pulsus paradoxus
● Taste sensitivity in microgravity.

The following UDP projects provide an example of the range of MPS expertise in Canada:

Beta Cell Encapsulation: Investigators from Canadian Astronautics Limited have experimented with methods which would eventually lead to encapsulating pancreatic beta cells in microgravity. Such techniques might be adapted for producing these insulin-producing cells and implanting them in the pancreas of diabetics

Alloy Formation: Scientists at Queen's University in Kingston are studying the formation of metal alloys in space. They hope to learn more about the role of gravity (by eliminating its' effects) so as to better understand the technology of manufacturing alloys on earth

Glassmaking Technology: In terrestrial glass processing, the main problems affecting the quality of the end products include devitrification, homogeneity, deformation, fining, impurities and limitation of heat treatment temperature. Scientists at B.M. Hi-Tech Inc. are using the microgravity environment of NASA's KC-135 aircraft to prepare speciality glasses which are not presently possible in earth's gravity

Water Droplet Dynamics: Spraying is a process that has application in many industries, for processes such as painting or crop protection. The effectiveness of spraying depends on the uniformity of the spray, the ability to control droplet size and to propel the droplets to the target. Investigators at the University of Western Ontario are using microgravity conditions to study the effectiveness of the charging process for electro-static spraying

The necessity to live and work in space is of considerable interest to Canadian scientists and engineers. The UDP is supporting SLS work in the fields of fluid physics, crystal growth and phase separation, in projects such as:

Microgravity Protein Crystallisation: Since the three-dimensional structure of human proteins determines to a large extent their function, the ability to grow protein crystals is important for the pharmaceutical industry. However, many crystals cannot be successfully grown in earth's gravity; it is difficult to obtain large, good-quality single crystals. Aastra Aerospace Inc. is studying the development of a protein crystallisation facility for future space-based crystal growth

Back Pain in Astronauts: Many astronauts report significant back pain during flight. This is thought to result from the lengthening of the human spine once beyond the effect of earth's gravity. A University of British Columbia team headed by Dr Peter Wing is investigating whether the range of motion of various joints, especially non-weight bearing joints, is increased in microgravity conditions. This may lead to a better understanding of human adaptation to space

Ocular Torsion: A team of scientists under Dr N. M. Kirienko of Toronto's St. Michael's Hospital is conducting research on the rotation of the human eye. A slight rotation of the eye in it's socket is observed as the axis of the visual field shifts. Such movements may be useful in predicting those individuals who will be susceptible to space motion sickness

International programmes

The Canadian space science programme strives to maintain a position of excellence for Canada in the world-wide scientific exploration of space. Space science experiments are complex and the environment of space requires development of facilities which are often beyond the resources of individual scientists, agencies or nations. International cooperation and collaboration are important to achieve common goals.

Canada is presently involved in collaborative space science ventures with France, Japan, the Soviet Union, Sweden and the USA Four areas of space science have been identified as encompassing the majority of Canadian interests and needs for the near future. These are: space plasma physics, upper atmosphere chemistry and physics, space astronomy, and microgravity science. At present in Canada, activities in the first two areas are dominant, but over the next few years, space astronomy and microgravity science are likely to become more active.

Table 9.2 gives a sample of international space science projects in which Canada is presently participating or which are under active consideration.

University space research

Much of the success of Canada's space programme is due to the expertise of Canadian university researchers and the excellent facilities which they have built up. At least eleven universities have space science and astronomy programmes of note. Seven of the most active have formed the Canadian Corporation for University Space Science (CCUSS). The purpose of CCUSS is to foster cooperative

Table 9.2
Canadian participation in international space science projects

Project	Description	International Partners
AURIO: Auroral Imaging	Instruments for a set of polar orbiting satellites being built in conjunction with the International Space Station. The Canadian instruments will help study in detail the processes causing the aurora (northern and southern lights). Canada will provide two instruments, a wide-field ultra-violet imager and a Doppler imaging interferometer.	European Space Agency
CANOPUS: Canadian Auroral Network for the Observation of Plasmas in the Upper Atmosphere and Space	Canada's contribution to the International Solar-Terrestrial Physics Programme. A set of instruments that sample, preprocess, transmit and analyse data from automated, unmanned remote sites to a central location	U.S.A (NASA) Japan
COBRA: Cosmic Background Radiation	This rocket payload included two instruments. First, a liquid helium cooled, two beam interferometer to measure part of the cosmic background spectrum. Secondly, a set of multiple electrostatic plasma probes to obtain electron density and temperature profiles.	USA
GEMINI: General Excitation Mechanisms in Nightglow	A sounding rocket experiment for airglow imaging. Main instrument will be an imaging spectrograph.	USA
FREJA	Canada will provide two instruments for the Freja mission, an ultraviolet imager and a Cold Plasma Analyser.	Sweden
INTERBALL ULTRAVIOLET AURORAL IMAGER (UVAI)	Canada will contribute the UVAI to the Soviet Interball project. Interball will study the explosive release of energy from the magnetotail region of the earth's atmosphere	U.S.S.R. (Russia)
MOPITT:Measurement of pollution in the troposphere	One of Canada's contributions to the NASA Earth Observing System (Eos) Polar Platform. The MOPITT Instrument measures carbon monoxide content in the tropospher	USA
RADIOASTRON	RadioAstron is a free-flying satellite that will function as a radio astronomy antenna in earth orbit and be used as part of a space-ground Very Long Baseline interferometry network. Canada will contribute Data Acquisition and Record Terminals (C-DARTs) and Playback and Correlation Systems (C-PCSs).	U.S.S.R. (Russia)
SMS: Suprathermal Ion Mass Spectrometer	A satellite-based instrument developed with Japan for the AKEBONO (Exos-D) satellite	Japan
WINDII: Wind Imaging Interferometer	A Canadian contribution to NASA's Upper Atmosphere Research Satellite (UARS) programme. Measures wind velocity and atmospheric temperatures, to aid in monitoring stratospheric ozone.	USA

research programmes in space science. Through CCUSS, the universities can collectively interact with industry and with government in joint undertakings in space science.

University space research spans an impressive range of topics. Given Canada's historical interest in the *aurora borealis*, a large number of scientists are involved in related studies. A growing number are involved with microgravity research in space life and materials science. University groups, such as those at the University of Toronto Institute of Aerospace Studies, have developed considerable expertise on the behaviour of different materials in the space environment. Such knowledge is essential for the construction of long duration space facilities such as the Space Station.

A number of universities are concerned with developing new techniques for airborne and space-based remote sensing and with applying remote sensing data to improving our understanding of atmospheric science and global climate change. Much of the basic research underlying Canada's Mobile Servicing System for Space Station is being conducted in universities. Researchers are studying problems of artificial intelligence, robotics, vision systems, sensors and speech recognition and synthesis. The development of new generations of communications satellites is calling for improvements in the understanding of microwave technology, microelectronics, power and mechanical systems; university scientists are active in all these fields.

Many of Canada's space science projects are conceived and managed by investigators based at universities. It is customary for university scientists to manage the scientific aspects of projects and to design space equipment. The manufacture of space hardware is generally contracted to Canadian industry.

Canadian space agency

For many years, responsibility for Canadian space efforts was divided among a number of government departments, each pursuing an area of interest related to it's own mandate. Principal players included the National Research Council (scientific research), Department of National Defence/Defence Research Board (military research), Energy, Mines and Resources (geophysical research), and so forth. In 1959, the National Research Council established the Associate Committee on Space Research (ACSR) to represent national interests in the non-military aspects of space research and to coordinate all space activities in government departments and agencies, as well as to coordinate participation in rocket programmes.[10]

Ten years later, in 1969, the Interdepartmental Committee on Space (ICS) was set up to coordinate all space activities in government departments and agencies. ACSR took on the role of the scientific subcommittee to ICS.

Over the years, the increase in activity led the government to consider establishing an agency which would be responsible for Canada's space activities. In it's first report of 1967, for example, the Science Council of Canada had recommended a Canadian space agency. In March 1989, the government announced the creation of a new Canadian Space Agency (CSA). CSA is headquartered in the Montreal suburb of St-Hubert, and, when fully staffed, will have approximately 300 employees. CSA will operate as a government department. The space activities previously carried out by several federal departments and agencies will be coordinated by the new Agency. CSA will directly manage the majority of space programme activities and will coordinate other Canadian space activities. At the same time it created CSA, the Government established a Space Advisory Board to advise the Minister responsible for space and to provide on-going appraisal of the Canadian Space Programme and the work of the Agency. The Board is made up of key representatives from the space science community, provincial governments and the private sector.

International space university

Canada was one of the earliest supporters of the International Space University (ISU), a ten week multi-disciplinary programme for graduate students and recent graduates in industry. The ISU programme is held in a different country each summer and annually attracts some 130 students from 30 countries. In 1990, Canada hosted the third ISU summer session in Toronto and Montreal. In 1991, a competition was opened for a permanent site of the International Space University. A number of cities from around the globe have submitted bids, and, as this chapter was being written, the city of Toronto had been selected on a short list of three finalists.

The Canadian Foundation for the International Space University (CFISU) was established in 1987 as a federally-incorporated non-profit charity, to act as Canada's official liaison to ISU. The Foundation was the first national liaison to affiliate itself with ISU. Since that time, many nations have adopted a liaison model similar to that of CFISU. CFISU's objective is to promote Canada's participation in the International Space University and to develop a programme of Canadian university space education. CFISU raises private and public funds to sponsor Canadian students chosen to attend ISU. Canadian students and faculty have been represented at every ISU session. Since 1988, CFISU has raised scholarship funds to allow almost 60 Canadian graduate students to attend the ISU summer programme. In 1992, CFISU sponsored the first Canadian university student space conference, which will be held in association with the 42nd Congress of the International Astronautical Federation. CFISU is governed by a board of Directors representing universities and industry.

Working under ISU guidelines, and with the assistance of the Association of Universities and Colleges of Canada, each fall CFISU solicits student applications from universities and colleges around the country. To qualify for ISU, students must be enrolled in or accepted to an accredited programme of graduate-level study. Individuals with a graduate-level degree who work in industry are also eligible to attend ISU, provided they have not been in the work force longer than five years.

The foundation convenes a panel of academic advisors to select a list of student semi-finalists. The panel meets in February and forwards it's list of semi-finalists to ISU for final selection. CFISU announces the final list of scholarship winners in March. The CFISU scholarship finances all travel and accommodation for Canadian students. Each scholarship is valued at approximately $15,000. Prior to each summer session, CFISU arranges a briefing about Canada's space programme for students departing to the ISU summer session. Following the session, CFISU organises a symposium at which students present the results of their work to the foundation's sponsors.

Conclusions

This chapter has described the extent of Canada's involvment in space,covering a wide range of areas from space science and astronomy to communications satellites and the Space Station. Canada's approach to this field of s&t is deeply rooted in it's history and international ties. Probably no other technology has had more effect on the country's development and to overcome the limitations of geography than space and communications research.

As this chapter is being written, the evolution of space s&t is being shaped by international shifts in commitments to space plans and domestic considerations about the future of the Canadian space industry and science. Public attention to this area is quite strong in Canada, fostered both by an attentive media and a society engaged by the possibilities (and limitations) of prospective investments in the final frontier. As a result, new roles for industry, governments and academe will be defined over the coming decade.

Notes

1 Kerwin L 1991 *The next century: Prospects for Space* The Charles Stark Draper Lecture given to the XLII International Conference on Astronautics Montreal
2 Jelly D H 1988 *Canada: 25 Years in Space* p14 Montreal Polyscience Publications Inc. and the National Museum of Science and Technology
3 *Ibid* p14
4 Hartz T R Paghis I 1982 *Spacebound* p52 Ottawa Department of Supply and Services
5 Jelly D H *op. cit.* p19
6 Senior pilots are known as 'commanders'. Commanders have overall responsibility for a Shuttle mission
7 Canadian Space Agency 1991 *Background Paper on the Canadian Space Programme* p9 Montreal
8 Canadian Space Agency *op. cit.* pp10–11
9 Canadian Space Agency *op. cit.* p20
10 Kelly D H *op. cit.* p100

Chapter 10

Information and telecommunications research and innovation

Roger Voyer and Peter MacKinnon

Introduction

In Canada, development of telecommunications and informatics have been driven in large part by geography, where there are vast territories with low population density except within a chain of cities stretching from the east to west coast of North America. As a result, Canada has set in place extensive world-class digital terrestrial and satellite telecommunications networks. In more recent times, with the advent of information technologies, these networks have become increasingly 'intelligent' while continuing to meet the demand for reliable and sophisticated services.

The fusion of telecommunications and informatics is leading to an ever more flexible Canadian communications system which is able to provide a wide range of services to users (see Figure 10.1). The development of this integrated communications system is placing many new demands and expectation on the related Canadian research system, including for instance the development of new high-speed digital networks for the r&d community.

Much of telecommunications research in Canada has taken place within government laboratories and within Bell Northern Research, the r&d wing of Northern Telecom, which in turn is the major supplier to Bell Canada. More recently, university and industrial consortia have launched a number of cooperative long-term telecommunications research initiatives.

In Canada, delivery of telecommunication services has traditionally been highly regulated. In recent years some facets of telecommunications have been opened to competition. In contrast to this concentrated and regulated environment, informatics developments in Canada have been largely performed by small companies operating in a largely unregulated environment.

The following pages present a historical, current and future perspective of telecommunications and informatics research in Canada with a particular emphasis on the convergence of these two aspects of information technology.

Historical perspective

Telecommunications

From telegraph to satellites
Telecommunications in Canada originated with the inventions of the telegraph

(1837) by Samuel Morse and the telephone(1876) by Alexander Graham Bell. These two technologies were milestones in the development of the Canadian telecommunications system, since they provided a means of communicating over the great distances that are synonymous with Canada.

In Canada, the first telegraph company, the Toronto, Hamilton and Niagara Electro-Magnetic Telegraph Company, was founded in 1846. The telegraph followed the development of Canada's railway system and by the 1930s, the Canadian National and Canadian Pacific railway companies had established themselves as the principal providers of telegraph services.

Figure 10.1: *The Convergence of Communications and Computing into IT*

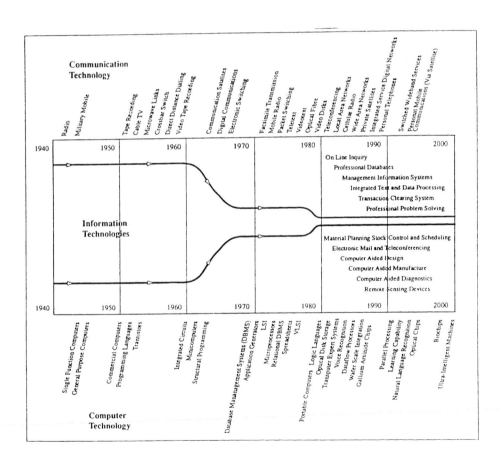

Bell's definitive tests of the telephone in Brantford, Ontario led to the inauguration of telephone service in Canada in 1877. The Canadian telephone system grew rapidly with the establishment of various competing companies across the country. The Bell Canada company emerged as the largest telephone company.

While the telegraph and telephone did set the basic structure for Canada's telecommunications system, the modern era of Canadian telecommunications is said to have begun with the launch of the Alouette I satellite in 1962, making Canada the third nation in space. As Chapter Nine discussed, Alouette I was designed and built by the Defence Research and Telecommunications Establishment (now the Communications Research Centre) and was used to study the ionosphere, which has a major effect on communications at the northern latitudes where Canada is situated. The Alouette I launch was followed by other related activities such as Canada's participation in ISIS (International Satellites for Ionospheric Studies), a joint project with the USA.

A major objective of such activities was to transfer the knowledge developed by government scientists to the private sector. At that time, RCA, DeHavilland Aircraft and Spar Aerospace were major participants.

By the late 1960s, the emphasis had shifted away from research *per se* to applications. In 1969, Telesat Canada was created by an Act of Parliament to operate Canada's domestic commercial satellite system. A series of communications satellites (Anik series) were launched in the 1970s.

In parallel, an experimental communication satellite system, named Hermes, was developed to open up the higher radio frequency 14/12 GHz band. Hermes was launched in 1976. It was the most powerful communications satellite at the time and the first to operate in the 14/12 GHz band. The Hermes programme included experiments in tele-health, tele-education, community communications, delivery of administrative services and scientific applications. Hermes was also used to demonstrate direct broadcasting in Peru, Australia and Papua New Guinea.

In keeping with the early objective of transferring technological capabilities to the Canadian private sector, SPAR Aerospace was designated as 'a chosen instrument' to develop communications satellite technology. Over time, Hughes Aircraft, which had been the prime contractor for the Anik series of satellites, transferred technology to SPAR to enable the latter to become the prime contractor for the Anik D satellite onward (see Figure 10.2).

The alliance between government, through it's principal research arm the Communications Research Centre, and the private sector resulted in the emergence of a sizeable space communications industry in Canada which employed more than 3,500 people by the late 1980s. Canadian firms had developed the capability to participate in international programmes such as those of the European Space Agency (ESA) where Canada is an associate member. In fact, Canadian firms have been very successful in their bids for ESA contracts; they have captured ninety-nine per cent of Canada's financial contribution to ESA.

The technology transfer objective of the Communications Research Centre (CRC) has led, over time, to the creation of a large number of firms as shown in Figure 10.3. In fact, it has been shown that the CRC has more than paid for itself through the taxes paid by the firms that have spun-off from this research establishment.

The telecommunications industry: the dominance of Northern Telecom
While government led the development of advanced telecommunications systems, Bell Canada's manufacturing arm Northern Telecom was, and continues to be, Canada's private sector flagship in telecommunications research and de-

velopment (r&d) generally. As can be seen from Table 10.1, more than eighty-five per cent of the r&d expenditures of the Canadian telecommunications industry is accounted for by Northern Telecom.

The development of r&d at Northern Telecom was the result of a 1956 anti-trust decision in the USA which severed the r&d links between Northern Electric (Northern Telecom's predecessor) and AT&T. This decision led to the establishment of separate research facilities at Northern Electric including Bell Northern Research (BNR) in Ottawa. As shown in Figure 10.4, BNR has been instrumental in spinning-off companies, some of which such as Mitel and its' own spin-offs,

Figure 10.2: *Satellite Communications Activities – Canadian Content*

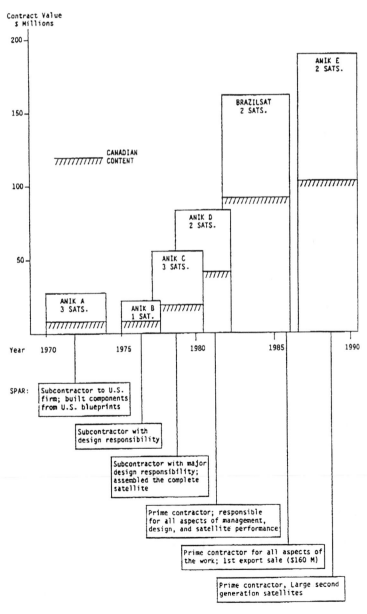

Figure 10.3: *A Partial List of Companies Whose Creation was Influenced by CRC*

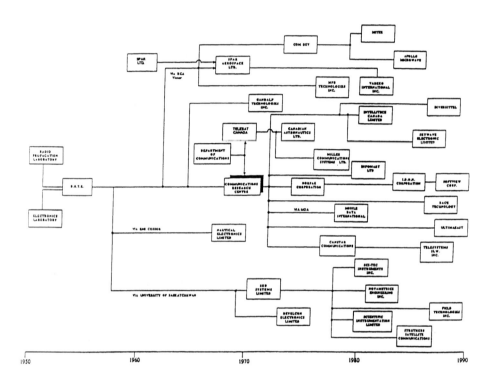

Newbridge and Corel, are now major players in the Canadian telecommunications and informatics sectors respectively.

Led by Northern Telecom, the Canadian telecommunications equipment industry is the most r&d intensive industrial sector in Canada (*i.e.* sixteen point nine per cent of total revenues were spent on r&d in 1987). The share of r&d carried out by firms in the industry was about seventeen per cent of total Canadian industrial r&d performance. As can be seen from Table 10.2, Northern Telecom spends relatively more on r&d as a percentage of sales than the average spent by the twelve top firms world-wide.

While the telecommunications equipment industry spends a lot on r&d, such is not the case for the telecommunications service firms such as Bell Canada, B.C. Tel, SaskTel and so on (*e.g.*, less than two per cent of sales are devoted to r&d). However, with the fusion of telecommunications with informatics and the increasing demand for value-added services, it is expected that the telecommunications service industry will be spending more on r&d in the future, especially as the field of personal communication emerges.

Informatics

From First Principles to World Class Software Products
Since the advent of the computer in the 1940s, Canadians have been active in the science, technology, products and services arising from the various technical, industrial and social transformations triggered by the digital computing engine.

Throughout the 1950s and early 1960s, computer science studies were scattered among various university departments such as mathematics and electrical engineering. In those early days, scientific interest focused on the principles of computation theory and associated domains in the pure and applied sciences. The University of Toronto was the first centre of higher learning in Canada to offer courses in computing science. This began in 1965. Over the next decade more than twenty Canadian universities formed departments for advanced training in computer related fields.

For purposes of this discussion, software is considered to be the instructions that make programmable machines work. In other words, software is the implementation medium that makes computer and communications hardware useful. It can also be thought of as the glue that holds computer and communication systems together. Informatics on the other hand is a combination of the methods and techniques for creating software as well as the software itself.

Table 10.1
1989 R&D Expenditures of Selected Canadian Companies
($ Million)

Company	Revenues[1]	Gross R&D Expenditures[2]	R&D Expenditures /Revenues (%)
Canadian Marconi Company[3]	$15.2	$1.7	11.2%
Develcon Electronics Ltd.	$11.3	$2.4	21.2%
Gandalf Technologies Inc.	$167.4	$17.9	10.7%
Glenayre Electronics Ltd.	$108.3	$8.0	7.4%
Memotec Data Inc.[4]	$72.4[5]	$11.9	17.7%
Mitel Corporation	$423.1	$55.3	12.8%
MUX LAB Inc.	$7.4	$0.8	10.9%
Newbridge Networks Corp.	$67.4	$11.7	17.4%
Northern Telecom Ltd.[6]	$7,161	$948.0	13.2%
SPAR Aerospace Ltd.	$233.2	$20.0	8.6%
SR Telecom Inc.	$29.8	$6.1	20.5%
TIE/Communications Cda Inc.[4]	$66.7	$3.4	5.0%
TOTALS:	$8,367	$1,088	13.0%

Source: Company annual reports

Notes: [1] Worldwide revenues
[2] Worldwide R&D expenditures (includes contracted R&D)
[3] Estimated amount for telecommunications (assumed to 5% of total)
[4] 1988 figures
[5] Product revenues only
[6] Converted to Canadian dollars at 1.1728

Figure 10.4: *Spin-offs from Bell Northern Research Laboratory*

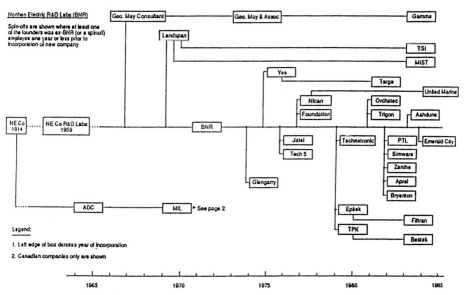

Source: Bell Northern Research

In the 1960s, many Canadian academics directed their research efforts towards new techniques for controlling and expressing operations of the computer (*e.g.*, software). This work focused on operating systems and languages. From the early days of computing, the sequencing of events and control of activities has been the responsibility of the operating system. Equally, the means of communicating with the computer has been handled by families of rules and procedures described as languages. Both of these broad fields have received considerable research attention in Canada. One of the most widely known contributions in the language area came with the research work on the WATFOR series of FORTRAN language compilers at the University of Waterloo.

Canadian researchers, particularly in academic circles, were among the first contributors to the multi-disciplinary field of artificial intelligence. As early as 1973, Canadian researchers established the first North American professional society for artificial intelligence, the Canadian Society for Computational Studies of Intelligence/Société Canadienne pour les Etudes d'Intelligence par Ordinateur. In fact, this society was created eight years before the USA formed such a group. Areas of expertise have evolved at a number of centres across the country such as University of British Columbia, University of Toronto and McGill Uni-

Figure 10.5:   *Spin-offs from Microsystems International Limited (MIL)*

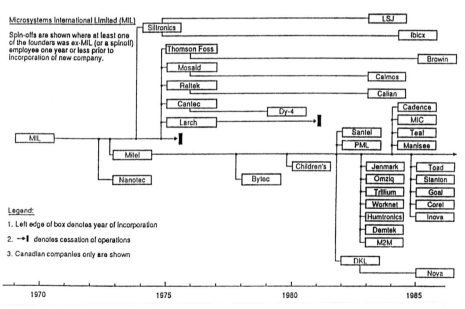

Source: Bell Northern Research

Figure 10.2
1988 R&D Expenditures of Selected Global Companies

Company	Country	R&D Expenditures ($US Billion)	R&D/Sales (%)
Siemens	West Germany	$3.5	10.5%
NEC	Japan	$3.4	15.9%
AT&T	U.S.A.	$2.6	7.3%
Hitachi	Japan	$2.5	6.5%
Fujitsu	Japan	$1.5	9.3%
Alcatel	France	$1.3	10.0%
Northern Telecom	Canada	$0.71	13.1%
Motorola	U.S.A.	$0.66	8.1%
LM Ericsson	Sweden	$0.58	11.3%
AEG	West Germany	$0.52	8.0%
Plessey	U.K.	$0.17	7.2%
Harris	U.S.A.	$0.12	5.7%
TOTAL		$17.6	9.4%

Source: Northern Telecom "Telecommunications Market: Operating Telephone Companies, Carriers, Interconnect and Manufacturers – Statistics", June 1989 and Northern Business Information.

versity. Among the speciality areas are image analysis (particularly satellite imagery), basic research and theorising in machine perception of certain kinds of photo-images (such as X-rays and metal castings), basic problems of knowledge representation, and basic research in the human-machine interface. There is also a strong presence in Canada in the area of logic programming. As well, Canadian researchers have pioneered a number of advances in machine translation, in particular the Q-systems approach which allows for a core part of the machine translation process to be natural language independent.[1]

Today there are a number of active research areas of artificial intelligence focused on problems of national interest (*e.g.*, aids for the management of commercial forests through the use of computer-based analysis of satellite imagery; and computer-aided analysis of X-ray images for medical research).

Canada, like a number of other countries in the 1960s, saw the national government attempt to stimulate development of a domestic computer industry. However, Canada only played a small role in efforts to commercialise computers. The best known example in this regard involved a partnership between the federal government and a subsidiary of the UK-based Ferranti company. In 1961, a commercial effort was launched in Canada to develop a mainframe Ferranti computer for general purpose markets. After several millions of dollars and three years' effort, the initiative was cancelled.

By the mid 1960s, industry and government began to adopt computers for business purposes. This coincided with the availability of general purpose computers from U.S. manufacturers. However, the widespread availability of machines was offset by a serious lack of qualified personnel to program these computers. A major effort was launched to find and train personnel. As a result of this effort, most of the on-the-job programmers of the 1960s and early 1970s were recruited from corporate cafeteria advertisements seeking staff looking for a job change.

Up to this time, Canada had little experience in the industrial production side of digital technologies. The telecommunications industry was not yet committed to digital technology, universities were only beginning to produce scientific and engineering graduates with computing skills and the domestic software industry was nearly non-existent.

Three defining factors changed this situation by the beginning of the 1970s. Firstly, the main players in the Canadian telecommunications industry committed to developing digital technologies. This triggered a chain of events ranging from efforts to stimulate domestic capability in mass production of microelectronic devices (*e.g.*, Microsystems International) to enhanced supply of qualified scientists and engineers in the digital hardware and software fields. It also led to the creation of an industrial support structure through both start-ups and spin-offs from the telecommunications manufacturers, notably Northern Telecom and it's r&d operation, Bell-Northern Research. Second, IBM Corporation in the USA made a strategic decision to unbundle the applications software environment from the hardware and operating system. This caused a market driven response to develop a packaged software industry. As the Canadian public and private sectors were already embracing the use of computers to automate traditional business tasks and to improve efficiency of operations, various Canadian businesses and entrepreneurs launched into developing software for re-sale.

The third and related event was the rapid growth in the number of computer installations and the need for software applications to satisfy business needs. With so many organisations ranging from banks to geophysical exploration companies buying into computing at the same time, there arose a major service industry to assist in the specification, development and operation of new computer-based applications. Thus, by 1980 the informatics industry in Canada could be characterised as having reached a mature organisational structure with clear sights on rapid growth tied to software developments for various hardware platforms and environments. Added to this, consulting firms began to differentiate their capabilities as both the domestic and international market places became more sophisticated. As a result, firms began to offer facilities management, real-time software engineering, software products and systems integration.

The arrival and general acceptance of the personal computer (PC) in the early 1980s saw yet another attempt in Canada to become a source of computer hardware. This time the effort was directed to the educational use of computers. By example, two provinces, Ontario and Quebec, each set out to foster the development of provincially based manufacturers to produce a PC targeted at the educational community within their jurisdiction. In the end, the proprietary operating environment of the machines made them uneconomical in the face of larger scale commercial successes emanating from the USA. Although the actual computer hardware for these educational machines was not a commercial success, the projects lead to further strengthening domestic capabilities in software; this was particularly true for the creation of software products.

By the 1970s, the Canadian telecommunications industry began a massive digitisation programme, including establishing one of the first, and certainly the largest, fully digital packet switching networks (DATAPAC). This activity accelerated both demand for and capabilities in real-time software. The strategic decision to introduce digital technology to telephony marked a major step in bringing communications and computing closer together.

By the late 1970s, Bell-Northern Research and the recently formed Mitel Corporation were among the largest employers in the world of real-time software specialists. Their work concentrated on imbedded software for central office and

PBX digital switches. In addition, many smaller firms were established as spin-offs and suppliers of specialised services.

Also during the 1980s, U.S. based firms, notably IBM and Digital Equipment Corporation, established major facilities in Canada. IBM expanded it's Canadian software development facility into a major world class laboratory. At the same time, Digital moved to establish a world product mandate for the manufacturer of a mid-range computer line. Since that time a number of other U.S. hardware manufacturers have established r&d facilities in Canada. For example, Sun Microsystems has a UNIX lab in Montreal and Hewlett-Packard undertakes software development for telecommunications in Toronto.

The 1970s also saw the emergence of large MIS (Management Information Systems) departments throughout government and within most of Canada's larger corporations. These groups were charged with providing custom built software applications to support business needs as well as procurement and management of corporate computing resources. Initially, the applications focused on 'back room' systems such as payroll, inventory management, and accounting systems. During this period the computing resources tended to concentrate around mainframe computers.

By the mid 1970s the sheer demand for systems placed MIS organisations under considerable pressure. As the corporate backlog for new application systems grew from months to years, many users within organisations turned to alternative means of securing their requirements. This included taking advantage of departmental scale computing in the form of mini computers. This was in contrast to the corporate-wide systems associated with mainframes. The lower cost mini's, coupled with localised MIS teams within a functional business group, the advent of high quality and price competitive third party professional services, and the emergence of packaged software tools and applications accelerated the decentralisation of corporate computing. In addition, this period experienced a major technical advance facilitating this transition from punch card based batch computing applications to terminal based time sharing computing. This represented one of the pivotal steps in bringing computing and communications closer together.

Although Canada's involvement in the technical developments of this period were relatively small, the 1970s saw Canadian organisations responsive and adaptive to the major international trends in business computing and related services. This would prove to be a valuable asset in preparation for the 1980s and beyond.

Current directions

Telecommunications

Canadians are still learning how to use the highly-featured products which were installed in the 1980s. The 1990s are bringing more user-driven products and services and, as costs fall, these products and services will multiply. The technological underpinnings to support this thrust include the following[2]:

- **microelectronics**: continued development of low-powered, high-density microelectronic systems
- **photonics**: evolution in power, electrical and optical characteristics
- **value-added products**: increasing intelligence in all aspects of the telecommunications system, from network switches to customer terminals will shape

user requirements
- **energy storage**: battery technology will be of growing importance to support the development of portable equipment for personal communications
- **learning systems**: increasing sophistication of software systems leading to user-friendly 'learning' systems
- **human computer interface**: continued evolution of display technology and input techniques
- **proving large systems**: new methods for testing large complicated systems will be needed
- **manufacturing technologies**: continued access to state of-the-art manufacturing technologies will be essential. This necessitates a capability to evaluate competing foreign technologies

These technological directions are being integrated in several current research programmes within firms, universities and government. Because of the increasing costs of undertaking research, several programmes are of a pre-competitive nature and involve a number of participants from both the private and public sectors. Examples include (some of these are detailed in the Appendix):

- The Institute for Telecommunications Research (ITR) established in 1988 under the $240 million federal government National Networks of Centres of Excellence programme. ITR received a five-year $14.7 million grant to conduct research in broadband and wireless communications
- Micronet, another federal Network of Centres of Excellence which received $10.8 million to study ultra-large scale integration (ULSI) of microelectronic systems
- The Telecommunications Research Institute of Ontario (TRIO), an Ontario government sponsored Centre of Excellence which undertakes research in digital networks, radar and satellite systems
- The Information Technology Research Centre (ITRC), another Ontario government Centre of Excellence studying artificial intelligence, microelectronics, communications mathematics, *etc.*
- The Wireless Communications Research Centre (WCRC), supported by a $14.5 million contribution from the government of British Columbia
- B.C. Advanced Systems Institute (ASI), which has an $8 million, five year, programme to study artificial intelligence, telecommunications and microelectronics
- National Optics Institute (NOI), in Québec which has a $35 million programme to study optical communications and related microelectronics
- Vision 2000, a private sector consortium which is developing portable personal communications technology
- CANARIE a pan-Canadian high-speed network supported by the federal government to stimulate the development of telecommunications applications (under development)

These programmes are in place across the country, mainly in universities and involve some 3000 researchers out of the 15,000 or so researchers, who are involved in IT research overall.

These initiatives are supported by the research activities of federal and provincial research institutions. For example, the Communications Research Centre continues to be a leader in the development of both terrestrial and satellite communications systems. At the provincial level, eight provinces have provincial research organisations (PROs) and other research bodies (*e.g.*, Alberta's TR Labs) whose programmes are in part related to telecommunications.

Canada has a number of well-articulated telecommunications r&d programmes across the country aimed at maintaining international leadership in this sector.

Informatics

By the 1980s, informatics was clearly segmented into three broad categories, custom software, packaged software and imbedded software. Also by this time, official statistics on the 'industry' began to be collected on a routine basis. In fact, by 1982 there were approximately 4,000 informatics firms or firms actively involved in informatics.[3] Many of these firms were small consultancies providing custom software services. In the same year, 60 firms reported performing software r&d.

In addition to these software industry suppliers, there exists a significant service sector providing both domestic and international consultancy advice. This 'software industry' is both supporting and independent of the in-house software environment (i.e. MIS). It is also useful to note that some MIS organisations also spin-off their customer applications for commercial gain. Figure 10.6 illustrates a structure to the domestic informatics industry along with representative examples of firms and applications.[4]

By 1988, the last year in which detailed national statistics on the informatics industry were officially collected, Canada had some 5,000 software related firms of which more than 2,400 were directly involved in software product development.

Like many other OECD countries, the bulk of activities in job creation, wealth creation and innovation are occurring in small to medium size firms. Canada is no exception. As an example, Alberta, a province in the western region of the country, has some 1,500 firms involved in computing and related service activities. In fact some 250 firms are considered to be directly involved in value-added aspects of software development.[5] Of these, some fifty-four per cent have six or less employees and only five per cent have more than 100 employees. The largest number of firms, some 115, are located in Calgary.

In terms of current employment patterns, some 150,000 people are engaged in the software sector as of 1990. Figure 10.7 illustrates the distribution of these jobs between the direct software industry and the in-house software sector. It is useful to note that the growth rate in the software industry is forecast at some fifteen per cent above in-house activities. On this basis, employment by 1995 is estimated to reach 280,000 jobs.

A general profile of the current Canadian informatics sector is illustrated in Figure 10.8. The human resource side of the software industry component is only fifty per cent of the size of the in-house software development component, however it represents both the largest growth opportunity and the highest volatility as new forums are established and unsuccessful ones fail. It is also important to note that the in-house skills set is rapidly being replaced by highly skilled software product developers.

The software products emanating from Canadian companies are backed by a large pool of research capability. Computer science and related fields are now taught and research performed at more than three dozen degree granting universities and colleges. Government labs (e.g., the Institute for Information Technology, National Research Council of Canada; the Canadian Workplace Automation Research Centre, Department of Communications) provide ongoing research programmes, facilities and technology transfer in support of public and private sector needs. In addition, all the larger software firms maintain active research departments. For example, Cognos, a $150 million a year software products firm, has some 175 people on it's r&d staff.

Figure 10.9 illustrates the trend in private sector r&d expenditures in support of software development. By 1988 some $225 million was being directed toward software r&d. Today, the estimate is in excess of $300 million.

This r&d community is actively involved in using computer networking as a means of communicating and conducting collaborative research. There are currently ten regional networks, one for each province, interlinked by a nation wide 56 kilobits/second backbone known as CA*net. This in turn is linked to the Internet network of r&d networks and thus provides world-wide coverage.

The advent of computer-to-computer networks has also fostered the development of research consortia. One of the largest Canadian pre-competitive consortia is known as PRECARN. It involves some three dozen firms, ranging from large resource and utility companies to technology intensive firms in aerospace and software. The consortia manages over $50 million in r&d funds for targeted multi-year research in artificial intelligence. PRECARN maintains close ties to the academic community, particularly to the Artificial Intelligence (AI) programme within the Canadian Institute for Advanced Research. This institute has established long-term research programs to exploit Canadian academic research excellence in a number of areas. The AI programme has been running since the Institute's inception in 1985.

Today the Canadian informatics industry, which is comprised of over 5,000 hardware, software, and service companies fills a domestic market valued at $15 billion and projects an eleven per cent annual growth.[9] Figure 10.10 illustrates the general growth trends in Canadian consumption of hardware, software

Figure 10.6: *Software in Canada*

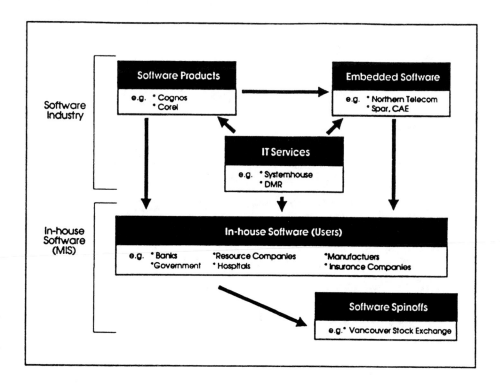

Figure 10.7: *Software Employment in Canada*

	SOFTWARE INDUSTRY	IN-HOUSE SOFTWARE
Current Employment:	50,000	100,000
Projected Employment Growth:	20%	5%
Ownership:	Canadian	30% are foreign owned
Size:	Under 20 employees	More than 100 employees
Training Provided:	Little	Some (often on-the-job)
Software:	Services & products designed & produced for sale	Designed & developed using applications & languages for internal business use: some spin-off software is sold externally
Results:	Employment shortfall	Skills obsolescence

But skills mismatch hinders mobility

products, professional services, and processing services from 1984 through to 1993.

Future Possibilities

Telecommunications

Canadian telecommunications r&d programmes are increasingly linked to international programmes, reflecting both the trend to globalisation and the high cost of doing r&d in this sector. Examples include:

- Canada's associated membership in the European Space Agency
- Canada's participation in the Space Station
- TRIO's participation in the Telepresence Project through Ontario's partnerships with the Four Motors for Europe (*i.e.* Rhône-Alpes, Baden-Wuerttemberg, Lombardy and Catalonia). This project aims at developing a user friendly international tele-conferencing system
- Canada has over 330 federal and provincial s&t agreements with other countries and regions many of which deal with telecommunications(See Chapter Seven)

Supporting such specific initiatives are federal government programmes and activities aimed at establishing closer links with other countries. These include, for example, Canada's interest in negotiating a s&t agreement with the European

Community and the Investment Canada strategic alliance programme aimed at establishing partnerships between Canadian and foreign firms and research organisations in advanced technology areas such as telecommunications, microelectronics and software. International partnerships are likely to keep growing in the future.

Informatics

Markets for products and services are increasingly becoming globalised as a result of common technologies, the emergence of major trading blocks and transformations in the flows of investment capital. As competitive pressures mount, technology and know-how are increasingly becoming a hallmark in differentiating competitors in the global market place. Leading Canadian firms in the informatics business are keenly aware of these trends, particularly in the realisation that much of these changes have been brought about because of the convergence of computing with communications. Central to this interaction is the emergence of new types of communications networks – high-speed electronic highways – linking the intellectual resources of nations in vast arrays of computers networked across local and distant geography.

Canada has traditionally and boldly spanned it's own geography with networks of waterways, rails, roads, airlines, telegraph and telephone lines, and national broadcast systems. Over the years, the rate of technological change has gradually challenged and altered our use of each of these elements of national infrastructure. Today, the rate of change in the globalisation of research, development, markets and economies is placing enormous demand on electronic

Figure 10.8: *Informatics Industry Profile*

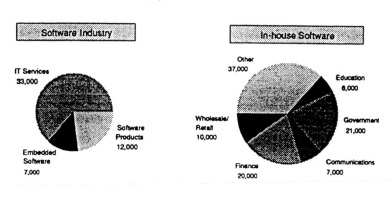

Total Employment - 150,000 in 1990

Projected Annual Growth - 20 %   Projected Annual Growth - 5 %

Total Employment - 270,000 in 1995

Figure 10.9: *Canadian Software R&D Expenditures*[8]

Inhouse R&D ($millions)

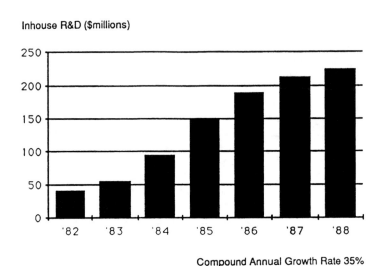

Compound Annual Growth Rate 35%

highways. This is further compounded by recent international trends in deregulation and decentralisation of communications.

Conventional networking services are being challenged by new demands for:

● keeping up with the exploding body of knowledge
● utilising more information to make better decisions
● developing better understanding of relationships between pieces of information
● providing better visualisation of massive amounts of data
● supporting the ability to capture and/or create data (*e.g.*, point-of-sale, satellite imagery)

All of these mean increased demand for greater speed and higher bandwidth, in other words, a need to go faster with more and wider lanes in order to pump ever increasing amounts of information in more diverse forms down the communications conduits. Indeed, networks are evolving to support full multimedia capability in the form of integrated voice, data, text, image, and full motion video.

This challenge must also be addressed from the perspective of international standards. It is essential that new Canadian IT products embrace the prevailing norms of the industry, such as the TCP/IP protocol. At the same time, they should be designed to evolve with emerging standards such as those defined by OSI, broadband ISDN, and SGML.

These new forms of computer networking and their associated standards provide the means to increase the value of new products and services. They also shorten the time to market through collaborative activities spread across diverse geography. In short, computer networks bring a new dimension to competitiveness and new capabilities to education.

Canadians from industry, government and academe are currently planning for a major upgrade to the existing CA*net backbone and associated regional networks. This upgrade, to be phased in over several years, is aimed at enhancing the overall national r&d infrastructure in order to:

● provide or assist in providing network-based services for Canadian industry, government, and the education communities
● serve as an educational and training resource to assist in the preparation of Canadians for the information age
● ensure that the network conforms to international standards and provides a test-bed for the development of technologically pre-eminent Canadian network products, applications and services
● stimulate and accelerate the formation and use of major national assets in the form of centralised and distributed databases
● serve as a show-piece of Canadian advanced information technologies and know-how
● establish state-of-the-art communications capabilities in support of the Canadian r&d community to stimulate and synergise collaborative r&d

It is anticipated that achieving these goals will come through the catalytic role played by both applications/service development by the IT industry and users.

This entire project possesses enormous opportunities for future r&d, which should lead to new technologies, products and services. It is also intended to bring many of the thousands of Canadian businesses and researchers closer together through pre-competitive and cooperative r&d activities.

Figure 10.10:   *Components of the Canadian Informatics Industry*

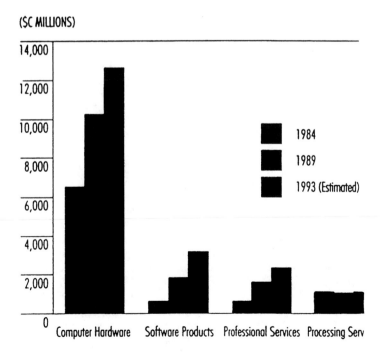

(\$C MILLIONS)

Another major area for further research lies in the field of reusable software. Today, most software is single purpose. In addition, the cost of developing software, whether it is in custom in-house applications or in the form of a software product, is much higher than it need be. Advances in software engineering and choices in software constructs will have enormous economic impact. To this end, a number of Canadian universities, firms, and government laboratories are exploring ways to enhance software development and re-use.

Finally, the role of informatics is continually increasing in support of international r&d. For example, the multi-national Human Genome Project and the Human Frontier Science Program, both require advanced software to handle and analyse the massive volumes of data these projects generate. Canadian researchers are actively seeking technical solutions to these types of data management problems. This includes research on such things as large scale data structures, data navigation and analysis procedures and visualisation of computer generated results. These types of needs are also present in handling data streams from satellites such as the Hubble Space Telescope, for which Canada is one of the primary archive sites.

Conclusions

Historically, the telecommunications and informatics sectors have developed through publicly-funded research; the former largely through the federal government, the latter through the universities. While the private sector has undertaken an increasing amount of research in recent years – to ensure that it remains internationally competitive – publicly-funded r&d continues to support leading-edge projects, such as CANARIE and TRIO's Telepresence Project. Since these activities are technology driven, it is expected that publicly-funded research will continue to be needed to lower the risk and share the cost with the private sector.

The convergence of telecommunications and informatics into the information technology (IT) sector is leading to an increasingly complex policy debate as to the directions of future research. The federal Department of Communications, for example, is restructuring its' research activities within the context of a new research institute which is intended to be to be more flexible and responsive to national research requirements as defined by industry needs. The federal Department of Industry, Science and Technology, for its part, is in the process of developing a comprehensive IT strategy as a framework for it's financial support of IT areas.

The provinces are also becoming more involved players in this policy debate as they position themselves to set in place the advanced technology-based infrastructures that will be needed to support IT development. Ontario's TRIO and ITRC, Quebec's NOI and B.C.'s WCRC and ASI are examples of the involvement of provinces in strengthening the advanced IT infrastructure. Ontario's ambitious and multi-sector strategy, *Telecommunications: Enabling Ontario's Future*, released in August 1992, is an example of recent attempts by the provincial governments to develop a more information-based economy capable of better competing internationally.

The policy debate is also influenced by the two industry associations most directly involved with IT; the Information Technology Association of Canada (ITAC) and the Canadian Advanced Technology Association. The former includes in it's membership the larger multinational IT firms while the latter represents largely smaller, Canadian-owned, firms in all fields of advanced technologies.

While the specific objectives of these two industry associations can differ, they agree on the need to strengthen Canada's IT sector since it is a key sector which enables productivity gains across the economy as a whole.

The central importance of IT to the well-being of Canadians is now accepted. The policy debate is therefore focused on the IT requirements to ensure that Canada remains internationally competitive into the 21st Century.

Notes

1   1984 *National Language Automated Processing and Artificial Intelligence: State of the Art* Vol 1 Cognos Inc. for the Government of Canada
2   For details, see NGL Consulting Ltd 1991 *A Proposal towards a Strategic Plan for the Canadian Telecommunications Equipment Industry*
3   1990 *Industry, Science and Technology Canada* Internal report on the IT sector
4   Employment and Immigration Canada 1992 *Software and National Competitiveness* p26
5   Albert Software Industry 1991 *Products and Capabilities Directory* Alberta Ministry of Technology, Research and Telecommunications
6   Employment and Immigration Canada 1992 *Software and National Competitiveness: Human Resource Issues and Opportunities* Executive Report p5
7   Employment and Immigration Canada 1992 *Software and National Competitiveness: Human Resource Issues and Opportunities* Executive Report p22
8   ISTC, Internal report on IT sector
9   Government of Canada 1991 *If Software is your Business* High Technology Opportunities Series
10   1990 Evans Research Corporation

Chapter 11

Medical and health services research

Douglas Angus and Patricia Lemay

Introduction

Health care, as in all industrialised countries, is big business. In Canada, this sector employs four per cent of the work force directly, represents more than $60 million annually in spending, and accounts for nine point two per cent of the total Gross National Product (GNP). By comparison, in 1980, health care spending accounted for seven point five per cent of GNP.

To support this level of expenditure, a research infrastructure for biomedical and health care research has evolved in Canadian universities, research institutes, hospitals and other treatment centres, federal and government departments, and in private industry. Funding for research flows in large part from federal and provincial sources, non-government voluntary organisations, and the pharmaceutical, biotechnology and related industries. An estimated $1.24 billion are spent on biomedical and health care research and associated overhead costs, with the majority directed to basic and applied medical projects. As governments at the federal, provincial and municipal levels struggle to keep health care spending under control, more research funds are expected to be channelled into health care research aimed at cost-effectiveness, medical technology assessment, program evaluation and service delivery efficiency.

Health care indicators

Population Demographics

Canada's population, while still relatively young when compared to many industrial countries, is ageing. In the latter part of the 1980s, for the first time in the country's history, one half the population was over thirty-one years of age. During the past twenty-five years, the proportion of the population sixty-five years of age and older has increased from eight per cent to over eleven per cent of the total population. In the same period, the proportion of Canadians under fifteen years of age has dropped from thirty-four per cent to about twenty per cent.

Health Status

From an international perspective, the health status of Canadians rates favourably when compared with the twenty-three other OECD countries. From the perspective of overall mortality, Canada has the fifth lowest age-standardised death rate. In terms of male infant mortality rates, Canada ranked ninth in the late 1980s. For females, Canada is in eighth place. (OECD, 1987)

Life expectancy at birth for Canadian males and females in the late 1980s was seventy-three years and eighty years, respectively. For the first time since data on life expectancy have been collècted in Canada, the gap in mean years of life between women and men has diminished. The reduction of the gap may reflect the increasing number of women in the labour force who are exposed to new lifestyles and pressures.

For both men and women, the two leading causes of death are heart disease and cancer. Whilst HIV infection and AIDS have emerged as the most recent epidemic, the leading causes of hospitalisation are mental disorders, heart disease, stroke, cancer, accidents, diseases of the nervous system and respiratory disease.

Health Delivery System

Health care costs as a proportion of Canada's Gross National Product (GNP) have increased since the late 1970s from seven point five to nine point two per cent in 1991. Few countries spend more than nine per cent of the GNP on health care services, with the exception of the United States that spends about twelve point four per cent of GNP. (HWC, 1992) Health expenditures per capita in Canada are exceeded only in the USA. In Canada, we dedicate more resources to health than any other country with a national health system. Countries such as Sweden and Japan with relatively older populations spend eight point seven and six point eight per cent of GNP respectively, and have higher measures of health status.

Institutional services, including hospitals and other institutions, account for almost half of all health care expenditure in Canada. In 1988, 1,250 public, private and federal hospitals were operating in addition to over 6,000 special care facilities such as nursing homes. The number of public hospital beds has been declining, whilst the number of long-term care beds has been increasing. The hospital bed-to-population ratio was about seven beds per 1,000 population in 1989. Canada has the seventh lowest number of inpatient medical beds per 1,000 population of the twenty-four OECD countries. (Statistics Canada, 1992)

Direct employment in Canada's health care sector accounts for four per cent of the total labour force including about 57,000 physicians and 250,000 nurses. The growth in the number of health and allied health professionals has, in most cases, exceeded the growth in the Canadian population. Nonetheless, there continue to exist distribution problems across regions and between urban and rural areas.

Use of Health Care Services

Data on hospital utilisation show that the rate of hospital admissions has increased only for the sixty-five plus age group in Canada. For all other age groups, the rate of admissions to hospitals has declined. The average length of stay (ALOS) in Canadian hospitals in the late 1980s was just over fourteen days. Compared to the twenty-four OECD countries, Canada has the ninth lowest ALOS.

A number of factors influence health service utilisation patterns. Factors such as the availability of facilities, equipment and trained personnel; population, age structure and health status of the population; and variations in medical practice all affect utilisation rates. For instance, data from the early 1980s comparing Cae-sarean-section rates in eighteen industrialised countries showed C-Section rates per 100 hospital deliveries varied from seven in Austria to sixteen in Canada and eighteen in the USA For other surgical procedures such as cholecystectomy, hys-terectomy, tonsillectomy and hernia repair, similar variations in surgical rates are found between and within countries.

Health technologies

The pace of technological change in health care in the developed world has been rapid over the past fifty years. Research has led to new generations of sophistic-ated pharmaceuticals, medical devices, surgical procedures and diagnostic tech-nologies being introduced almost daily. Anticipated health care technologies will present society with a complex array of challenges in terms of research and planning for health services, human resource planning, funding and evaluation of health care outcomes.

Modern Health Care in Canada

Canadians enjoy a health care system that provides universal and comprehens-ive medical and hospital services to all Canadians. This is generally known as a universal system. This chapter will review how Canada's system evolved, and in so doing will establish a basis for considering questions related to medical and health services research in Canada.

Prior to the mid-1950s, when health care was rendered largely in the home set-ting, the hospital was a unique environment reserved for treating the acutely ill. The cost of treatment was borne entirely by the patient. This led to two conditions:

● the patient often did not seek treatment because of lack of personal resources
● most hospitals, lacking a stable funding source, were not able to meet basic standards of care, or, to be a viable economic unit

In the late 1950s, the federal government decided to finance the major portion of hospital costs from the public purse, with the result that access to hospital care was equalised and hospital funding was stabilised. By the 1960s, Canadians de-cided to put even more public money into the health care system by insuring a range of medical services provided both in hospital community settings. Thus, government participation in the funding and provision of what have been deemed "essential health care services" has been a significant cornerstone of Ca-nadian social policy over the past thirty years. (Angus, 1990)

As a result of the wording of Canada's Constitution, the ten provinces and two territories are responsible for delivering health care to their constituents. While health is primarily an area of provincial and territorial jurisdiction, there has been a long and necessary partnership between the federal and provincial gov-ernments in terms of funding and provision of insured hospital and medical ser-vices. The *Hospital Insurance and Diagnostic Services Act* of 1957 and the *Medical Care Act* of 1966 helped to establish a framework between the federal and pro-vincial governments. This partnership implies that provinces are the planners, managers and administrators of the system and that the federal government provides partial financing of the cost of services and, in so doing, establishes conditions respecting the transfer of funds.

Until 1977, every dollar spent by the provinces on health care was matched by the federal government. As provincial health expenditures grew, so did federal spending. The federal government had little control over spending under this arrangement and in 1977, the *Federal-Provincial Fiscal Arrangements and Established Programs Financing Act* (EPF) was enacted in response. The EPF changed previous dollar-for-dollar funding arrangements between the provincial and federal governments to lump sum transfers based on a one-time transfer of tax points and *per capita* block payments. At the same time, conditions governing transfers became more flexible and funds could be used for new programmes such as drug benefit plans. These new arrangements placed increased pressure on the provinces to effectively manage their health resources.

By the early 1980s, primarily as a result of a major recession and an increased federal deficit, the federal government decided to phase out it's financial support of EPF. Between the mid-1970s and the late 1980s, the federal share (in cash and tax point transfers) of provincial government health care expenditures for insured hospital and physician services declined by slightly more than three per cent. Cash contributions from the federal government increased from $2.1 billion to $5.7 billion from 1977 to 1989. However, the rate of increase decreased significantly from a high of sixteen per cent to zero point three per cent in 1991.

While primarily government based, health care expenditures in Canada are a mix of the public and private sectors' contributions. (Table 11.1)

The Canada Health Act

A major legislative event took place in 1984 with the passage of the *Canada Health Act*. The Act was introduced largely to curb extra charges for health services being levied by some provinces and physicians. It was designed to ensure that every Canadian received medical and hospital treatment paid by general taxes or compulsory health insurance. While many provinces viewed this Act as a direct affront to their sovereignty, it nevertheless received unanimous support from all political parties at the federal level.

The Act reasserted five principles upon which the Canadian health system is based and which are considered fundamental and essential to its operation.

Public Administration

Provincial governments are the sole insurers for listed insured services. Since each province is responsible for administering its own program, there actually are ten 'Medicare' plans.

Governments at any level in Canada are rarely the direct provider of health services. Rather, they are insurers of services and as such have little control over

Table 11.1   Sources of Health Care Funds

● Government	{ Provincial	42%
	{ Federal	30%
● Private		25%
● Residual (Workers)		3%
Compensation and local government)		
Total		**100%**

Source: National Health Expenditures in Canada, 1975–1987

the mechanisms of delivery and utilisation. In this respect Canada's health care system is quite different from the systems in Great Britain and Scandinavia for example, where governments own services and employ their own providers.

Comprehensive Coverage

A broad range of medical services are rendered by medical practitioners in hospitals, clinics, or doctors' offices at standard ward level. Health services are upgraded to private or semi-private if medically necessary. Diagnostic tests and a broad range of outpatient services are provided. Under this system the patient is free to choose his or her physician or surgeon. Depending on identified needs and priorities, this coverage can vary from province to province.

Universality

All legal residents of a province are eligible to receive health services.

Accessibility

Access to all required health services is available to everyone; doctors determine whether or not services are required.

Portability

Canadian residents receive health services in any province or territory.

Federal-Provincial Issues

Reductions in federal transfers for health care under EPF combined with restrictions implicit in the Canada Health Act, have been a source of federal-provincial tension. The provinces are forced to respond to growing demands for health care but with limited taxing authority and little flexibility to introduce alternative services or revenue generating programmes such as user fees or health insurance premiums as a means to supplement shrinking resources. In light of recent federal budgets which have resulted in freezes and cuts in federal transfers for health care (and post-secondary education), the provinces are increasingly hard pressed to maintain current levels of medical services.

In view of a high federal deficit and health care costs that continue to increase out of proportion to the growth of the population and the national wealth, it is reasonable to expect fundamental changes in the health care system. Restructuring will be required if a universally accessible and affordable system of health care is to be maintained.

Medical and health services research

In addressing the topic of health research in Canada, there are two main components:

(i) biomedical research which relates to research in the areas pertaining to bio-

logical organs, organ systems, and organisms
(ii) health care research which embraces health care delivery, administration, and the socio-psychological influences on health and health care

Overall Canadian Funding of Health Research

In Canada, health research is supported by the federal government, provincial governments, National Voluntary Health Organisations (NVHOs) and private industry. Statistics Canada estimates that total r&d expenditure in the health field in Canada in 1991 amounted to $1.24 billion (a six per cent increase from the previous year's figure of $1.17 billion). This figure includes government and industrial in-house research and an estimate of research-related university infrastructure costs as well as extra mural funding available to researchers on a competitive basis. Canada accounts for one point six per cent of the total world's spending on biomedical research. By comparison, total health r&d expenditure in the UK in 1989/90 amounted to $3.15 billion, slightly more than Canada on a per capita basis. The corresponding figure for expenditure in 1989 in the USA was $20.6 billion, indicating a considerably higher per capita. (Heacock, 1992)

The Association of Canadian Medical Colleges estimates that extra mural funding to universities for biomedical research was $577 million in 1989/1990. This represented a real rate of growth of greater than thirteen per cent over funding in the previous years. Funding for biomedical research in universities has increased in real terms every year for the past decade, although the rates of increase vary widely from a low of two per cent to a high of over thirteen per cent. (Ryten, 1991)

Federal sources now account for about thirty-seven per cent of total funding down from forty-five per cent in 1984/85. The single most important source is the Medical Research Council which provided thirty per cent of funds to universities in 1989/90, down from over thirty-nine per cent in 1984/85. Provincial governments now provide about ninteen per cent of funding for university biomedical research, down from twenty-three per cent in 1984/85. The percentage of funding from National Voluntary Health Organisations has fallen from a high of eighteen point three per cent of total funding for university biomedical research in 1982/83 to fifteen point two per cent in 1989/90. (Table 11.2)

The most striking increase has come from private sector industry that provided eight point three per cent of funds in 1989/90, up from approximately two per cent in 1981/82. The other major growth areas include funds from internal university sources and from local sources. (Ryten, 1991)

About forty per cent of total research funds available to faculties of medicine is spent on basic biomedical science, fifty-two per cent on clinical or applied research, four per cent on community health and 'related' research and approximately four per cent on other fields. (Table 11.3). 'Related' programmes refer to activities in areas which influence health rather than sickness.*i.e.*, behavioural sciences, environmental health, epidemiology, health administration, medical humanities, occupational health, public health and social and preventive medicine.

Federal Funding

Reflecting the dual nature of health research in Canada, there are two principal federal funding sources: the Medical Research Council of Canada (MRC) and

Table 11.2   Expenditures for Biomedical Research in Canadian Faculties of Medicines, 1989/90

Source	($000)	% of Total
Federal Sources, Total	**212,633**	**36.9**
MRC	174,036	30.2
Health and Welfare	21,930	3.8
National Research Council	666	0.1
Natural Sciences and Engineering Research Council	8,264	1.5
Other Federal Agencies	7,737	1.3
Provincial Sources	**111,565**	**19.3**
Voluntary Health Orgs', Total	**97,485**	**16.9**
National	87,705	15.2
Provincial	9,780	1.7
Other, Total	**154,961**	**26.9**
Private Industry	47,762	8.3
Local Sources	29,879	5.2
Internal University Sources	37,310	6.5
Hospitals & Universities	1,313	0.2
Foreign Sources	38,097	6.6
Miscellaneous	608	0.1
Total, All Sources	**576,652**	**100.0**

Source: Ryten, Eva (1991), ''The Funding of Research Conducted by Canadian Faculties of Medicine'', *ACMC Forum*, Vol. XXIV, No. 3, 1–6.

Health and Welfare Canada's National Health Research and Development Programme (NHRDP). MRC was established by Parliament in 1969 to promote, assist and undertake basic, applied and clinical research in Canada in the health sciences and to advise the Minister of National Health and Welfare about health/medical research. Since the 1960s, there have been a number of efforts to re-direct research from purely biomedical to health care delivery, which culminated in the establishment in 1975 of NHRDP.

Table 11.3   Research Expenditures by Canadian Faculties of Medicine, by Field of Research, 1986/87

Field of Research	Expenditures $('000)	% of Total
Basic Sciences	158,672	40.4
Clinical Research	206,031	52.3
Biomedical Engineering	2,508	0.6
Community Health & Related	15,114	4.1
Continuing Education	770	0.2
History of Medicine	244	0.1
Nursing	386	0.1
Nutrition & Dietetics	2,181	0.6
Miscellaneous	6,441	1.6
Total	**392,347**	**100.0**

Source: Ryten, Eva (1989), ''Financing the Research Mission of Canadian Faculties of Medicine, Comparisons & Trends'', *ACMC Forum*, Vol XXII, No. 3.

The Medical Research Council of Canada (MRC)

The MRC's mission is 'to improve the health of Canadians through the promotion and support of excellent basic, clinical and applied research in the health sciences' (MRC, 1990/91). Major recipients of MRC research funding have been universities or university-based institutions such as teaching hospitals. In 1990/91, the MRC provided $235 million to support 2,249 operating, equipment, group or programme grants and 1,519 awards for research personnel.

MRC has a full-time president, twenty-one volunteer council members representing the scientific and lay community, and three associate members representing Health and Welfare Canada and other federal granting agencies. The Council also has 39 committees which review applications for research projects and awards, with input into these committees provided on a part-time, unpaid basis by some 350 researchers.

MRC funds research under the following programmes:

● Research Funding Programme – support for projects by investigators in Canadian universities, institutions and other health science schools and departments. Support is provided through operating, equipment and development grants and through general research grants to university deans. Under this programme, collaborative research funds are provided to multi-disciplinary research projects involving groups of scientists from different areas of expertise working on specific projects. The Council administers four federal Networks of Centres of Excellence programmes related to biomedical research in the areas of Bacterial Diseases; Genetic Basis of Human Health; Neural Regeneration and Functional Recovery; and Respiratory Health. (NCE Programme)
● Personnel Support Programmes – support for research personnel under three categories: (a) salary support programmes, *e.g.*, research associate, career investigator and MRC scientist; (b) research training programmes, *e.g.*, fellowships; and (c) travel and exchange programmes.
● University-Industry Programmes – support for collaboration between Canadian companies and researchers conducting research in Canadian universities. Costs are shared with the private sector
● Other Programmes – under this programme a small percentage of Council funds are allocated to projects such as the President's fund.

For the 1990/91 fiscal year, MRC provided a total of $235 million for biomedical and health care research of which $192 million was allocated to the grants programme and $36 million to personnel support programmes (Table 11.4).

A breakdown of MRC funding for operating grants by area of research is presented in Table 11.5.

National Health Research Development Program (NHRDP)

NHRDP operates under authority of the Department of National Health and Welfare, and incorporates the provisions of the former National Health and Public Health Research Grants. It's mandate is: '... to enable the Department of National Health and Welfare to obtain information and to evaluate and develop innovative options pertinent to the achievement of broad departmental objectives which embrace promotion, protection, maintenance and restoration of the health of the residents of Canada' (HWC,1988).

NHRDP is the only federal programme that supports public health and health

Table 11.4   Medical Research Council Expenditures, 1990/91

Item	$000
Research Funding Program	192,270
Personnel Support Program	35,529
University-Industry Program	5,746
Other Activities	1,876
Total	**$235,421**

Source: Medical Research Council of Canada (1991), *Report of the President, 1990/91*, Supply and Services Canada, Ottawa

services research. Recent priorities of the NHRDP include organisation and delivery of health care, risk assessment, health promotion, rehabilitation, immune status and communicable disease control, native health and AIDS.

Funding from NHRDP
NHRDP funds the following programmes: (a) research projects and study awards including formulation grants and demonstration projects; (b) personnel support programmes, including research training, career development and student fellowships; and (c) conferences and workshops. A selection committee structure, composed primarily of university-based researchers evaluate projects for funding.

In 1990/91, NHRDP provided a total of $30.4 million in research grants, of which about three-quarters was for the grants programme and another twenty per cent was for personnel support programmes. Based on a detailed breakdown of 1987/88 data, about thirty per cent of NHRDP funding went to 'health care delivery' research, followed by grants for AIDS projects at ninteen per cent. (Table 11.6). (HWC, 1988)

It has been suggested that NHRDP, established to fill the gap created by

Table 11.5   Medical Research Council (MRC) Operating Grants by Research Area, 1990/91

Classification	Funding $('000)	% of Total
Neuroscience	26,466	16.5
Biochemistry	22,580	14.1
Cardiovascular	11,363	7.1
Transplantation	8,099	5.0
Reproductive	7,752	4.8
Endocrinology	7,402	4.6
G.I. and Liver	7,111	4.4
Cancer	6,927	4.3
Metabolism	6,617	4.1
Cell Biology	6,116	3.8
Respiration	6,071	3.8
Genetics	5,611	3.5
Other	38,516	23.9
Total	**160,631**	**100.0**

Source: Medical Research Council of Canada (1991), *Report of the President, 1990/91*, Supply and Services, Canada, Ottawa.

MRC's emphasis on biomedical research, increasingly reflects the funding pattern of the MRC. NHRDP has funded some projects in biomedical research areas and in particular has provided support for biomedical researchers through it's post-doctoral fellowship, scholar, and workshop programmes. During the three year period 1986–1988, about sixty-six per cent of post-doctoral fellowships were awarded to applicants with biomedical interests.

Provincial Funding

Total expenditures on health research by provincial agencies are estimated at $94 million in 1990. Provincial research agencies that fund at least $100,000 in research annually include: (NHRDP, 1992)

- Alcoholism and Drug Addiction Research Foundation (Ontario)
- Alberta Cancer Board
- Alberta Foundation for Nursing Research
- British Columbia Health Research Foundation
- Fonds de la Recherche en Santé de Quebec
- Manitoba Health Research Council
- Ministère de la santé et des Services sociaux, Service de la coordination de la recherche (Quebec)
- Ontario Ministry of Health
- Alberta Mental Health Research Foundation
- Saskatchewan Health Research Board
- Alberta Heritage Research Foundation for Medical Research

Based on a review of health research supported by provincial health research programmes during a recent twelve month period, about seventy-five per cent of funding goes to the biomedical/clinical research sector and the remainder to public health/health service research. (Heacock, 1992)

Table 11.6   National Health Research Development (NHRDP) Grants by Health Field, 1987/88

Health Field	$('000)	%
Grants, Research, Demonstration Projects, Total	**11,916**	**76.3**
Health Care Delivery	4,679	29.9
Risk Assessment	1,419	9.1
Prevention/Promotion	819	5.2
Population Immunity	78	0.5
Human Biology	806	5.2
Rehabilitation	527	3.4
Aboriginal Peoples Projects	620	4.0
AIDS Projects	2,967	19.0
Personnel Support, Total	**3,490**	**22.3**
Training Awards	953	6.1
Career	2,537	16.2
Other		
Conferences	218	1.4
Total	**15,624**	**100.0**

Source: Health and Welfare Canada, unpublished.

Research funding by provincial ministries of health, however, are targeted increasingly towards health services research. This is not surprising given that approximately one-third of provincial budgets are now spent on health care and there is an urgent need for work in the areas of cost-effectiveness, medical technology assessment, program evaluation and service delivery efficiency.

National Voluntary Health Organisation and Foundations (NVHOFs) funding

NVHOFs accounted for an estimated $142 million of research spending or about thirty per cent of all non-industry funding sources. Research funded by these organisations and foundations is for the most part basic or applied research in an area of special interest. Organisations such as the National Cancer Institute of Canada, Canadian Red Cross Society, Kidney Foundation, and Canadian Cystic Fibrosis Foundation are examples of 56 of Canada's largest NVHOFs listed by the federal government in its Reference Guide of Funding Sources for Health Research in Canada. (NHRDP, 1992). Of the 56 agencies listed, eighty-five per cent fund biomedical and clinical or applied research; the remaining fifteen per cent represent eight organisations that fund health services research.

Industry Funding

Canada spends a smaller proportion of it's GNP on r&d than any other large industrialised nation. Spending on biomedical research follows this general trend. (Cruess, 1988) Traditionally, biomedical research has been carried out by large multi-national companies at corporate headquarters and in major markets. Increasingly, these companies are decentralising and specialising r&d operations to capitalise on local, scientific expertise and facilities, research infrastructure and local incentives. Canadian subsidiaries face stiff competition from other developed countries for a share of global r&d funds. This is especially true for the pharmaceutical industry although not as prevalent in the less mature biotechnology sector.

Biotechnology

A recent Science Council of Canada document characterised biotechnology as a broad term used to describe the production of innovative products, devices, and organisms through the use of biological processes. (Science Council, 1992) It is not an industry. It is described as 'a set of biological techniques, developed through decades of basic research, that are now being applied to research and product development in several existing industrial sectors'. (Congress, 1991) It is an enabling technology in the sense that it has the capacity to affect a wide variety of processes and organisms and to bring about advances in a wide range of applications.

Biotechnology has a huge economic potential. The Canadian industry had estimated sales of $1.3 billion and r&d spending of about $500 million in 1991. The industry is research intensive and highly competitive and virtually all western countries have taken steps to ensure a strong biotechnology sector. Canada's National Biotechnology Strategy, in place since 1983, has moved from supporting some scientific research in government labs and universities to fostering a supportive environment for commercialisation and technology transfer. The National Biotechnology Advisory Committee, which advises the federal Minister for Science, has promoted an international competitive position for Canada in biotechnology and has developed a business strategy for biotechnology. (Science Council, 1992)

The Canadian biotechnology industry encompasses about 290 companies actively involved in r&d to create new products and processes based on biotechnology. This is up from 220 in 1989. Biotechnology activity has grown fastest in the health care sector. There are seventy-eight companies currently identified with the human and/or animal health sectors as developers of diagnostic agents, pharmaceuticals, drug delivery systems, transgenic products, or genetic therapy. A recent survey of biotechnology firms in Canada showed that biotechnology firms involved in health care are further advanced and faster growing that firms in other sectors. These firms report the highest average r&d expenditures, the highest r&d/sales ratio, the highest average number of biotechnology employees, the highest number of employees with bioscience qualifications, and the highest spending projections on r&d facilities. (Ernst & Young, 1992)

The same survey showed that 78 of 290 biotechnology firms are involved in health care. Companies in this sector spend an average $3.3 million per firm on r&d, a reflection of the significant 'discovery' costs in developing new diagnostics, therapeutics and vaccines. About twenty-four per cent of r&d spending in the health care sector is government funded. (Ernst & Young, 1992)

The biotechnology industry in Canada employs a total of 7,175 people, up slightly from about 6,500 in 1989 but well short of forecasts made in a 1989 survey of biotech companies. About 2,500 people are employed in the health care sector; approximately seventy-five per cent of these employees have a background in the bioscience or chemistry. (Ernst & Young, 1992)

A number of federal government departments and agencies are involved in medical biotechnology.

Health and Welfare Canada
This Department regulates biopharmaceuticals and other biotechnology-based products; evaluates the safety of biotechnology products for environmental and industrial application, pesticides and consumer products; provides a national advisory service on safe operating procedures in biotechnology labs. HWC also administers a Biotechnology Network, BIONET, in the area of human and animal health care products. The Biotechnology Networks have been established to facilitate communications and cooperation between performers and users of research. Other networks are established in the areas of aquaculture, agriculture, forestry, minerals, and waste treatment. (ISTC, 1992).

National Research Council of Canada Biotechnology Programme
This organisation undertakes research with application to health and medical fields at the Biotechnology Research Institute in Montreal and the Institute for Biological Sciences in Ottawa. NRC also encourages expansion of Canadian biotechnology capabilities through collaboration with NRC researchers, financial support through cost-sharing agreements; contracting; and guest worker arrangements for corporate and university personnel. (ISTC,1992)

Pharmaceutical Research

The pharmaceutical industry in Canada is composed of 148 drug manufacturers and distributors, of which over eighty per cent are foreign owned. Almost all companies are located along the Quebec City-Windsor corridor of Central Canada. In 1990, prescription and non-prescription drugs were worth $4.2 billion in manufacturer's sales and represented about two per cent of the world market for pharmaceuticals. (PMAC, 1992) Annual after-tax profits on capital employed averaged nineteen point three per cent for the five years ending in 1985, placing this industry among the most profitable manufacturing sectors in the country. Prescription and non-prescription drugs represent the fastest growing segment of health care expenditures in Canada. In 1990, pharmaceuticals accounted for fourteen point one per cent of total health care expenditures, considerably higher than in 1980 when this sector accounted for eight point nine per cent of health care spending. (Figure 11.1)

The pharmaceutical industry is highly competitive and fragmented with no company holding more than five per cent of total dollar sales. It is composed of an innovative, research-based sector represented by the Pharmaceutical Manufacturers Association of Canada (PMAC). Most companies in this sector are owned by multi-national companies with head offices in the USA or Europe. A generic sector, represented by the Canadian Drug Manufacturers Association (CDMA), is dominated by two Canadian owned companies which together account for eighty per cent of generic drug revenues and prescriptions dispensed.

It is estimated that Canada is responsible for less than one per cent of pharmaceutical r&d spending world-wide. A 1990 study estimated it costs US$230 million and an average twelve years to bring one new chemical entity from concept to the market, well beyond the resource capability of most Canadian firms (Dimasi et al 1990). Indeed, the cost of pharmaceutical r&d, and the need to bring together special facilities and personnel for cost-effective research programmes have, in part, been responsible for a number of corporate mergers, joint ventures and acquisitions world-wide. Canada's research-based sector competes for r&d funds within this global marketplace.

Arguably the piece of legislation that has most influenced pharmaceutical r&d in Canada over the past twenty-five years was an amendment to the Patent Act in 1969 that provided for 'compulsory licensing'. Under this legislation, a compulsory license may be granted to import patented pharmaceutical products into Canada. The licensee pays a royalty of four per cent to the Canadian patent holder as the licensee's share of the cost of research leading to the invention. Canada is the only major developed country that permits compulsory licensing of pharmaceutical patents as a normal commercial practice.

This legislation helped spawn a generic drug industry in Canada that now accounts for twenty-five per cent of all prescriptions dispensed and represents eight to the per cent of total sales. Competition introduced by the generic industry has meant lower drug costs for Canadians. A landmark review of the Canadian pharmaceutical industry was conducted by the Commission of Inquiry on the Pharmaceutical Industry (Eastman, 1985). In it's final report, the Commission noted the growth of the generic industry in the post-1969 period and also concluded that compulsory licensing legislation had not adversely affected the innovator sector in terms of profits, rate of growth, value of shipments and employment. In the opinion of the innovator sector however, this was one factor that created a 'negative investment climate' for pharmaceutical r&d in Canada.

This 'climate' also extended to investment in r&d. Expenditure on r&d as a percentage of revenues stayed between four point six and five per cent from 1967 to 1987 well below world levels of approximately twelve point five per cent of sales (Eastman, 1985; Brogan, 1990)

In 1987, the Patent Act was changed again with the introduction of Bill C-22. This legislation left compulsory licensing intact but guaranteed a minimum

Figure 11.1: *Health Expenditures in Canada, 1980 and 1990 (% of total)*

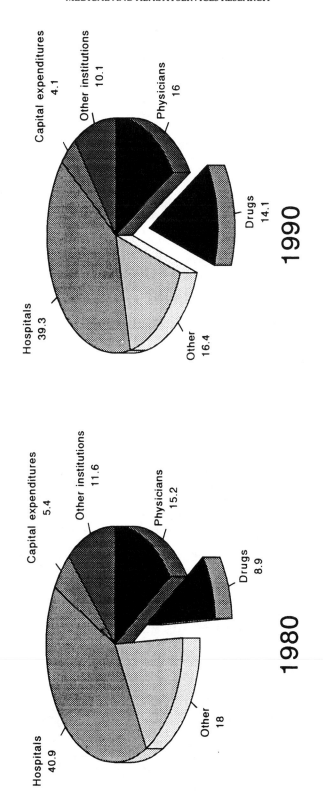

Sources: Health & Welfare, Canada, (1990), National Health Expenditures; Health & Welfare, Canada, (1992), Health Expenditures in Canada, Fact Sheets

period of market exclusivity of seven to ten years. This falls short of other G-7 countries where an average sixteen to twenty years of market exclusivity is provided. Nevertheless, in return for improved patent protection the pharmaceutical industry committed to spend ten per cent of it's revenues on r&d by 1996, up from a pre-1987 baseline of four point nine per cent. The Patented Medicine Prices Review Board (PMPRB) was established in conjunction with passage of Bill C-22 to monitor the prices of patented drug products and to record expenditure on pharmaceutical r&d by patentee companies and the pharmaceutical industry overall. PMPRB reports show that by 1991, r&d as a percentage of sales had reached nine point seven per cent, up from six point one per cent in 1988. (PMPRB, 1989, 1991)

Further legislation was tabled in the Canadian parliament in 1992 that will eliminate compulsory licensing and extend patent exclusivity to twenty years. The pharmaceutical industry has guaranteed further increases in spending on research and development as a percentage of sales.

The National Advisory Council on Pharmaceutical Research stated in it's 1992 final report to the Minister for Health and Welfare Canada that 'this sector is so competitive that only countries with highly qualified manpower, top quality research teams and infrastructure, and universities and hospitals receptive to clinical research will be the targets of heavy investment by the multinationals and will benefit from the favourable economic side-effects such investments inevitably bring with them'. (NACPR, 1992) The Council's report identified key factors the pharmaceutical industry considers when choosing a location for its r&d investment dollars. These include:

Patent Protection
Until recent legislation, intellectual property rights for pharmaceuticals have not been protected to the same extent as Canada's major trading partners.

Drug Review Process
The Food and Drugs Act and Regulations govern the safety, efficacy and quality of drugs marketed in Canada. The Act is administered by the Drugs Directorate, Health and Welfare Canada; approximately 19,000 drug products have been approved for use in Canada.

Federal standards for drug approval are considered to be among the most rigorous in the world. However, the drug approval process has been the subject of a number of internal and external reviews mostly related to the backlog of submissions. In response, a number of internal changes have reduced the backlog for most categories of drugs over the last three years. In addition, the federal health department has proposed a number of operational and policy changes including 'drug product licensing' that will require continuous post-market reporting on marketed drugs and a streamlined drug approval process; harmonisation of post-market reporting with other countries notably the European community; and use of expertise in universities and hospitals to evaluate new drug therapies.

Provincial regulations
Canada's provinces and territories add additional layers of regulations to those already imposed by the federal government. Many regulations are aimed at cost-containment as the budgets for provincial drug benefit plans have increased significantly over the past decade but they also address safety and efficacy, bioequivalency, substitution, availability and distribution issues. Of concern to the pharmaceutical industry is the lack of coordination

and harmonisation between provinces. Independent provincial policies create costly duplication of federal drug approval processes, weaken the domestic market place, erode effective patent life of tradename products, and discourage investment in r&d. Harmonisation of provincial policies has been recommended as a priority by NACPR. (NACPR, 1992)

Income Tax Incentives
Revenue Canada applies a narrow definition of Scientific Research and Experimental Development (SR&ED) under Canada's Income Tax Act and the Patented Medicine Prices Review Board uses this interpretation in reporting r&d expenditure in Canada. The interpretation of SR&ED limits what the industry can claim as tax credits. (See Chapter Five).

Basic and applied research as well as development research[1] are included in the definition of SR&ED. However, post-market research, quality control, routine testing, or research in the social sciences and humanities, routine data collection, and training of graduate students and researchers, university chairs and educational grants are not eligible. Limits on research training may contribute to a shortage of scientists in a number of biomedical disciplines. (NACPR, 1992)

Infrastructure for research
The infrastructure for pharmaceutical research includes facilities, skilled researchers and scientists, funding, and the intellectual property and regulatory environments already discussed. Canada offers a number of strengths. The health care system offers a standardised environment for clinical trials. Canadian universities and colleges provide a high level of education and training to scientists, researchers and technicians in a broad range of specialities. The research community in federal and provincial laboratories, research institutes, hospitals, universities and industry constitute a considerable pool of expertise. Researchers from key laboratories and different sectors and specialities have been brought together under collaborative programmes such as the National Networks of Centres of Excellence, university-industry granting programmes at the Medical Research Council, National Research Council, Natural Sciences and Engineering Research Council, and federal and provincial government departments.

The National Advisory Council on Pharmaceutical Research identified a number of areas of research strengths in Canada, including pharmaceutical formulation; clinical trials; clinical epidemiology; clinical pharmacology; genetic and molecular biology. In addition, Canada has significant expertise in specialities such as information systems, artificial intelligence, telecommunications and robotics that are complementary to research in a high-technology sector such as pharmaceuticals.

Shortages of skilled scientists in specific disciplines (*e.g.*, pharmacology, epidemiology and biostatistics) as well as a shortage of scientists and clinicians with the skills to manage complex, inter-disciplinary research projects are anticipated as university programs are forced to cut programmes due to funding constraints (Science Council, 1991). The drug approval process itself, despite operational problems already mentioned, is viewed by the industry in a positive light in terms of it's reputation as an excellent review programme. This is an advantage when drugs in Canada are marketed abroad.

Local incentives
Ontario and Quebec together account for about ninety per cent of total expenditure for pharmaceutical r&d by patentees. Most in-house and approx-

imately three-quarters of extra-mural pharmaceutical research spending occurs in these two provinces. Whilst spending decisions are influenced by factors such as company location and proximity to research expertise, local provincial and regional incentives play a role.

R&d facts and figures

The Patented Medicine Prices Review Board reported r&d expenditure of $376 million in 1991, an increase of twenty-three per cent over 1990.[2] This figure includes spending by 66 patentee companies and represents the bulk of intramural and extra mural r&d pharmaceutical expenditure in Canada. In 1991, $201 million, or fifty-seven per cent of r&d expenditures were intra-mural. A further $154 million was spent on r&d in universities, hospitals, and other institutes and companies. (Table 11.7) (PMPRB, 1991)

In contrast to other advanced technology sectors, the pharmaceutical industry shows little dependence on direct r&d subsidies from governments. More than ninety-six per cent of 1990 research investment was financed from corporate working capital. Of the $376 million in total r&d expenditures, federal and provincial governments supplied $11.1 million.

Statistics Canada reports that pharmaceutical r&d expenditures increased by an average of seventeen point six per cent per year between 1983 and 1988. This figure conceals wide variations in year-to-year spending including significant increases in spending in the late 1980s. These increases are linked to commitments made by the innovative sector of the pharmaceutical industry at the time Bill C-22 was passed.

PMAC companies reported that 925 new jobs were created in medical r&d between 1987 and 1990. In excess of twenty-six per cent of r&d spending in 1990 was for basic research, mostly conducted in Québec, an increase from nineteen per cent recorded in 1988. Applied research, mostly tied to regulatory requirements for pre-market clinical trials, accounts for the largest share of r&d, with over sixty per cent of total expenditures for clinical and pre-clinical drug trials. (PMPRB, 1991) Table 11.2 shows the distribution by type of research.

Health policy research

Funding of health service research in Canada has lagged behind that of biomedical research. While there have been continuing recommendations over the years for changes in health policy, available funding continues to be directed to biomedical investigator-initiated research. This trend has persisted despite publicly-articulated policy goals aimed at improvements in quality of care, effective delivery of health services and management of resources. In 1974, a land-mark report, *A New Perspective on the Health of Canadians*, the Minister of National Health and Welfare pledged to give studies into human biology, the environ-

Table 11.7   Current R&D Expenditures* by R&D Performers, 1989–1991

R&D Performer	$ million (% of total)		
	1989	*1990*	*1991*
Patentee Pharmaceutical Company	134 (59)	134 (50)	201 (57)
University & Hospital	55 (24)	68 (26)	85 (24)
Other	40 (17)	64 (24)	70 (19)
Total	**229 (100)**	**266 (100)**	**376 (100)**

*excluding capital and depreciation expenses

Figure 11.2: *Pharmaceutical R&D Expenditures* by Type of Research, 1988–1990*

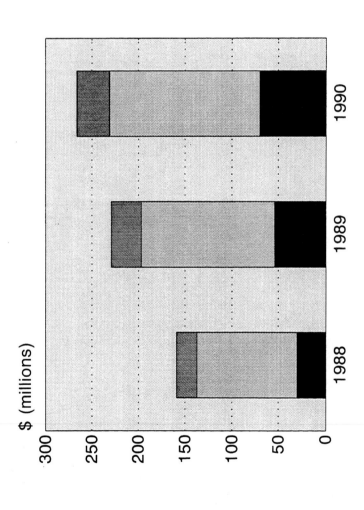

* excludes capital and depreciation expenses
Source: PMPBR second and third Annual Report

ment and life style as much attention as it had to financing the health care organisation. (Lalonde, 1974) The document recognised that factors other than health care contribute significantly to the health of populations. In 1986, the federal health department released *Achieving Health for All: A Framework for Health Promotion*, that recognised the need for health promotion along side conventional health care modalities. (HWC, 1986)

Despite these efforts, there continue to be inadequacies in terms of the lack of attention to the promotion of positive health and prevention of illness as well as in the quality, availability and accessibility of diagnostic and treatment services for some geographically and socially isolated groups when illness does occur. The paucity of data, the shortage of qualified personnel to carry out research and evaluation studies in the health care field and the relatively small amount of research effort which has been devoted to the more applied aspects of health care and health care delivery are all part of a more general phenomenon in Canada. Canada has valued and emphasised basic and clinical research and has spent proportionately less on health care research.

Heacock (1992) traced the evolution of the budgets of the two major federal programmes for funding biomedical/clinical research, on the one hand, and public health/health services research, on the other. In 1974/75 the MRC's budget was $40 million and NHRDP's budget was $10 million (one quarter of the MRC amount). In 1990/91 the MRC budget had grown to $235 million, while expenditures under the NHRDP had increased to only $30.4 million (one-seventh of the MRC amount). About eighty-seven per cent of federal funding in 1990/91 went to support biomedical/clinical research. During the period 1974/75 to 1990/91, the federal government provided about $2 billion for biomedical/clinical research, while at the same time providing $275 million for public health/health services research.

Changes are occurring in the economic, social, cultural, communication, transportation, technological, housing **and** political environments which are modifying the Canadian pattern of living and, with it, its health care system. By not taking steps now to forecast and examine these issues, many of the 'solutions' which have been offered in developing to meet current difficulties are being rendered obsolete and irrelevant by the arrival of a largely new set of conditions. (Institute, 1988)

Notes

1 Development research is defined as the use of basic and applied research for the purpose of creating or improving existing material, devices, products or processes
2 Reported r&d expenditures include only those eligible for an investment tax credit under provisions of the Income Tax Act in effect on December 1, 1987

References

Angus D E 1990 An Option for Discussion: The Canadian Health Care System", in Meyer R (ed.) *Who Pays for a Healthy America? Options for Health Care in the 1990s* p97–109 University of Kansas
Brogan T 1990 *Analysis of Investment in Canadian-Based Pharmaceutical Research and Development*Report Prepared for Industry, Science and Technology Canada
Congress of the United States, Office of Technology Assessment *Biotechnology in a Global Economy* Washington, D.C U.S. Government Printing Office

Cruess R L 1988 *The Current State in Funding Targets for Medical Research* University Research and the Future of Canada, Proceedings of a National Conference held in Edmonton Alberta, April 1988.

Dimsai J A. 1990 Press Briefing, Investigation into the Cost of New Drug Development, Centre for the Study of Drug Development, Tufts University, Boston, Massachusetts

Eastman H C 1985 *The Report of the Commission of Inquiry on the Pharmaceutical Industry* Minister of Supply and Services Canada, Ottawa, Canada

Ernst & Young in collaboration with Industry Science Technology Canada (1989) *Canadian Biotech '89: On the Threshold* Minister of Supply and Services, Canada

Ernst & Young 1992 *Canadian Biotech '92, Towards Realization* Survey of the Canadian Biotechnology Industry Toronto Canada

Heacock R.A 1992 Health and Health Care Research: An Investment in Future Well-Being paper presented at the CSCI/CMA/CHR Conference *The Future of Health and Health Research in Canada: What Are We Heading For?* Ottawa, February 27

Health and Welfare Canada 1986 *Achieving Health for All: A Framework for Health Promotion* Ottawa Health and Welfare Canada, Minister of Supply and Services

Health and Welfare Canada 1988 *National Health Research Development Program Inventory, 1986–1988* Ottawa

Health and Welfare Canada 1991 *National Health Expenditures in Canada 1975–1987* Ottawa Minister of Supply and Services Canada

Health and Welfare Canada 1992 *Health Expenditures in Canada* Fact Sheets, prepared by Policy, Planning and Information Branch, HWC, Ottawa

Industry, Science, Technology Canada 1992 *Partnerships in Biotechnology* Ottawa Minister of Supply and Services Canada,

Institute for Health Care Facilities of the Future 1988 *Future Health: A View of the Horizon* Ottawa,

Lalonde M 1974 *A New Perspective on the Health of Canadians: A Working Document* Ottawa Government of Canada

Medical Research Council of Canada 1991 *Report of the President, 1990/91* Ottawa Supply and Services Canada,

National Advisory Council on Pharmaceutical Research 1992 *Time to Act, A Strategy for the Development of a Growing Sector: Pharmaceutical Research* Ottawa Health and Welfare Canada

National Health Research and Development Program, Health and Welfare Canada 1992 *Reference Guide of Funding Sources for Health Research in Canada* Ottawa Minister of Supply and Services Canada

NCE Program Backgrounder *Networks of Centres of Excellence* Government of Canada.

OECD 1987 *Financing and Delivering Health Care, A Comparative Analysis of OECD Countries* Paris OECD

Patented Medicine Prices Review Board 1989 *First Annual Report, 1989* Ottawa Minister of Supply and Services Canada

Patented Medicine Prices Review Board 1991 *Second Annual Report, 1990* Ottawa Minister of Supply and Services Canada

Patented Medicine Prices Review Board 1991 *Third Annual Report, For the Year Ended December 31, 1990* Ottawa Minister of Supply and Services Canada

Patented Medicine Prices Review Board 1992 *Fourth Annual Report* For the Year Ended December 31, 1991, Ottawa Minister of Supply and Services Canada,

PMAC 1992 *The Canadian Pharmaceutical Industry, A Backgrounder* prepared by the Pharmaceutical Manufacturer's Association of Canada, Ottawa,

Ryten E 1989 Financing the Research Mission of Canadian Faculties of Medicine: Comparisons and Trends *ACMC Forum* **22**(3)

Ryten Eva 199 The Funding of Research Conducted by Canadian Faculties of Medicine *ACMC Forum* **24**(3)

Science Council of Canada 1992 *Reaching for Tomorrow, Science and Technology Policy in Canada 1991* Ottawa Minister of Supply and Services Canada

Statistics Canada 1992 *Canada Year Book 1992* Chapter 4 Ottawa Minister of Supply and Services Canada

Chapter 12

Science, technology, and society in Canada

André Lemelin and Camille Limoges

Introduction

That science is, or ought to be, an autonomous societal sub-system is a view still current today, mainly among scientists. It perpetuates the notion of a 'republic of science'. Science as the disinterested quest for truth would best be warranted by society at large granting self-governance to the practitioners in the scientific enterprise. Society itself would be served best by that attitude: the applicability of science, the source of all technologies and the very foundation of the improved level and quality of living, would be predicated upon it's objectivity, requiring, in turn, it's freedom and seclusion from social pressures.

In this view, science and society are two different entities that relate optimally one to the other by being carefully kept apart. One of the main tasks of the 'public understanding of science' is to make people conscious of this requirement and of the benefits flowing from meeting it.

However, most analysts and decision-makers nowadays consider this conception as highly abstract, idealised and even misleading. What is at issue is to better conceptualise and manage the role of science in society, or even of science itself as a social process. Indeed, the interconnections between s&t are not seen any more as unidirectional, and many have come to talk of 'technoscience', emphasising the basic unity of s&t, as scientific discovery appears grounded in the potentialities of instrumentation, and as the differentiation between scientific results and innovation has become highly questionable in many fields . Moreover, decision-makers as well as concerned citizens have come to debate societal issues which are seen as having essential and intrinsic technoscientific dimensions. Keeping apart science and society in that regard appears unrewarding.

One has to recall here how, after unprecedented decades of sustained economic growth, the energy crisis of the 1970s, the two petroleum shocks, two-digit inflation, rising unemployment, monetarist and restrictive policies, the system – of which technoscience was a pillar – though not collapsing was considered seriously 'ill', and in need of a cure. New instabilities appeared in the context of the extension of free trade, exacerbated competition, displacement of production units in new industrial countries, automation, more lay-offs, and zero new jobs growth. In that economy which was said to be adjusting, the so-called anti-science movement emerged, and triggered a renewal of preoccupation with 'public understanding of science', and 'science awareness programmes'. Also, during the last two decades, the ecological movement achieved political significance. Development was no longer necessarily synonymous with social progress. The feasible and the desirable were no more seen as the two sides of the same equation. Governments responded with new regulations, and with the creation of new fields of expertise, such as technology assessment and risk assessment. This opened new avenues for research, and on the university scene, traditional fields of scholarship, such as the history of s&t, expanded into social studies of science, or science, technology and society programmes.

These developments are symptomatic of a growing interest and concern with technoscientific development and of the need for better societal management. In the 1990s, as the notion that science is not only a driving force behind the economy, but is permeating all dimensions of social life, this trend is gaining strong impetus. This chapter focuses on the description of the institutions and activities through which the promotion and management of a scientifically informed societal culture is pursued.

Section one will review some changes through which scientific culture becomes a recognised major issue on the science policy agenda in Canada. Section two will examine the goals and means for the popularisation of s&t. The last section will review some management devices aiming at a better interface between technoscience and the drive to innovate on the one hand and, on the other hand, the preservation of the environment and the betterment of the quality of life.

Constructing 'science in society' as a policy object

A newcomer to the policy agenda

In Canada, as elsewhere, the setting and role of science in society, for a long time, were taken to be unproblematic and not an object of special concern. Science policy was focused almost exclusively on patterns and levels relevant to the funding of the research enterprise.

Of course, the less comforting aspects of technoscientific development were not totally ignored, even then. For instance, the most extensive review of Canadian science policy in the late 1960s and early 1970s – the work of the Senate Special Committee headed by Maurice Lamontagne – published a detailed picture of the scientific revolution and of it's stunning results; the material improvement of life. However, the report also noted 'a destructive side of technology that was not foreseen by the prophets and founders of the scientific revolution'[1]. It further pointed out that 'the dark side of technology, the impending breakdown in man's relation to nature, and the affluence gap are three major problems of which we are forewarned by current trends and projections.'[2] However, it is fair to say that in Canada at the time, the sense of urgency was still quite low, and in policy terms, few practical actions pertaining to the diffusion of scientific culture were contemplated.[3] It is indeed highly symptomatic that technology assessment was dealt with in that report in some fifteen lines.[4]

However, substantial changes have occurred during the last decade. The Canadian government – and those of some of the provinces as well – have moved from a classical approach to science policy to policy for innovation. This is perhaps best exemplified by the federal government's regrouping of most of it's actions in s&t under the umbrella programme called *InnovAction*. It is not expected anymore that simply supporting scientific research warrants economic development, social progress and a better quality of life. Indeed, it is the innovation processes themselves that become targeted in the new policy for innovation. Within that perspective, science is embodied in society, it is no longer conceived of as a separate activity. In brief, it comes to be perceived, even more than before, as a condition for development and for the solution of societal issues. On the other hand, science is also seen as sharing part of the responsibilities for generating the negative externalities of technological development. This is why a policy for the support of innovation, much more than a science policy, calls attention to governmental responsibilities towards the assessment of technologies; economic, social, and cultural as well.

This is reflected in the new policy arena. Not only research, but also the broad diffusion of a s&t culture in a society competing on global markets, and in which most societal issues come to embody technoscientific dimensions, becomes a priority. This aspect will be examined in the next section of this chapter. Moreover, the negative externalities have to be forecasted, prevented or mitigated; the processes and institutions through which this is done will be examined in section 3.

InnovAction, the federal umbrella action plan in s&t, has identified the promotion of a more science-oriented culture as one of it's five major thrusts. It is on that base that it has increased the efforts of it's public science awareness programme, to which we will come back later in this section.

This new vision of s&t, of their closer mutual relationships, and of their increasingly crucial societal role, has generated the need for new types of experts, new forms of governmental support, and created opportunities for a stronger involvement of actors from the private sector. These are the three points that this section shall explore.

Research and training in the analysis of technoscience in society

One of the results of the trend described above is that science has come to be perceived as a much more complex activity than it used to be; not only in it's management of scientists, but also in the development of new competencies, skills, and tools.

This has been reflected, in Canada as in most other advanced industrial countries, in transformations affecting the analysis of technoscientific development, as well as those providing the knowledge base for policy studies.

Indeed, during the past two decades, academic research on the historical and sociological aspects of scientific development has undergone a transformation in depth and scope. The history of science, which had largely focused on the understanding of the genesis of scientific theories, has been complemented by sociological analysis of research communities and institutions, by ethnographic approaches, economic analysis, policy and evaluation research, including scientometrics. Research and training in the broadened field of what is now called 'science studies', 'social studies of science' or 'science, technology and society' [STS] has become a highly interdisciplinary affair. It is now cultivated in Canada in a vast spectrum of programmes and organisations, still currently in development.

The Institute for the History and Philosophy of Science and Technology, created at the University of Toronto at the end of the 1960s, was the first Canadian graduate studies unit essentially devoted to research and training in that field. It has played a pioneering role in the scholarly study of the history of Canadian s&t. In the field of the history of medicine, the Department of the History and Social Studies in Medicine of McGill University has played a similar role in training historians of medicine and more recently also in the sociology and anthropology of medicine. The Institut d'histoire et de sociopolitique des sciences, at the Université de Montréal, created in 1973 and disbanded in 1985, had been the only programme to have included the policy studies and the economics of r&d, with the history and the sociology of science in the training of graduate students. During that period, it was the only major producer in Canada of specialists in the STS field.

However, many other universities offer graduate courses pertinent to STS studies and provide opportunities for students to write doctoral dissertations on STS topics. At the undergraduate level, most universities offer a few courses in that field, and some make it possible to major in 'science and society'. The 'Science and Human Affairs' programme of Concordia University in Montré al is the oldest undergraduate programme in Canada. Since 1987, the Université du Québec à Montréal [UQAM] has offered a complete Bachelor degree in Science, Technology and Society studies and is currently developing a PhD programme in STS.

Though formal programmes are not yet numerous, recent years have seen an increase in the number of formally institutionalised centres where some form of STS is pursued with continuity by teams of researchers. These centres often provide loci for the training of graduate students, in a research environment, even where no formal training programme exists.

Many of these recent initiatives are to be accounted for by the rise of the new accentuation of science policy as policy for innovation, and are focused on the management of technology issues. Examples include the Management of Technology Institute [MTI] at McMaster University, the National Centre for Management of Research and Development [NCMRD] at the University of Western Ontario, the Research Centre for High Technology Management at Carleton University and the Research Centre for the Management of New Technology at Wilfrid Laurier University.[5]

In June 1989, the Natural Sciences and Engineering Research Council and the Social Sciences and Humanities Research Council jointly launched a programme for the creation of Chairs in the Management of Technological Change. These chairs are funded in partnership with business, industrial firms, or other private sector entities, while universities and the provincial governments are encouraged to participate in the funding.[6] Among other goals, the programme aims to 'improve the development of public policy and public understanding related to technological change'.

In Quebec, a special governmental programme,[7] designed to create new poles of development in innovation related research in universities, has identified the 'social mastery of technology'[8] as one of it's priorities. New research centres have emerged on the basis of grants provided through that programme. The Centre for Research on the Social Assessment of Technology [CREST],[9] brings together researchers from the Université du Québec à Montréal, the Ecole Polytechnique de Montréal and McGill University, the Centre for Research on Development in Industry and Technology [CREDIT],[10] which has members from the Université du Québec à Montréal, the Université de Montréal, the Université de Sherbrooke and Université Laval, and the Research Group on Law and Information Technologies [GRID][11] whose members come from the Université du Québec à Montréal and the Université de Montréal. The increasing interlinkages between these three interuniversity centres has created a critical mass of researchers in different aspects of the social studies of s&t in the Montreal area.

Risk research, a field which emerged in the 1970s, and which is taking on increasing importance as more and more technologies and projects induce public controversies, is also taking root in the academic context, mainly through the work of the Risk Research Institute at the University of Waterloo, in Ontario, and of the Centre for Policy Studies on Science and Technology, at Simon Fraser University, in British Columbia.

Of course, and this is a trend more evident since the 1970s, the rapid development of concerns for the environment has generated a large amount of research and training activities in Canadian universities. Though these studies do not take science *per se* as their object, but rather make use of it, they are remarkable in that more often than elsewhere they associate experts from the natural sciences, from engineering and the social sciences. Environmental planning and design, social impact assessment and environmental assessment have indeed become areas in which studies are pursued in a highly interdisciplinary context.

Governmental actions

Beyond those mentioned earlier, a number of new governmental programmes have recently accelerated the development of training and research in science studies. One of these new initiatives has been a joint programme of the Natural Sciences and Engineering Research Council and of the Social Sciences and Humanities Research Council, providing for Master's scholarships in Science Policy. The two Councils state among the objectives of the programme that: 'as issues in s&t increasingly affect people in their day-to-day lives, Canada has an urgent requirement for trained personnel to help direct and oversee policy-making in the field of science. Those responsible for future science policy decisions must recognise the essential role of the social sciences and humanities in the development of science policy, and understand the scientific dimensions of public policy'.[12]

In recent years, much of the research in science studies in Canada has been supported by the Social Sciences and Humanities Research Council through it's Strategic Grants Programme. Indeed, many themes of that programme, such as 'Science and Technology Policy in Canada', 'Applied Ethics' or 'Managing for Global Competitiveness' have encouraged the pursuit of research on a wide spectrum of problems.

In the domain of environmental research, the federal government has launched an ambitious new programme that is likely to have some influence on the development of science studies research in Canada. Under the federal Green Plan for the environment, the Eco-Research Programme for research on the ecosystems, is jointly managed by the three federal granting councils ($50 million over the next six years).[13] The programme has three components: doctoral fellowships, research chairs, and research grants. Though it aims mainly at a better understanding of the Canadian ecosystems, it is of interest here, because the programme also intends to promote interdisciplinary research approaches, with the management of ecosystem, policy considerations, decision-making, and consensus building given a high priority.[14] It is likely that s&t policy research, risk studies and the sociology of public controversies research, among others, will benefit significantly from this programme.

Finally, it should be emphasised that some of the new responsibilities assumed by ministries will help foster the development of research in the STS domain. Again, it is obviously the case in environmental matters where, as public concern increases, controversies multiply and regulations require more formal impact assessment reviews, government departments and agencies will need to understand better the social dynamics pertaining to the information of the public, the role and weight of scientific expertise and the conditions for good management of issues having technoscientific dimensions. The same situation obtains in regard to the regulatory processes themselves, not only in the realm of the environment, but also in the registration of food products, of drugs, pesticides, *etc.* Moreover, it probably is worth underlining, as symptomatic of this new trend, that mandates for 'insuring the social mastery of s&t' have appeared in the laws constituting the Ministry for Higher Education and Science and the Ministry for Industry, Trade and Technology, in the Québec government. This might indicate that the assessment and management of technoscientific development is indeed taking a new turn, which will call for the intensification of STS research and training.

The private sector

But it is not just governments and university-based researchers that have shown concern for a better understanding of the development, role and significance of technoscience in society. Indeed, labour unions and councils have shown an active interest in assessing the impact of technological change on employment, job skills, the organisation of firms, the quality of working life and management control, and they have conducted or supported a substantial amount of research on these issues.[15] From the private sector side, organisations such as the Canadian Manufacturers Association have also often intervened publicly, for instance on innovation policy issues.

Canada's National Academy, the Royal Society of Canada, has also intensified it's public presence over the past few years. Its *Corporate Plan* includes activities aimed at the 'Public Awareness of Science'. Moreover, with the support of disciplinary societies, umbrella organisations and other institutions, such as the National Museum of Science and Technology, the Society supports a Committee of Parliamentarians, Scientists and Engineers [COPSE], which includes a programme of luncheon addresses and symposia. It also links with writers and media specialists to encourage better understanding of science, and promotes the involvement of scientists in public awareness activities.[16] In addition, a guide has been published to help scientists become more efficient in communications science.[17]

With the support of the Royal Society and private sector organisations, a new association, the Federation of Associations for Science and Technology (FAST) has been recently established that will facilitate enhanced communications amongst s&t societies and other organisations to increase public awareness, understanding and support of science in Canada.

In 1990, the Association canadienne-française pour l'avancement des sciences [ACFAS], modernised it's statutes to identify as one of it's major objectives to intervene publicly in debates on issues having a science or technology dimension, with the purpose of providing better information to the public. Created in the 1920s, it's membership now includes some 7000 French speaking scientists throughout Canada.

Finally, it should be pointed out that small firms have emerged during the last ten years that have expertise in the social dimensions of scientific and technological development and which specialise in contract work for industry and government. The 'Impact Group' from Toronto, for instance, has prepared the communication strategy of Industry, Science and Technology Canada's public awareness programme.[18] In Quebec, 'Média-Science' does similar work. Moreover, there has been a recent trend in corporate donations from financial to 'in kind' contributions, and recommendations have been made to increase this sort of involvement and to help industry associations raise the awareness of their members, for instance through newsletter material.[19]

Conclusion

Despite the undeniable progress made in enhanced training, research and public awareness, there are serious reasons for concern. First, there is a notable lack in Canada of institutions dedicated to the public assessment of s&t. Canada has no equivalent to the U.S. Office of Technology Assessment, or to the parliamentary offices for scientific and technological choices that have been created in recent years in Europe. Moreover, this gap is aggravated by the loss of two distinguished institutions recently abolished by a decision of the federal government (the Science Council of Canada, and the Economic Council of Canada). Both had been a major source of public studies and public advice over the past three decades.

Diffusing science in society

Introduction

Over the past several decades, democratic countries have engaged in supporting mass education, no doubt partly as a response to the requirements of mass economic production initiated by the industrial revolution. Complementary state-sponsored enterprises of popular education were added to educational institutions, covering issues such as public health and hygiene, or new agricultural techniques, embodying some notions of s&t.

However, teaching, as well as popularisation of science, have generally been predicated upon the notion that it is a special subject. Indeed, for many people, science remains something quite inaccessible. At a time when it has become conventional wisdom among decision-makers that industrial competitiveness is grounded in technoscientific innovative capability (even in a country like Canada, where traditionally economic success was based on the export of natural resources and a huge financial sector), this issue of public science has become a matter of serious concern. Calls become more and more frequent for a new industrial culture centred upon an understanding of s&t. Indeed, scientific culture has come to be seen as holding the key to future economic success.[20] As the Chairman of the Science Council of Canada wrote in one of the Council's last documents, the quality of culture in society, on which r&d activities are dependent, is certainly as important as r&d itself.[21]

On this score, there seems to be good reason for concern in Canada. Canada has only 30 r&d scientists and engineers per 1000 workers, whereas it's partner in the Free Trade Agreement, the USA, has 65. Canadian high school students get poor marks in international comparative surveys of science and mathematics literacy. The drop-out rate of Canadian high school students is very high at twenty-five to thirty per cent compared to a mere five per cent in Germany. The enrolment in undergraduate science and engineering programmes is steadily declining. Canada is facing important personnel shortages in key s&t sectors. Finally, another major problem is that the enrolment of women in science and engineering programmes remains still far too low, despite a significant growth during the past two decades.

On this last point, it is worth emphasising that the training of women alone will not suffice. Unfortunately, it seems that the market-place is still not ready to welcome them. A recent report of the Canadian Committee on Women in Engineering paints a picture of 'a profoundly sexist profession that has begun accepting women slowly and painfully – principally because it needs their numbers'.[22] Industry, Science and Technology Canada has hired the coordinator of that study, with a view to help implement the 29 recommendations of that report. From the data compiled by ISTC, it seems that the situation in the science disciplines is still very similar to that in engineering.[23]

Opinion polls, which have multiplied in recent years, give few reasons for further comfort. Some of these studies indicate disquieting trends among young Canadians: they show little interest in science and often entertain negative stereotypes towards s&t.[24] Although the perceptions of the adult population are generally positive, the overall level of information remains rather low.[25]

All these facts are a cause for concern in a period of rapid technological change. Nevertheless, according to the report already mentioned from the Science Council of Canada, there is no coordinated strategy to face these problems at the present time.[26] Nonetheless, there are numerous programmes and initiatives intended to convince Canadians of the importance of education, to promote science education, and to improve the public awareness of s&t development. We will review some of these in the following pages.

Governmental actions

The resources invested by the federal government into public awareness remain of modest size. Nor have some of it's programmes proved very innovative. It has been its tendency, it would appear, to pick up some interesting initiatives from the provinces and to adapt them into national programmes.

The federal Public Awareness Programme for s&t was introduced in 1984, redefined in 1987, and renamed Science Culture Canada, with a budget of $2.5 million a year. It's objectives are to enhance the public awareness of scientific and technological realisations, in particular those carried out in Canada or to be applied in the country; to stimulate the public's interest for and comprehension of the role and impact of s&t in today's society; to improve the communications between scientists and non-scientists, between those who develop and those who use technology, between those who are responsible for the technological changes and those who are subjected to them; to stimulate and maintain the interest of young Canadians for s&t, and make them sensitive to this question.

Another initiative of the federal government has been the Canada Scholarships Programme, intended to encourage students in science and engineering undergraduate studies, and in college level technology programmes. It had a budget of $80 million for the 1988–1992 period. Based solely on academic performance, the scholarships provide $2500 per year for up to four years. At least fifty per cent of the first-year scholarships are awarded to women. In a typical year, about 2500 scholarships are awarded to first-year undergraduate students. (See Chapter Six). Women are also a specific target of the Public Awareness Campaign, a $2 million a year marketing effort.

The action of the federal government in these matters has led to the production of a considerable amount of promotional material, and it has designed promotional campaigns using the techniques of the advertising industry.[27] The results of this strategy are difficult to assess.

Another recent initiative is the National Science and Technology Week, inspired by similar events organised in British Columbia and in Québec. Since 1990, it has spurred quite an enthusiastic response from coast to coast, and triggered many local initiatives, with considerable provincial involvement.

Recently, a new programme called 'Innovators-in-the-School' was created by mutual agreement of the ministers responsible for s&t. Adapted from a British

Columbia experiment, it's formula is known in many countries: practising or retired scientists pay a visit to an elementary school or high school to explain what is science, and the interesting life and opportunities that scientific careers offer. Each province has its own list of volunteers from industry, academia or the public sector. The federal ministry conducts coordinating and consulting work with local correspondents – not necessarily provincial governments. This has been the federal approach: not trying to reinvent the wheel, but instead to take advantage of what already exists, not only in the provinces, but in other federal departments as well. Industry, Science and Technology Canada is currently compiling a data base of all services available from federal ministries and agencies; it will be of great assistance to it's clientele across Canada.

The space available here does not permit discussion of the programmes of all ten provinces and of the two territories. It should be pointed out however that most provinces are very active in this area.[28]

Of all the provinces, Québec has provided much of the inspiration to the federal organisation preoccupied with these matters. According to the words of some observers, 'most provinces have a history in public awareness similar to that of the Federal Government. Québec outstrips all others, for it has long had a structure, science leisure system and policy of financial support for public awareness activities'.[29] It may be worth noting some of these initiatives considered to have exemplary value.

Indeed, one should recall here that Québec has taken leadership in science policy, together with Ottawa, as described in Chapter Three. Quebec was not only the first Canadian province to issue it's own science policy plan, in 1980, but also the first government to set specific goals in the policy for the diffusion of s&t culture, and to support democratic public debates on important issues.[30] The Quebec government annually awards it's prestigious *Prix du Québec*, the *Prix Marie-Victorain* (for natural and engineering sciences) and the *Prix Léon-Gérin* (for the humanities and social sciences). Efforts are made to publicise the events and to make the laureates career models. The Ministère de l'Enseignement supérieur et de la Science (MESS) has a fairly broad range of peer-reviewed, financial aid programmes. It helps non-profit organisations, including universities, to carry out all kinds of projects in the public understanding or awareness of s&t. It funds a few major science magazines, offers grants to science magazines for publication of special projects, and supports the writing of original undergraduate textbooks in the French language. MESS also helps to set temporary or travelling science exhibitions. The total annual budget for these programmes is about $2.5 million, and has been steadily rising in recent years.

The ministry also supports non-profit organisations such as the Association canadienne-française pour l'avancement des sciences (ACFAS), the Société pour la promotion de la science et de la technologie (SPST), and the Agence Science-Presse (ASP). ACFAS, similar to BAAS and AAAS, has no Canadian equivalent. Founded in 1923, it unites the largest part of the francophone research community throughout Canada, holds an important annual congress and publishes the high standard *Interface* science magazine. SPST organises the annual 'Quinzaine des Sciences' in October, a twenty day fair encompassing hundreds of activities, notably an international science film festival, and the Prix du Québec awards ceremony, now broadcasted. ASP, among other things, operates a science news service, mainly for weekly newspapers, and publishes the highly successful club-magazine *Les débrouillards*, for children from nine to twelve years old. This magazine has been adapted and translated in several countries and ASP now coproduces a television series by the same name. The agency also issues the monthly bulletin *Science Express,* aimed at supporting science teachers at the elementary and high school levels.

MESS also supports the production of television programmes, and covers the costs of the pedagogical design of major science museums, such as the Insectarium and Biodome in Montreal.

Science clubs

Another ministry, the Ministère du Loisir, de la Chasse et de la Pêche, also supports many non-profit organisations which are active in the field of the public understanding or awareness of science. There are some 90 scientific leisure groups throughout Québec, federated under the Conseil du loisir scientifique du Québec, which has existed under different names since 1968.

The tradition of scientific leisure clubs dates back to 1931, when ACFAS, which has always been deeply involved in supporting the popularisation of science, created the 'Cercles des Jeunes naturalistes'. In 1954, there were already 815 local circles, as well as 290 4-H Clubs, devoted to forest conservation. Science clubs are to be found across Canada, but they have been particularly successful in Québec, where they tend to be a mass movement. Elsewhere in Canada, it is more elitist, with a clear emphasis on experimentation and excellence.

For years, the activities promoted by the Science Fair Foundation (1966) have encountered much success in English speaking Canada. In Québec, participation in these fairs has mushroomed to nearly 10,000. The real success of this approach resides not so much in the big 'national' science fair (the Pan-Quebécois), but rather in regional and local fairs.

The media

Science acquired a higher status with the media during the 1970s. The most important newspapers and general interest magazines have hired specialists. There are now some high quality television programmes such as the Canadian Broadcasting Corporation's [CBC] *The Nature of Things,* with David Suzuki, or its counterpart on the CBC French speaking network *Découverte.*

Magazines dedicated to the popularisation of science have multiplied, especially in Québec. *Québec Science* magazine has been very influential, since its first appearance under the title *Le jeune naturaliste,* in 1951. It underwent an interesting evolution from natural history to a broader conception of science, and from a readership exclusively composed of high school students to a publication aimed at a broader general public audience. It has been, and still is, a springboard for careers of budding science writers.

Many specialised magazines have since been introduced, either devoted to the environment (*Forêt-Conservation, Franc-vert*) or to natural sciences in a more traditional way (*Québec-Oiseaux, Astronomie Québec*).

There is no English Canadian equivalent to the general public, multidisciplinary science magazine *Québec Science.* There are, however, more specialised magazines, like *Equinox,* a Canadian competitor to the *National Geographic Magazine* and *Geo.* A national science magazine was introduced in 1985, with English and French versions, *Science and Technology Dimensions* and *Dimension science et technologie,* published simultaneously in Toronto and Montréal, but it eventually collapsed.

Similarly, the *Canadian Science News Service,* a syndicated news service based in Toronto and reaching more than 200 Canadian newspapers and even more radio stations, unfortunately died in 1992.

Science policy magazines called *Science Forum* and *Canadian Research*, begun during the halcyon days of the Lamontagne Committee hearings, also folded. However, a number of new science policy newsletters have emerged on the scene. These include *Research Money, Science and Government Bulletin* and *Canadian R&D Manager*.

If Québec appears as the Canadian leader in the popularisation of science, it is largely because it has the paradoxical advantage of it's minority status in North America, due to the use of the French language. To a certain extent, this shields it from the overwhelming influence and competition from the USA.

Science writers and other media professionals

Science writing is not yet a discipline benefiting from university training. Contrary to the experience of some American universities, science writers in residence do not exist in Canada. There are no graduate studies in science journalism and the schools of journalism do not include science in their curricula.[31] The field has been mostly defined by self-made professionals, often science graduates with a special knack for writing, who switched from college journals to new-born science magazines, thence to the electronic media. Many were eventually regrouped in the Canadian Science Writers Association [CSWA] in the late 1960s. The CSWA has some 300 members, working in newspapers, magazines, radio, television, and in the film industry. The CSWA has inspired the creation of it's French speaking counterpart, the Association des communicateurs scientifiques du Québec in 1977, the year when the forerunner of science communication in Quebec, the late Fernand Séguin, a biochemist turned science writer and TV producer, was awarded the UNESCO Kalinga prize.[32]

The museums

It is often believed that Canadian scientific museology was non-existent before the inauguration of the Ontario Science Centre in 1969.[33] This is not true, but symptomatic of the considerable impact this institution has had. The first of its kind in Canada, it opened it's doors in 1969, the same year as the San Francisco Exploratorium, which gave impetus to a wave of similar centres throughout the USA.

Science centres are what Hudson calls Phase III of the evolution of science museums.[34] In Phase I museums, collection and conservation are top priorities, the past being the main object of discourse and exhibition. McGill University's Redpath Museum (est. 1882) is a perfect example of this kind of museology. In Phase II museums, the emphasis is on education. Past technological innovations are explained with the help of functional models, as in the case of many Canadian interpretation centres, for example, those located in hydroelectric powerhouses. Phase III museums emphasise participation and interaction. This hands-on museology, born in Paris' Palais de la Découverte (est. 1937), focuses on the basic principles and applications of today's science.

Today, it is not very clear which institutions are museums and which are science centres. The development of science centres has had a considerable impact on traditional institutions, which gradually incorporated the participative approach into their practices. As a consequence, scientific museology has evolved dramatically during the past twenty years, and along with it, the whole field of museology. As a matter of fact, visits to science museums, everywhere, are the most numerous, longest, and most frequent of all. Science museums now play a primary role in the diffusion of science, and have become the focus point for all sorts of related activities.

Some 350 museums and science centres have been identified in Canada.[35] Needless to say, most of them are local institutions, and their drawing power is rather limited. With large urban centres, Ontario and Alberta, for example, are better off in that respect. Toronto is the home of the Ontario Science Centre, which welcomes more than a million visitors each year. Canada's capital, Ottawa, is also very well equipped with the National Museum of Science and Technology, the National Aviation Museum, and the Agriculture Museum. Science North, in Sudbury, is focused on the geology of this rich mining region of Northern Ontario. Alberta is especially proud of the Royal Tyrrell Museum of Palaeontology, in Drumheller, one of the richest dinosaur sites in the world. It also is the home of the Edmonton Space and Science Centre, and the Fort McMurray Oil Sands Interpretative Centre.

Quite surprisingly, Québec, the second biggest province in terms of population, has not a single large science centre, at least not of the type defined here. In the mid 1980s, an important provincial project, the Maison des sciences et des techniques de Montréal, was aborted for budgetary reasons. Another initiative, a federal-provincial venture called the Musée des sciences et des techniques de Montréal, has been under study for several years, yet was no assurance of realisation. In recent years, however, the huge Montréal Botanical Garden (now ranked in the top three in the world), has come to house an extremely popular Insectarium, close to the recently opened Biodome. The latter is the first of its kind in the world. Neither zoo, nor museum nor botanical garden, it contains under one roof not less than four functional ecosystems, inviting the visitor to an initiation to the diversified ecology of the Americas. However, scientific museology in Québec remains characterised by an overwhelming emphasis on the natural sciences, with little yet to offer in the physical sciences and industrial technologies.

Problematising, interfacing and regulating science

Introduction

There is much ambivalence in citizens' perceptions of s&t: the same knowledge that could heal and feed humankind and make for a happier future could also destroy any hope for a future. Science has become a force so potent that it ought not to be left to private interests, to bureaucrats and even to duly elected politicians. It is nowadays widely felt that the evaluation of the benefits and impacts of technoscience and of it's utilisation ought to be openly discussed, and that it's governance ought to be made in accordance with public policies widely established.

Problematical technoscience

That there is a need for management of technology not only at the level of the firm, but at the societal level as well, has become in the past two decades a widely shared opinion. The new public management of technology is not just a management of technology for innovation and competition on world markets, but also – and often mainly – an attempt at managing the consequences of the development and implementation of technologies. This is especially the case in areas such as the environment, safety in the workplace, employment and quality of work, and telecommunications.

In relation to these relatively new interests, a number of organisations and programmes have been established, either to advance research or to perform research on the use and impacts of technologies. Such examples include the Ontario Quality of Working Life Centre, or in Québec, the Institut de recherché appliquée sur le travail [IRAT], the Institut de recherché en santé et sécurité du travail [IRSST], the Canadian Centre for Research on Information Technology [CCRIT], or the Technology Impact Program of Labour Canada for investigating issues of relevance in the work place, some of those involving technology.[36]

However, the concerns for assessing the significance of technoscientific development and for better managing it socially are perhaps best expressed in two sets of issues which have taken unprecedented importance in recent years: ethical issues, and those pertaining to public controversies.

Ethical issues

The nature and the impact of human action through s&t are such that traditional prescriptions may prove of little use when it comes to deciding whether an act is morally acceptable or not. This is no doubt a major reason why the field of ethics, and specially of 'applied ethics' has developed considerably during the past two decades in Canada (as mentioned earlier in this chapter, applied ethics has been recognised as a priority in the context of the strategic programs of the Social Sciences and Humanities Research Council). In this regard, Canada no doubt has been strongly influenced by the surge of interest for ethical issues in the USA since the late 1960s.[37] In any case, the main areas of interest in applied ethics are the same in Canada as in the USA: bioethics, professional and business ethics, and environmental ethics.

In Canada,[38] it is in bioethics that research and teaching are more developed and better structured. Graduate studies are now emerging, though it is mainly at the undergraduate level that most teaching is conducted. The training of future doctors and nurses in clinical ethics is based on case histories, includes legal and deontological considerations and is given by health care professionals and philosophers/theologians. Continuous education is also available in many hospitals. Other, more theoretical courses are offered in some humanities departments. They are intended to respond to the public demand on questions like access to medical files, the right to decline treatment, to die in dignity, to have a child, *etc.* They also stem from concerns related to the meaning and limits of technological development, and to the implementation of social and legal norms.

The presence of ethicists has become widespread in hospitals. They are often hired as consultants on urgent and delicate clinical cases. Multidisciplinary ethics committees, composed of clinicians and researchers, are also common now.[39]

Research in applied ethics focuses on issues such as abortion, or the reproductive technologies. It is often of an interdisciplinary nature, involving mainly sociologists and lawyers, and is supported by the Social Sciences and Humanities Research Council (SSHRC) and the Medical Research Council (MRC). Several research centres have emerged in recent years, such as the Centre de bioéthique de Montréal, the McGill Centre for Medical Ethics and Law, the Westminster Institute in London, Ontario, and the Medical Bioethics Group in Calgary (Alberta).

Environmental ethics is taught in some humanities departments. Many other units integrate it's notions and concepts in technical courses, to support reflection on topics such as nuclear safety and reliability, care of laboratory animals, fauna management, *etc.* Much of the consulting, and part of the research, are conducted outside the university scene, by all sorts of associations, such as Friends of the Earth, GREEN, *etc.* Their main concern is to inform the public and to encourage participation in public debates. The concept of the environment has become so comprehensive, and the eco-ideology so pervasive, that environmental ethics and bioethics are now beginning to merge into one unified, broad field.

Other fields as well are now adding ethical preoccupations to their traditional, deontological rules. This is rarely the case in the teaching of professional schools in Canada. However, business ethics has began to be taught in some engineering schools. At the graduate level, some programmes have recently emerged where research and teaching attempt to broaden the perspectives and to help professionals reflect on the impact of their work on society. Such programmes are offered by the group ETHOS in Rimouski (Québec), the Manitoba Centre for Professional and Applied Ethics, the University of Waterloo and York University, in Ontario. The transdisciplinary broadening of the perspective on the ethical and social aspects of s&t development indeed seems to be developing. The significance of this can be illustrated by two international conferences, one held at the University of Guelph in 1989 on *Ethical Choices in the Age of Pervasive Technology*[40], and the other one in Québec City in 1990, *Les pratiques de l'évaluation sociale des technologies*[41]

Public debates and their management

Nowhere perhaps does it become as clear that technoscience has become deeply embedded in the daily life of our society than when public controversies erupt. Increasingly, it has become obvious that most controversial issues have scientific and technological dimensions. The environment, particularly since the end of the 1960s, has provided numerous occasions for constructing interfaces between science and the public.

This is reflected in numerous media series or programmes featuring 'nature', pollution problems, ecological catastrophes, but also in the creation of numerous environmental groups which have become active promoters for the diffusion of scientific knowledge, of – admittedly – uneven quality.

Significant social innovations have given prominence and legitimacy to these movements. New types of legislation and regulations now require environmental and even social impact assessment for a broad range of projects, making obligatory public disclosure of information. Most of the time these have a strong technoscientific content, and provide more room than ever before for public participation in the discussions conducive to decision-making.

An example was the creation of the Federal Environmental Assessment Review Office, in 1973, responsible for the administration of the Environmental Assessment and Review Process on behalf of the Minister of the Environment, and 'for producing policy and providing guidelines, procedures and administrative advice to participating agencies and *to the public*' on the operation of that process.[42] Such institutions create new fora for public debate on scientific issues and call for a better diffusion of sound 'popularised' technical knowledge.[43] Similar initiatives at the provincial (such as the Bureau d'audiences publiques in Québec) or even the municipal levels (for instance, the Bureau de consultation de Montréal), are clear indications of a trend: more and more issues having technoscientific dimensions are to be discussed, whatever their territorial significance, by the public having a stake in them.

One of the specificities of these new institutions is that they operate continuously and keep these issues permanently under the public eye. This makes it more likely that more citizens will themselves get a better grasp of what too often before seemed irreducibly arcane, and will take a more active role in their discussion and resolution.

Of course, more traditional institutions such as parliamentary committees, or royal commissions continue to be used. Indeed, at the present time, a Royal Commission on New Reproductive Technologies is generating heated controversy, even though it's final report has not yet even been made public. Another commission is studying proposals for the disposal of nuclear power plant wastes.

The use of nuclear energy, which is regulated in Canada by the Atomic Energy Control Board[44], has not generated in Canada as much controversy as in many European countries, or in the USA, though nuclear power plants are numerous in Ontario.[45] Hydro-electricity, which remains an option for several provinces in Canada however, has been at the centre of much controversy in provinces such as Manitoba and Québec.

On the whole, and contrary to what occurs in the USA, where controversies rage over the assessment of risks related to research in biotechnology or the release of genetically modified organisms, or over pesticides and diverse carcinogenic substances, in Canada debates not directly related to the environmental effects of large projects have been rare and have not taken on national proportions. Issues such as establishing standard thresholds for carcinogens, or lead in the environment, have been either more or less limited to experts and regulators, or have been of a rather local interest.[46]

Because of it's globalism, however, environmentalism with it's concern for ecosystems, and other living beings and health, may well lead in the short term to a higher level of public mobilisation. It would be supported also by increasingly well-organised public interest groups which have been acquiring experience, expertise and financial means through actions conducted during the past ten to fifteen years. Since the drive for public participation in such issues is generally seen as a progress for the political process, this cannot be countered in an authoritarian fashion. It will make ever more important the social learning process concerning the management of public controversies. This indeed may prove to be the most challenging aspect of the increasing presence of technoscience *in* society.

Notes

1 Information Canada 1972 *A Science Policy for Canada, Report of the Senate Special Committee on Science Policy* Vol. 2 p347 Ottawa
2 Information Canada 1972 *A Science Policy for Canada, Report of the Senate Special Committee on Science Policy* Vol. 2 p365 Ottawa
3 Indeed, the theme of 'cultural enrichment' was limited to the support for basic research. See *Ibid* p375–6
4 *Ibid* p379
5 Salter L 1990 Overview of the field and its researchers in Salter L Wolfe D (eds) *Managing Technology. Social Sciences Perspectives* p17–43 Toronto Garamond Press
6 NSERC/SSHRC 1989 *A Joint Programme of Chairs in the Management of Technological Change* Ottawa
7 The 'Programme des Actions structurantes' of the Ministry of Higher Education and Science
8 'Maitrise sociale des technologies'
9 Centre de recherche en evaluation sociale des technologies
10 Centre de recherche en developpement industriel et technologique

11 Group de recherche en informatique et droit
12 Social Sciences and Humanities Research Council of Canada 1990 *SSHRC Fellowships: Guide for Applicants* p18 Ottawa SSHRC
13 The Medical Research Council, the Social Sciences and Humanities Research Council and the Natural Sciences and Engineering Research Council.
14 1991 *SSHRC News* 4(2) p1
15 Wolfe D 1990 Innovation and the labour process in Salter L Wolfe D (eds) *Managing Technology: Social Science Perspectives* p89–111 Ottawa Garamond Press
16 Royal Society of Canada 1989 *Corporate Plan 1989* p21 Ottawa SRC
17 Royal Society of Canada 1990 *Communicating Science: How and Why* Ottawa SRC
18 Ministère d'Etat (Sciences et Technologie)/ The Impact Group 1988 *Une action concertée. Résumé de la strategie de communication concernant la sensibilisation du public aux science et à la technologie* Ottawa Ministère d'Etat (Sciences et Technologie) Canada
19 PAI-Public Affairs International 1989 *Involving the Private Sector in the Promotion of Public Awareness of Science and Technology in Canada* Ottawa PAI-Public Affairs International
20 See Conseil de la science et de la technologie 1988 *Les enjeux: vers une nouvelle culture industrielle* Québec
21 Conseil de Sciences de Canada, 1992 *Prendre les devants*
22 Mitchell A 1992 Sexism in Engineering Decried *The Globe and Mail* April 8 p5
23 ISTC *Women in Science and Engineering* Vol. 1: Universities 1991 Vol 2: Colleges 1992 Vol 3 Workplace 1992
24 Décima 1988 *Etude des attitudes des Québécois et des Québécoises à l'égard de la science et de la technologies et évaluations de l'impact de la publicité* Montréal Decima. *Also 1988 A study of Canadians' attitudes towards s&t and advertising effectiveness for the Ministry of State for Science and Technology* Montréal Decima. Also Corneille S 1989 *Exploration des perceptions et attitudes des jeunes face à la science et aux technologies* Montréal CREATEC.
25 Tremblay V Roy J 1985 *Sondage d'opinion en matiére de science et de technologie* Québec Ministére de l'Ensignement supérieur et de la Science et de la Technologie. Filiatrault J Ducharme J 1990 *Le développement des sciences et de la technologie au Québec: Perceptions de la population* Montréal. Esédel E F *Scientific literacy: A Survey of Adult Canadians.*
26 Conseil de Sciences de Canada *Prendre les devants* p43 Ottawa Science Council of Canada.
27 See for instance ISTC *Science and Technology Public Awareness Campaign, 1989–1990 Communications Plan* Ottawa
28 See for instance ISTC *Science and Technology Public Awareness Campaign, Provincial and Territorial S&T Awareness Activities for 1988–89* Ottawa
29 Royal Society of Canada 1988 *Science and the Public, report on a conference of National Scientific and Engineering Representatives* compiled by Neale Ward E R Ottawa RSC
30 Gouvernement du Québec 1980 *Un projet collectif. Enoncé d'orientations et plan d'action pour la mise en oeuvre d'une politique Québécoise de la recherche scientifique* p21 Québec. *(For the situation at the present time see Ministére de l'Enseignement supérieur et de la Science 1992) Le développment scientifique au Québec* p36–41 Québec Gouvernement du Québec
31 Crelinsten J Impact Group ongoing research
32 See the forthcoming biography by Jean-Marc Carpentier and Danielle Ouellette, Editions Libre Expression Montréal
33 We thank Bernard Schiele, CREST for his comments on this section
34 Hudson K 1987 *Museums of Influence* Cambridge University Press
35 Royal Society of Canada *Science and the Public* p11
36 Salter L *Overview of the field and its researchers* p32
37 Conseil de recherches en sciences humaines du Canada 1989 L'éthique appliquée au Canada' in *Comités d'éthique à travers le monde, recherches en cours 1988* Paris Tierce-Médecine/INSERM
38 For a complete description of this field, see McDonald M Parizeau M H Pullman D

1989 *Towards a Canadian Research Strategy for Applied Ethics. Report to the Canadian Federation for the Humanities to the Social sciences and Humanities Research Council* Ottawa

39 See for instance Groupe de recherche en éthique médicale (Université Laval) 1991 *Les comités d'éthiques au Québec. Guide des ressources* Québec Ministére de la santé et des services sociaux

40 Nef J Vederkop J Wisemen H 1989/1990 *Ethics and Technology. Ethical choices in the age of Pervasive Technology. Proceedings of the conference held at the University of Guelph, October 25–29, 1989* Toronto Thompson

41 Limoges C (ed) 1991 *Les practiques de l'évaluation sociale des technologies* Québec Conseil de la Science at de la Technologie. See for instance in the proceedings of that conference Rolleston F *Public Policies in the Ethics of Research* p101–106

42 Federal Environmental Assessment Review Office 1986 *Environmental Assessment in Canada. Directory of University Teaching and Research 1985–1986* p2 Ottawa Minister of Supply and Services Canada

43 Bourgon J M 1991 Le PEEE, un processus qui questionne l'évolution technologique in Limoges C (ed) 1991 *Les practiques de l'évaluation sociale des technologiques* p143–146 Québec Conseil de la Science et de la Technologie

44 Lévesque R J A 1991 Le rôle de la Commission de contrôle de l'energie atomique du Canada in Limoges C (ed) 1991 *Les practiques de l'évaluation sociale des technologiques* p57–60 Québec de la Science et de la Technologie

45 There have been. however, in the recent past public enquiries relative to the nuclear industry, for instance on the mining of uranium (Cluff Lake in Saskatchewan), on the construction of a nuclear power plant in New Brunswick (Pointe Lepreau), and on the nuclear power plants that might be required to meet the energy demand in Ontario (Porter Commission). The controversies occasioned by these enquiries do not seem to have been long lasting.

46 Salter S 1988 *Mandated Science. Science and Scientists in the Making of Standards* Dordrecht Kluwer Academic Publishers.

Chapter 13

The end of geography: science, technology and the future of Canada

John de la Mothe, Paul Dufour and Richard Lipsey

As this volume opened, it was suggested that both the Canadian mind and the Canadian experience are deeply technological. To a considerable extent, Canadian thinkers as diverse as George Grant, Harold Innis, Marshall McLuhan, Northrop Frye and Margaret Atwood have all helped to frame science, technology and the global society for the modern world. This ability grows out of an historical and physical reality which dictated that, from it's very beginning, Canada would be, by definition, a place of 'otherness': a place that was, by design, culturally different than both the U.S.A and Europe. This was only made possible through the development and deployment of space binding innovations. As a result, the history of Canada has been littered with innovations which tie its citizens east to west and north from the 49th parallel. Only by innovating in such a spatial way could Canadians fulfil a fledgling dream of nationhood.

Thus, the achievement of a transcontinental railway gave meaning to the national motto *ad mari usque ad mare* (a country from sea to sea). Other innovations followed, many of which were both enabling and symbolic. These ranged from the Canadian Broadcasting Corporation, to Anik 1, to the invention of the variable pitch propeller, to Canadian capabilities in remote sensing.

But just as Canada has fundamentally understood space, so too has it's innovative capacity been challenged in recent years. This challenge has come from two fronts: one internal and one external. The internal challenge can be succinctly typified by the observation that;

> *Canadian public opinion has two dangerous propensities...to accept the American myth that with enough science you can reach and sustain an ideal state of society, and to accept the European myth that with enough policy of any kind, you can become as rich and powerful as the United States.*[1]

Clearly, this reveals the flip side of Canada's deeply technological understanding of itself. This inferiority complex, which is not uncommon amongst ex-colonial nations - colours much of its debate over the future course of s&t today and conditions the style of response that Canada will take.

This leads us to the external challenge regarding science, technology and the future of Canada. While Canada has long understood, at an intuitive level, the spatial qualities and potential of technology (and – along with them – communications, trade, finance, knowledge and technology transfer, and so on...) the rest of the industrialised world has rapidly become articulately aware of the competitive fact that trans-national and sub-regional spaces are superseding the politico-economic relevance of national spaces. Canada's technological future thus lies at the end of geography. Caught between it's history and physical space, between it's intuition and a restructuring global space, Canada is coming to realise that it has to mobilise it's understanding of s&t into an aggressive 'propensity to prosper'. The role of science, technology and innovation have become central to this debate.

One thing that makes and conditions Canada unique response to this debate is the fact that while it is a small country in terms of it's 'production' of s&t, it tends to consider itself a big country geo-politically (as well as geographically). As this volume has made clear, Canada *is* a small country in some respects. It's domestic market is confined to just over 27 million potential consumers, versus 270 million in the USA, 500 million in the new Europe, and 123 million in Japan. Sixty five per cent of all it's industrial r&d is performed by 100 firms in six industries, and roughly half of this is performed by foreign multinationals. (Crown corporations represent a significant portion of industrial r&d as well.) Canada's gross expenditures on research and development (GERD) as a percentage of gross domestic product (GDP) have consistently fluctuated around one point three per cent as compared to roughly double that ratio for the USA It's share of research scientists and engineers in the labour force is the lowest of the G-7 nations. It's contribution to the pool of world science (it varies from field to field) is estimated as being about four per cent. And concerns about the 'appropriate' balance of concentration in r&d expenditures has led to the creation of national Networks of Centres of Excellence, as well as funding for such 'big science' projects as the Sudbury Neutrino.Observatory.

However, just as Canada can be thought of as a small nation, so too can it be asserted that Canada falls within the ranks of the advanced, industrialised, countries. But there are several factors in addition to overall economic performance (such as per capita GNP, overall GNP, gross expenditures on r&d per GNP, overall level of health care, *etc.*) which make size an important limiting factor in the technologically-driven competitiveness of nations.

To begin with, small countries such as Canada tend to have small domestic markets. This means that firms in Canada must export to foreign markets in order to gain economies of scale. Increasingly, they also need to justify heavy expenditure on r&d for, as technology increases the sophistication of products and processes, so too do the technological and industrial resources and skills for development become more costly and complex. Furthermore, small countries like Canada tend to have proportionally less money to spend on r&d, they tend to have fewer research personnel *per capita*, and there is a considerable pressure on the researchers from smaller nations to emigrate. In addition, small countries – and this is especially true for Canada – tend to have problems in concentrating their r&d resources. While they recognise the need to fund strategic areas of research, they also need to allocate funding across the spectrum of scientific research for educational purposes. But, in so doing, they tend to fail to define their own technological niche areas – opting, instead, for buying into, or falling behind, the research agendas of larger countries. As a result, and although it may be a dubious exercise to generalise about smaller nations, technologically successful small countries seem to adopt an r&d specialisation strategy. They tend to seek market niches in which larger firms (and even the smaller firms of larger countries) do not bother to compete, while the industries in which they tend to focus are those in which they already have strong multi-technology transnational corporations (such as Bell Northern Research and Alcan International). Thus, the effective rate of growth in a small nation like Canada is highly influenced by the existing levels of economic development (which is high), technological production (which is low), technological sophistication (which is high), the growth of innovative activity (which is low), and the factors influencing the innovative and knowledge processes, as these play a decisive role in the long-run performance of an economy. Indeed, contrary to popular belief, neither the growth of relative unit labour costs nor differences in welfare activity seem to have strong effects on the balance of payments constraint and growth, whereas 'brain power' – the ability to identify, create, adapt, and diffuse technology – clearly do. Exactly what the relationship is between science, technology, institutional change and economic growth is a topic which is now being carefully explored in a major five year study by the Canadian Institute for Advanced Research.

The object of this study, which is being monitored by policy makers, policy analysts and academics alike, is to shed light on the long-run determinants of economic growth. Among the major views which inform the approach of this project are the following:

- technological change is one of the key driving forces determining long run economic growth
- competition is a critical creating force for technological change and innovation. Firms use innovative techniques and activities as their strategic variables
- many decisions relating to r&d and other innovative activities deal with such uncertainties that the paradigm of 'full information and global maximisation' is really not applicable to the firm
- growth needs to be understood in its historical and institutional contexts
- technological change takes place in waves. In the early stages, products, processes and organisations which are based on both old and new paradigms compete with each other while the infrastructure that is required for the efficient operation of the new is not yet fully developed or in place. This has the effect of causing a slowdown in macro growth rates and a widening in the dispersion of incomes followed by periods of 'boom' in later stages when new paradigms reach a more mature stage
- the economic order is undergoing a widespread transition away from the old paradigm of mass-production to flexible, global, information intensive, and lean production. The emphasis is moving towards intangible assets (skills and know-how), intangible investments (such as non-plant and equipment, offshore investments in research and development, networks, *etc.*), both up-and-down stream diversification and value added, and, finally, towards 'invisible' enemies in which competition is generated by non-traditional adversaries
- a major determinant of economic growth is the institutional structure of an economy. Institutions created by government policy, such as limited liability companies, are important, as are the institutional structures of firms and groups of firms
- patterns of comparative advantage are key determinants of growth and are largely created by the activities of firms dealing at the micro-economic level of activity
- the inward and outward flows of foreign direct investment, along with the behaviour and ownership of transnational corporations, have important potential effects on growth rates

This research project, coupled with it's starting premises, have strong implications for Canadian policy makers in that it both underlines the future importance of technology, investment and education/training policy for trade competitiveness and seeks to reveal some of the key relationships which will be of increasing concern.

While this research exercise unfolds, however, the federal government has put into place a so-called 'Prosperity Initiative'. Through a series of 186 'community talks' which have taken place in every region of the country and which have covered an extremely wide array of issues, the Initiative has offered a useful snapshot of the general public's concerns regarding Canada's economic performance. It has also drawn on a wealth of policy analysis that had been conducted by various think tanks (such as the now-defunct Economic Council of Canada, the Science Council of Canada, as well as the C.D. Howe Institute). The Initiative tried to address four main themes: 'competitiveness' (ranging from tax regimes, government regulation, trade and investment to technology, infrastructure, human resources), learning (school system and literacy), social issues (aboriginal issues, population health, poverty) and the environment.

Within the competitiveness element, a 'Task Force on Science, Technology and Related Skills' was established to look into the possibility of offering an action plan for Canadians in this area. It's report, published in the Fall of 1992 by the Conference Board of Canada, offers a host of recommendations on how Canada needs to change course in it's approach to innovation. The Task Force, in less than eight months, put together a report that drew on much of the past work dealing with innovation policy and set a new course for prospects in s&t. Among the key elements of this action plan are a nation-wide education initiative aimed at creating marked improvements in the educational standards achieved by all Canadians, including enhanced teaching in science and mathematics; an enhanced innovation environment; the promotion of technological diffusion; a proposed overhaul of the role and management of government in s&t, particularly through a re-definition of the role of government laboratories in it's linkages to the productive sectors of the economy; and a greater differentiation of the roles and missions of universities and technical/community colleges.

Quite naturally, this Task Force report has complemented and fed into the Initiative's larger report on competitiveness issued by the Steering Group *Inventing the future* which also underscores the need for building an innovative society through, for example, an improved investment climate (involving a review of duty rates, tax incentives for exporters, the replacement of anti-dumping trade remedy rules by a more articulate competition policy and so on); an improved trade performance; enhancement of the technology absorption rates within the private sector; and so on.

These recommendations and other broad-based (consultative) and research-based initiatives have served to reinvigorate the national innovation policy debate in Canada. It is now a question of in what form, and under what conditions, will the responses to recommendations take place.

Clearly, as Canada uncomfortably adjusts to compete at the end of geography, the next Canadian century will have to face the complex reality that knowledge of all kinds will help shape the country as it seeks to establish it's place in the restructured world economy. A new generation of stakeholders will need to be nurtured, trained and educated in the sciences, engineering, medicine, management and mathematics. Business and government leaders will need a stronger grasp of pitfalls present in ignoring s&t (as a cultural force as well as an economic force). Research professionals will need to develop a much more articulate appreciation of the nature, potential and limits of their craft. Science, technology and civic duty will need to be made more deeply ingrained in the fabric of the this, still, very young nation. The challenge is Canada's.

Notes

1 Johnson H G 1971 Address at the Canadian Economics Association annual meeting.

Appendix 1

Selected Bibliography

Adjusting to Win: Report of the Advisory Council on Adjustment. (Ottawa, March 1989).

Association of Universities and Colleges of Canada (1991). *Commission of Inquiry on Canadian University Education, Report (Stuart Smith, Commissioner, Ottawa: Association of Universities and Colleges of Canada).*

Babbitt, J.R. (ed.) (1965). *Science in Canada: Selections from the Speeches of E.W.R. Steacie* (Toronto: University of Toronto Press).

Betcherman, G. and K. McMullen (1986). *Working with Technology: A Survey of Automation in Canada* (Ottawa, Economic Council of Canada).

Bliss, Michael, (1982). *The Evolution of Industrial Policies in Canada: An Historical Survey* (Economic Council of Canada, Discussion paper no. 218).

Boismenu, Gérard et Ducatenzeiler, Graciela (1985). "Le Canada dans la circulation internationale de la technologie", in Duncan Cameron et Francois Houle, *Le Canada et la nouvelle division internationale du travail* (Ottawa: Editions de l'Université d'Ottawa).

Boismenu, Gérard et Ducatenzeiler, Graciela (1984). *Technologie et politique au Canada: Bibliographie 1963–1983* (Montréal: ACFAS).

Bonneau, L.P. and Corry, J.A (1972). *Quest for the Optimum: Research Policy in the Universities of Canada* Association of Universities and Colleges of Canada.

Bothwell, Robert, (1988). *Nucleus: The History of Atomic Energy of Canada Ltd.* (Toronto: University of Toronto Press)

Bourgault, Pierre L. (1972) *[Special Study No. 23]: Innovation and the Structure of Canadian Industry* Science Council of Canada, Information Canada.

Britton, John N.H., and James M. Gilmour (1978). *The Weakest Link: A Technological Perspective on Canadian Industrial Underdevelopment* (Ottawa: Supply and Services of Canada).

Brown, J.J. (1967). *Ideas in Exile: A History of Canadian Invention* (Toronto: McClelland and Stewart).

Canada, Parliament. Standing Committee on Industry, Science and Technology, Regional and Northern Development (1990). *Canada Must Compete* (Ottawa).

Canada, Royal Commission on Government Organization. (1963) Vol. 4 *Special Areas of Administration* (Ottawa: Queen's Printer).

Canada, Royal Commission on the Economic Union and Development Prospects for Canada (1985). *Report Volumes 1–3* (Ottawa: Supply and Services).

Canada, Senate Special Committee on Science Policy (1970). A Science Policy for Canada, Vol. 1; *A Critical Review: Past and Present* 1970, *Vol. 2; Targets and Strategies for the Seventies* 1972, Vol. 3; *A Government Organization for the Seventies* 1973; Vol. 4, *Progress and Unfinished Business* 1977, Information Canada, Ottawa.

Canada, Senate Standing Committee on National Finance (1984). *Federal Government Support for Technology Advancement: An Overview* (Ottawa: Queen's Printer).

Canada, Task Force on Federal Policies and Programs for Technology Development (1984). *A Report to the Honourable Edward Lumley, Minister of State for Sci-*

ence and Technology (Ottawa: Ministry of Supply and Services).

Canadian Chamber of Commerce (1988) *Focus 2000: Report of the Task Force on Technology* Ottawa.

Canadian Institute for Advanced Research (1989). *Innovation and Canada's Prosperity: The Transforming Power of Science, Engineering and Technology* (Toronto).

Canadian Space Agency, (1992) Background Paper on the Canadian Space Program (Montreal).

Clark, T.C. (1980). "Federal Governments and University Science Research: A Comparison of Practices in the United States and Canada, 1970–1979", *Canadian Public Policy*; VI:no.2, 324–351.

Conseil de la Science et de la Technologie (1988). *Les avantages fiscaux associés aux activités de recherche et de dévelopment* (Québec: Gouvernement du Québec).

Conseil de la Science et de la Technologie (1988). *Science et Technologie.*

Conjoncture 1988, (Québec: Gouvernement du Québec).

Conseil de la Science et de la Technologie (1986). *L'organisation de la politique scientifique et technologique* (Québec: Gouvernement du Québec).

Cordell, Arthur J. (1971). *The Multinational Firm, Foreign Direct Investment and Canadian Science Policy* Science Council of Canada Special Study No. 22, (Ottawa: Information Canada).

Council of Science and Technology Ministers of Canada (1987). *The National Science and Technology Policy* (Ottawa)

Crane, David, (1992) *The Next Canadian Century: Building A Competitive Economy* (Toronto: Stoddart Publishers)

Dalpé, R. (1988). "Innovation and Technology Policy in a Small Open Economy: The Canadian Case" in C. Freeman and B. Lundvall (Eds.), *Small Countries Facing the Technological Revolution* (London: Pinter).

Dalpé, R. and Réjean Landry. (1993). *La Politique Technologique au Québec* (Montréal: Presses de l'Université de Montréal).

Davis, Charles H. and J. Alexander (1991). "The Political Incorporation of Innovation Systems: Collective Action and Canadian Science and Technology Policy," *Reseaux, Nos. 56–57-58*

de Bresson, Christian (1988). *Understanding Technological Change* (Montreal: Black Rose Books).

de la Mothe, J. (1989). Special Issue on Science and Technology Policy Under Free Trade, *Technology in Society*, Vol. 11, no.2.

de la Mothe, J. and L.M. Ducharme (1989). " Towards an Innovation Policy Framework", *Futures* April.

de la Mothe, J. and Louis Marc Ducharme (Eds.), (1991). *Science, Technology and Free Trade* (London: Pinter).

de la Mothe, John, and Paul Dufour (1990). "Engineering the Canadian Comparative Advantage: Technology, Trade, and Investment in a Small, Open Economy", *Technology in Society* 12, 369–396.

de la Mothe, John, and Paul Dufour (1991). "The New Geopolitics of Science and Technology", *Technology in Society* 13, 179 - 187.

de la Mothe, John, (1992) "A Dollar Short and a Day Late: A Note on the Demise of the Science Council of Canada," *Queen's Quarterly* (Autumn).

de Vecchi, Vittorio (1984–85) "Science and Scientists in Government, 1878 - 1896," parts 1–2, *Scientia Canadensis* Nos. 27–29.

Doern, G. Bruce (1970). "The National Research Council: the Causes of Goal Displacement", *Canadian Public Administration* v.XIII, no. 2.

Doern, G. Bruce (1970). "Science Policy Making: the Transformation of Power in the Canadian Scientific Community," *Journal of Canadian Studies* V.4. 23–35.

Doern, G. Bruce (1971). "The Senate Report on Science Policy: A Political Assessment", *Journal of Canadian Studies* 6:42–51.

Doern, G. Bruce (1972). *Science & Politics in Canada* (Montreal: McGill - Queen's University Press).

Duchesne, Raymond (1978). *La science et le pouvoir au Québec* (Québec: Editeur officiel).

Duchesne, Raymond (1981). "Historiographie des sciences et des techniques au Canada," *Revue d'histoire de l'amérique francaise* vol.35, 2, 193–215.

Dufour, P. and Y. Gingras (1988). "Development of Canadian Science and Technology Policy", in John de la Mothe (ed.), Science and Technology in Canada (Special issue of) *Science and Public Policy*, Vol. 15, 13–18.

Economic Council of Canada (1983). *The Bottom Line: Trade, Technology and Income Growth* (Ottawa: Supply and Services).

Economic Council of Canada (1987). *Making Technology Work: Innovation and Jobs in Canada* (Ottawa).

Economic Council of Canada (1992). *A Lot to Learn* (Ottawa).

Eggleston, Wilfrid, (1978) *National Research in Canada: The NRC 1916–1966* (Toronto, Clarke, Irwin and Co. Ltd.).

Einsiedel, Edna, (1992). "Framing Science and Technology in the Canadian Press," *Public Understanding of Science* Vol.1, 1, 89–101

Fawcett, Ruth,(1991) "The Intense Neutron Generator and the Development of Science Policy in Canada," *Queen's Quarterly* vol.98, no.2, 409–23

Franklin, Ursula (1990) *The Real World of Technology* (Toronto: CBC)

Garique, Philippe (1972). "Science Policy in Canada", The Private Planning Association of Canada.

Gilmour, James (1978), "Industrialization and Technological Backwardness: The Canadian Dilemma," *Canadian Public Policy* IV: 1, 20–33

Gilpin, R. (1971). *" Science Policy for What?'*, paper presented at the Conference on Science Policy and Political Science, Science Council of Canada (Ottawa).

Gingras, Yves (1991). *Physics and the Rise of Scientific Research in Canada* (Montreal: McGill-Queen's University Press).

Grant, George, (1969). *Technology and Empire: Perspectives on North America* (Toronto: House of Anansi)

Grove, J.W. (1990). *In Defence of Science* (Toronto: University of Toronto Press)

Harris, Richard G. (1985). *Trade, Industrial Policy and International Competition* (Toronto: University of Toronto Press).

Hayes, F. Ronald (1973). The Chaining of Prometheus: Evolution of a Power Structure for Canadian Science (Toronto: University of Toronto Press).

Herzberg, Gerhard (1972). "Bureaucracy and the Republic of Science", *Impact of Science on Society* V.XXII.

Hill, C. (1989). "National Technology Strategies Under Free Trade: Some Implications of the US-Canada Free Trade Agreement", in John de la Mothe (ed.), *Technology in Society* Vol. 11, no.7, 161–180.

Hudson, J. (1989). "Canada as a Small Country: Creating A Science Policy for Competitiveness?", *Insight* National Science Foundation, March.

Hull, J.P. and P.C. Enros (1989). "Demythologizing Canadian Science and Technology: The History of Industrial R&D", *Canadian Studies.* Vol. 24, no.1, 1–22.

Industry, Science and Technology Canada (1988). *National Conference on Technology and Innovation* January 13–15, 1988: Proceedings (Ottawa).

Industry, Science and Technology Canada (1992). *An Overview of Selected Studies on Canada's Prosperity and Competitiveness* (Ottawa)

Innis, Harold, (1972). *Empire and Communications* (Toronto: University of Toronto Press).

Irvine, John, (1989). "Research Foresight in Canada," in John Irvine and Ben R. Martin, *Research Foresight* (London: Pinter)

Jarrell, Richard A. and Yves Gingras (eds.)(1992). *Building Canadian Science: The Role of the National Research Council of Canada* (Ottawa: CSTHA).

Jarrell, Richard A. and Ball, Norman R. (eds.) (1978). *Science, Technology and Canadian History: the first Conference on the Study of the History of Canadian Science and Technology* (Waterloo: Wilfrid Laurier University Press,).

Jarrell, Richard A. (1986). "Government, Technology and Industry: The Emergence of the Canadian Pattern," *Atkinson Review of Canadian Studies* Vol. 2, Spring- Summer, 27–36

Jarrell, Richard A., (1988) *The Cold Light of Dawn: A History of Canadian Astronomy* (Toronto: University of Toronto Press)

· Johnson, H.G. (1965). "The Economics of the 'Brain Drain': The Canadian Case," *Minerva* Spring.

Kenney-Wallace, Geraldine A. and J. Fraser Mustard,(1988) "From Paradox to Paradigm: the Evolution of Science and Technology in Canada", *Daedalus* Fall, 191- 214.

Kroker, Arthur, (1984). *Technology and the Canadian Mind: Innis/McLuhan/Grant* (Montreal: New World Perspectives)

Lacroix, Robert and Fernand Martin, (1988). "Government and the Decentralization of R and D", *Research Policy* 17, 363–373.

Lakoff, Sanford A. (1973). "Science Policy for the 1970's: Canada Debates the Options", *Science* 179:151–3.

Landry, Réjean (1987). "Science politique et politiques technologiques" in Paul Bernard et Edouard Cloutier, réd., *Sciences sociales et transformations technologiques* (Québec: Conseil de la science et de la technologie), 282–325.

Landry, Réjean (1984). *Les priorités de la politique scientifique et technologique de Québec, Rapport, (Qúbec: Conseil de la Science et de la Technologie).*

Le Bourdias, D.M. (1959). *Canada and the Atomic Revolution* (Toronto: McClelland and Stewart Ltd.).

Leiss, William (1971). *The Domination of Nature* (Montreal: McGill-Queen's University Press)

Limoges, Camille (1989). "De la politique des sciences à la politique de l'innovation", in Michel Leclerc, (ed.), *Les enjeux actuels des politiques scientifiques et technologiques* (Montréal: Presses de l'Université du Québec).

McFetridge, Donald (1993). "National Systems of Innovation in Canada, "in Richard Nelson (ed.) *National Systems of Innovation* (New York: Oxford University Press)

McLuhan, Marshall, (1964). *Understanding Media: the Extension of Man* (New York: McGraw-Hill).

Middleton, W.E.K. (1979). *Physics at the National Research Council of Canada 1929–1952,* (Waterloo: Wilfred Laurier Press).

Ministry of State for Science and Technology Canada (1979). *A Rationale for Federal Funding of University Research,* MOSST Background Paper No. 3 (Ottawa).

Ministry of State for Science and Technology (1983). *Towards 1990: Technology Development for Canada* (Ottawa).

Ministry of State for Science and Technology (1983). *Canada Tomorrow Conference* Summary, November 6–9. (Ottawa).

National Advisory Board on Science and Technology (1989). Private Sector Challenge Committee, *Keeping Canada Competitive: The Innovation Imperative*

National Advisory Board on Science and Technology (1990). Big Science Committee *Report on the KAON Project.*

National Advisory Board on Science and Technology, (1991). Committee on the Financing of Industrial Innovation. *Final Report* (Ottawa).

National Advisory Board on Science and Technology (1991). *Science and Technology, Innovation and National Prosperity: The Need for Canada to Change Course.*

National Advisory Board on Science and Technology. Human Resource Development Committee to the Prime Minister of Canada (1991). *Learning to Win: Education, Training and National Prosperity* (Ottawa).

National Biotechnology Advisory Committee (1991). *National Biotechnology Business Strategy: Capturing Competitive Advantage for Canada* (Ottawa: Industry, Science and Technology).

National Forum of Science and Technology Advisory Councils (1989). *The Halifax Declaration: A Call to Action* (Halifax).

Niosi, J. and M. Bergeron (1991). *Technical Alliances in the Canadian Electronics Industry* (Montréal: Centre de recherche en développement industriel et technologique, Université du Québec á Montréal).

Niosi, J. (forthcoming) "Technology in the Canadian Corporation," in D. Glenday et al. (eds.), *Canadian Society* (Toronto: McClellend & Stewart).

Ontario Premier's Council Report (1990). *People and Skills in the New Global Economy* Toronto.

Ontario, Ministry of Industry, Trade and Technology (1988). *A Commitment to Research and Development: An Action Plan* (Toronto: Queen's Printer for Ontario).

Ontario, Ministry of Industry and Trade (1986). *The Technology Challenge: Ontario Faces the Future* (Toronto)

Ontario Premier's Council on Science and Technology (1988). *Competing in the New Global Economy* Vols. 1–3 (Toronto: The Queen's Printer for Ontario).

O.E.C.D. (1974). *The research system; comparative survey of the organisation and financing of fundamental research. Vol. 3 - Canada, USA, General Conclusions* (Paris: Organisation for Economic Co-operation and Development).

O.E.C.D. (1969). Review of National Science Policy: Canada (Paris: OECD).

Organization for Economic Co-operation and Development (1988). *Innovation Policy: Western Provinces of Canada* (Paris: OECD).

Ostry, Sylvia,(1990) *Governments and Corporations in a Shrinking World: Trade and Innovation Policies in the United States, Europe and Japan* (New York: Council on Foreign Relations)

Palda, K.S. and Pazderka, B. (1982). *Approaches to an International Comparison of Canada's R&D Expenditures* (Ottawa: Economic Council of Canada)

Phillipson, Donald J.C.,(1983). "The Steacie Myth and the Institutions of Industrial Research: *HSTC Bulletin* 25, 117–134

Phillipson, Donald J.C. (1991), "The National Research Council of Canada: Its Historiography, Its Chronology, Its Bibliography," in Jarrell, Richard A. and Yves Gingras, Building Canadian Science: The Role of the National Research Council, *Scientia Canadensis* 15, 2

Polanyi, John (1992). "The Scientist as Citizen: Freedom and Responsibility in Science," *Queen's Quarterly* 99, 1, Spring.

Porter, Michael E., and the Monitor Company (1991). *Canada at the Crossroads: The Reality of a New Competitive Environment* (Ottawa: Business Council on National Issues and Minister of Supply and Services).

Prosperity Secretariat (1991). *Learning Well ... Living Well* (Ottawa: Minister of Supply and Services).

Quebec,(1988) *Technology in Quebec: Meeting the Challenge of the Future: Plan of Action, 1988–1992* Background paper, Quebec.

Richardson, Robert et al, (1989) *Technology and Innovation in Canada: The Case for National Action*. Report of the national and regional Conferences on Technology and Innovation, Ottawa.

Rugman, A.M., and J.R. D'Cruz (1991). *Fast Forward: Improving Canada's International Competitiveness* (Toronto: Kodak Canada Inc.).

Rugman, Alan M. and Joseph R. D'Cruz (1992). *New Compacts for Canadian Competitiveness* (Toronto: Kodak Canada Inc.).

Schroeder-Gudehus, Brigitte, "Les Relations Culturelles, Scientifiques et Techniques, " in *From Mackenzie King to Pierre Trudeau: Forty Years of Canadian Diplomacy 1945- 1985* Paul Painchaud (ed.), (Quebec: Presses de l'Universite Laval), 581–608

Science Council of Canada (1971). *Innovation in a Cold Climate: the Dilemma of Canadian Manufacturing* Information Canada.

Science Council of Canada (1979). *University Research in Jeopardy: The Threat of Declining Enrolment* Science Council Report No. 31, (Ottawa: Supply and Services Canada).

Science Council of Canada (1984). *Science for Every Student: Educating Canadians for Tomorrow's World* (Ottawa: Supply and Services).

Science Council of Canada (1988). *Winning in a World Economy: University- Industry Interaction and Economic Renewal in Canada* (Ottawa: Supply and Services, Canada).

Science Council of Canada (1989). *Enabling Technologies: Springboard for a Competitive Future* (Ottawa: Supply and Services, Canada).

Science Council of Canada and Canadian Advanced Technology Association (1990). *Firing Up the Technology Engine: Strategies for Community Economic Development* (Ottawa).

Science Council of Canada, (1991) *Science, Technology and Constitutional Change* (Ottawa: Ministry of Supply and Services) Science Council of Canada (1968). *Towards a Science Policy for Canada* Report No. 4, (Ottawa: Queen's Printer).

Science Council of Canada (1991). *Northern Science For Northern Society: Building Economic Self-Reliance* (Ottawa: Supply and Services, Canada).

Science Council of Canada (1992). *Sectoral Technology Strategy Series Reports:*
Sectoral Technology Strategy Series Reports No. 1 The Canadian Telecommunications Sector, No. 2, The Canadian Automotive-Parts Sector No. 3 The Canadian Iron and Steel Sector, No. 4, The Canadian Automotive-Vehicle Sector, No. 5, The Canadian Oil and Gas Sector, No. 6, The Canadian Petrochemicals and Resins Sector, No. 7, The Canadian Banking Sector, No. 8, The Canadian Nonferrous-Metals Sector, No. 9, The Canadian Forest-Products Sector, No. 10, The Canadian Machinery Sector, No. 11, The Canadian Food and Beverage Sector, No. 12, The Canadian Electric-Power Sector, No. 13, The Canadian Consulting-Engineering Sector, No. 14, The Canadian Electronics Sector, No. 15, The Canadian Computer Software and Services Sector

Science Council of Canada (1992). *Reaching for Tomorrow: Science and Technology Policy in Canada 1991* (Ottawa: Supply and Services Canada).

Sinclair, Bruce, (1979) "Canadian Technology: British Traditions and American Influences", *Technology and Culture* V. 20, 108–123.

Solandt, O.M. (1971). "Science Policy and Canadian Manufacturing Industries", *Nature*: v.234.

Solandt, O.M. (1968). "The Utilization of Scientific and Technical resources in Canada," in *Applied Science and World Economy* A Compilation of Papers Prepared for the Ninth Meeting of the Panel on Science and Technology. U.S. House of Representatives Committee on Science and Astronautics.

Spurgeon, David (1974). "Science Policy Changes in Canada", *Nature*; 248:189–190.

Steed, G. (1982). *Threshold Firms: Backing Canada's Winners* Background study No. 48 (Ottawa: Science Council of Canada).

Steed, G. (1989). *Not a Long Shot: Canadian Industrial Science and Technology Policy* (Ottawa: Science Council of Canada).

Steed, Guy and Scott Tiffin (1986). *A National Consultation on Emerging Technology* (Ottawa: Science Council of Canada).

Tarasofsky, A. (1984). *The Subsidization of Innovation Projects by the Government of Canada* (Ottawa: Supply and Services Canada).

Thistle, M. (1966). *The Inner Ring; The Early History of the National Research Council of Canada* (Toronto: University of Toronto Press).

Tiffin, Scott and Mary Wallis, (1989) "Negative Myth-Making: Canada's Self-Image and Its Implications for Scientific and Technological Development", *Journal of Canadian Studies* V.24(1), Spring, 32–49.

Tory, H.M. (ed) (1939). *A History of Science in Canada* (Toronto: Ryerson Press).

Uffen, Robert J. (1972). "How Science Policy is Made in Canada," *Science Forum* December, 3–8.

Venture Capital in Canada: A Guide and Sources (1989) Industry, Science and Technology Canada and Association of Venture Capital Companies, Ottawa.

Walsh, V. (1987). "Technology, Competitiveness and the Special Problems of Small Countries", in *STI Review* 2, OECD.

Walsh, V. (1988). "Technology and the Competitiveness of Small Countries: A Review", in C. Freeman and B.-A. Lundvall, (Eds.), *Small Countries Facing the Technological Revolution* (London: Pinter).

Wilson, Andrew H. (1974). "Canadian Science Policy: Report Number Four Revisited", *Research Policy* 3:202–215.

Wilson, Andrew H. (1974). "Innovation in a Federal State", *Research Policy* 2:365–379.

Wilson, Andrew H. (1971). *Governments and Innovation* Science Council of Canada. Special Study No. 26 (Ottawa: Information Canada).

Zaslow, Morris (1975). *Reading the Rocks: The Story of the Geological Survey of Canada* (Toronto: MacMillan).

Zeller, Suzanne (1989). *Inventing Canada: Early Victorian Science and the Idea of a Transcontinental Nation* (Toronto: University of Toronto Press).

Statistical Appendix

What follows is a selection of tables from the Services, Science and Technology Division of Statistics Canada. We are indebted to Bert Plaus for providing us with this sampling. Statistics Canada is responsible for maintaining and collecting national S&T statistics for Canada, and covers the broad spectrum of science and technology data for all funders and performers. Its regular publications are available for purchase by mail order at:

Publications Sales
Statistics Canada
Ottawa, Ont.
K1A 0T6

Telephone 1-800-267-6677
Fax: 1-(613)-951-1584

TABLE 1

Gross Domestic Expenditures on R&D (GERD), by Performing Sector and Funding Sector, 1971-1992

TABLEAU 1

Dépenses intérieures brutes en R-D (DIRD), selon le secteur d'exécution et le secteur de financement, 1971-1992

Year / Année	Federal government / Administration fédérale	Provincial government / Administrations provinciales	Business enterprise / Entreprises commerciales	Higher education / Enseignement supérieur	Private non-profit / Organismes privés sans but lucratif	Foreign / Étrangers	Total
	in millions of dollars - en millions de dollars						
Performing sector - Secteur d'exécution:							
1971	368	43	430	436	10	...	1,287
1972	399	50	462	434	12	...	1,357
1973	430	55	503	449	13	...	1,450
1974	485	68	613	485	15	...	1,666
1975	520	72	700	568	16	...	1,876
1976'	565	82	755	624	17	...	2,043
1977'	606	93	857	713	21	...	2,290
1978'	678	98	1,006	769	25	...	2,576
1979'	682	113	1,266	921	27	...	3,009
1980'	733	140	1,571	1,055	30	...	3,529
1981'	859	162	2,124	1,177	36	...	4,358
1982'	1,033	194	2,489	1,373	39	...	5,128
1983'	1,145	201	2,585	1,452	47	...	5,430
1984'	1,303	206	2,994	1,537	57	...	6,097
1985'	1,270	213	3,610	1,641	66	...	6,800
1986'	1,319	217	4,001	1,753	65	...	7,355
1987'	1,292	228	4,307	1,849	67	...	7,743
1988'	1,322	234	4,610	1,998	85	...	8,249
1989'	1,428	269	4,777	2,213	90	...	8,777
1990'	1,547	303	5,105	2,453	103	...	9,511
1991ᵖ	1,593	322	5,184	2,527	111	...	9,737
1992ᵖ	1,682	345	5,265	2,603	120	...	10,015
Funding sector - Secteur de financement:							
1971	574	76	348	226	38	25	1,287
1972	610	91	373	212	42	29	1,357
1973	653	99	409	210	47	32	1,450
1974	721	115	511	231	53	35	1,666
1975	768	126	582	294	57	49	1,876
1976'	827	149	625	332	60	50	2,043
1977'	904	171	702	383	66	64	2,290
1978'	995	189	839	390	88	75	2,576
1979'	1,029	211	1,104	507	76	82	3,009
1980'	1,149	249	1,392	558	80	101	3,529
1981'	1,416	303	1,800	573	97	169	4,358
1982'	1,705	372	1,971	707	95	278	5,128
1983'	1,904	381	1,901	685	114	445	5,430
1984'	2,182	402	2,176	684	123	530	6,097
1985'	2,198	429	2,747	760	136	530	6,800
1986'	2,281	462	3,077	828	143	564	7,355
1987'	2,321	476	3,203	824	172	747	7,743
1988'	2,430	530	3,415	811	209	854	8,249
1989'	2,545	591	3,666	941	204	830	8,777
1990'	2,782	664	3,929	965	239	932	9,511
1991ᵖ	2,855	696	3,994	995	250	947	9,737
1992ᵖ	2,979	729	4,061	1,025	260	961	10,015

TABLE 2.

Estimated Provincial Distribution of the GERD, 1979 to 1990

TABLEAU 2.

Estimations de la répartition provinciale de la DIRD, 1979 à 1990

Year / Année	B.C. C.-B.	Alta. Alb.	Sask.	Man.	Ont.	Que. Qué.	N.B.	N.S. N.-É.	P.E.I. Î.-P.-É.	Nfld. T.-N.	Canada[1]
in millions of dollars - en millions de dollars										
1979'	163	271	66	86	1,335	594	62	74	5	24	2,693
1980'	202	343	81	117	1,584	681	35	77	5	28	3,170
1981'	261	461	85	136	1,985	835	37	84	7	38	3,963
1982'	293	507	119	162	2,386	971	47	100	7	48	4,658
1983'	325	454	124	190	2,567	1,004	43	136	7	70	4,934
1984'	369	493	135	204	2,867	1,225	49	150	10	59	5,579
1985'	478	599	174	200	3,432	1,577	91	159	9	69	6,800
1986'	513	616	176	199	3,872	1,637	83	169	25	61	7,355
1987'	474	560	169	189	4,166	1,852	89	158	13	71	7,743
1988'	518	605	165	207	4,349	1,961	138	200	13	81	8,249
1989'	565	637	191	217	4,590	2,129	146	193	14	88	8,777
1990	684	710	193	249	4,893	2,353	116	208	15	86	9,511
as a percentage of the Canada total - en pourcentage du total canadien										
1979'	6	10	3	3	50	22	2	3	--	1	100
1980'	6	11	3	4	50	22	1	2	--	1	100
1981'	7	12	2	4	50	21	1	2	--	1	100
1982'	6	11	3	4	51	21	1	2	--	1	100
1983'	7	9	3	4	52	20	1	3	--	1	100
1984'	7	9	2	4	51	22	1	3	--	1	100
1985'	7	9	3	3	51	23	1	2	--	1	100
1986'	7	9	2	3	53	22	1	2	--	1	100
1987'	6	7	2	3	54	24	1	2	--	1	100
1988'	6	7	2	3	53	24	2	2	--	1	100
1989'	7	7	2	3	52	24	2	2	--	1	100
1990	7	8	2	3	51	25	1	2	--	1	100
as a percentage of PGDP - en pourcentage du PIBP										
1979'	0.49	0.76	0.63	0.83	1.28	0.91	1.15	1.21	0.64	0.61	0.97
1980'	0.53	0.79	0.65	1.04	1.38	0.94	0.70	1.22	0.59	0.68	1.02
1981'	0.58	0.92	0.59	1.03	1.50	1.02	0.62	1.14	0.69	0.82	1.11
1982'	0.63	0.96	0.81	1.16	1.74	1.13	0.72	1.18	0.67	0.95	1.24
1983'	0.67	0.82	0.81	1.27	1.69	1.09	0.57	1.41	0.60	1.28	1.22
1984'	0.72	0.84	0.82	1.23	1.67	1.21	0.58	1.40	0.77	0.99	1.25
1985'	0.88	0.92	1.00	1.13	1.87	1.46	1.01	1.33	0.68	1.08	1.42
1986'	0.89	1.07	1.03	1.08	1.91	1.39	0.82	1.30	1.67	0.90	1.46
1987'	0.75	0.94	0.98	0.97	1.86	1.43	0.82	1.13	0.82	0.96	1.41
1988'	0.74	0.97	0.91	0.96	1.72	1.38	1.17	1.33	0.73	1.02	1.36
1989'	0.73	0.97	0.98	0.95	1.69	1.42	1.15	1.20	0.74	1.04	1.35
1990	0.85	1.01	0.95	1.05	1.76	1.53	0.87	1.22	0.75	0.98	1.42

[1] Includes the Yukon and Northwest Territories.   [1] Comprends le Yukon et les Territoires du Nord-Ouest.

TABLE 3

Federal Expenditures on Science and Technology, by Region, 1985-86 to 1989-90

TABLEAU 3

Dépenses fédérales au titre des sciences et de la technologie, selon la région, 1985-86 à 1989-90

Region - Région	Year - Année				
	1985-86	1986-87	1987-88	1988-89	1989-90
	millions of dollars - millions de dollars				
Yukon and N.W.T. - Yukon et T.N.-O.	25	24	29	33	29
British Columbia - Colombie-Britannique	286	281	301	301	320
Alberta	170	166	190	202	246
Saskatchewan	82	81	91	88	96
Manitoba	143	151	139	152	157
Ontario	1,825	2,027	2,097	2,250	2,401
Québec	626	659	777	813	866
New Brunswick - Nouveau-Brunswick	80	66	56	80	71
Nova Scotia - Nouvelle Écosse	158	158	183	196	208
Prince Edward Island - Île-du-Prince-Édouard	12	16	16	15	14
Newfoundland - Terre-Neuve	65	66	69	75	90
Canada	**3,472**	**3,695**	**3,948**	**4,205**	**4,498**

TABLE 4.   Total Expenditures of Provincial Governments on Scientific Activities, 1985-86 to 1991-92*

TABLEAU 4   Dépenses totales des administrations provinciales au titre des activités scientifiques, 1985-1986 à 1991-1992*

Province	1985-86	1986-87	1987-88	1988-89	1989-90	1990-91	1991-92*
	in thousands of dollars - en milliers de dollars						
Newfoundland - Terre-Neuve	24,039	24,823	27,674	32,689	33,408'	27,649	27,258
Nova Scotia - Nouvelle-Écosse	..	..	37,683	47,460	48,800	49,128	53,362
New Brunswick - Nouveau-Brunswick	22,088	20,952	22,715	26,422	24,612	30,037	29,957
Québec	..	..	..	430,326	545,377'	565,953	584,398
Ontario	262,365	362,388	358,557	411,745	477,145	521,096	535,423
Manitoba	38,119	44,907	37,149'	34,887'	36,051'	37,594	38,296
Saskatchewan	46,583	53,426	47,141	44,326	58,576	70,981	61,090
Alberta	272,378	333,427	254,173	248,658	247,262	268,163	251,952
British Columbia - Colombie-Britannique	85,614	99,118	116,758	129,977	156,811	193,318	208,687

Note: Social science data collected for New Brunswick as of 1987-88.
Nota: Commençant en 1987-1988, les dépenses des sciences sociales sont inclus pour le Nouveau-Brunswick.

TABLE 5   Total Expenditures of Provincial Governments on R&D, 1985-86 to 1991-92*

TABLEAU 5   Dépenses totales des administrations provinciales au titre de la R-D, 1985-1986 à 1991-1992*

Province	1985-86	1986-87	1987-88	1988-89	1989-90	1990-91	1991-92*
	in thousands of dollars - en milliers de dollars						
Newfoundland - Terre-Neuve	2,013	712	5,969	9,693	8,658	6,767	6,734
Nova Scotia - Nouvelle-Écosse	..	..	6,707'	9,955'	10,271'	11,725	13,412
New Brunswick - Nouveau-Brunswick	10,637	8,607	5,804	5,089	5,938	6,927	7,293
Québec	..	..	..	136,043'	156,930'	212,242	216,958
Ontario	110,242	184,504	176,650	226,950	271,180	294,784	322,482
Manitoba	9,722	7,943	5,266	4,749	6,324	6,476	8,051
Saskatchewan	18,147	24,616	26,023	22,965	29,699	36,359	29,958
Alberta	179,367	251,922	172,696	150,117	148,241	168,318	153,312
British Columbia - Colombie-Britannique	26,287	26,632	33,168	45,520	60,692	68,101	76,870

Note: Social science data collected for New Brunswick as of 1987-88.
Nota: Commençant en 1987-1988, les dépenses des sciences sociales sont inclus pour le Nouveau-Brunswick.

TABLE 6.   Estimated Costs of R&D in the Higher Education Sector, by Province, 1979-1980 to 1990-1991

TABLEAU 6.   Coûts estimatifs des dépenses de R-D effectuées dans le secteur de l'enseignement supérieur, selon la province, 1979-1980 à 1990-1991

Year Année	Province										Canada
	Nfld. T.-N.	P.E.I. Î.-P.-É.	N.S. N.-É.	N.B. N.-B.	Qué. Que.	Ont.	Man.	Sask.	Alta Alb.	B.C. C.-B.	
	millions of dollars - millions de dollars										
1979-80'	16.8	0.8	25.8	13.6	274.8	330.7	41.3	38.8	100.1	77.9	920.4
1980-81'	19.5	0.8	30.8	16.1	304.9	373.7	44.5	48.7	118.1	97.2	1,054.3
1981-82'	21.1	0.9	35.6	12.1	331.7	426.5	52.0	42.6	144.7	109.3	1,176.3
1982-83'	23.9	1.3	35.7	19.4	372.2	497.3	61.7	54.6	187.8	118.8	1,372.7
1983-84'	25.1	1.1	40.2	21.8	384.8	546.5	68.9	57.5	185.6	120.7	1,452.2
1984-85'	25.3	1.2	45.1	22.8	416.2	572.4	72.3	62.7	194.0	124.9	1,537.1
1985-86'	28.3	1.5	49.1	24.6	451.5	610.4	73.9	67.6	207.7	126.3	1,640.8
1986-87'	30.6	5.1	50.7	25.6	466.6	659.5	78.4	71.2	239.3	126.4	1,753.5
1987-88'	35.2	2.7	53.4	28.8	511.4	702.2	80.8	71.4	226.3	136.4	1,848.7
1988-89'	39.2	2.7	59.9	29.2	559.0	766.6	84.6	75.7	225.8	155.5	1,998.2
1989-90'	40.7	3.4	66.3	30.8	629.2	857.7	89.8	91.5	236.7	166.7	2,212.8
1990 91	43.9	3.4	93.4	33.1	690.5	957.8	97.3	87.1	249.8	197.3	2,453.5

TABLE 7.   Estimated Costs of R&D in the Higher Education Sector, by Source of Funds, 1979-1980 to 1990-1991

TABLEAU 7.   Coûts estimatifs des dépenses de R-D effectuées dans le secteur de l'enseignement supérieur, selon la source de financement, 1979-1980 à 1990-1991

Year Année	Federal government Administration fédérale	Provincial governments Administrations provinciales	Business enterprise Entreprises commerciales	Higher education Établissements d'enseigne- ment supérieur	Private non-profit Organismes privés sans but lucratif	Foreign Étranger	Total
			millions of dollars - millions de dollars				
1979-80'	233.8	75.9	36.4	506.9	60.0	7.4	920.4
1980-81'	287.2	96.3	41.3	557.8	63.7	8.0	1,054.3
1981-82'	353.5	114.7	47.5	573.6	78.2	8.9	1,176.3
1982-83'	393.1	141.8	45.4	706.8	75.0	10.5	1,372.7
1983-84'	457.3	153.1	55.7	685.1	89.9	11.1	1,452.2
1984-85'	517.3	168.4	60.4	684.0	96.0	11.0	1,537.1
1985-86'	515.0	178.1	69.9	759.5	110.1	8.3	1,640.8
1986-87'	522.9	206.2	72.5	827.9	112.9	11.1	1,753.5
1987-88'	560.3	217.8	93.8	823.7	141.5	11.6	1,848.7
1988-89'	624.9	261.2	115.1	810.9	172.8	13.2	1,998.2
1989-90'	669.4	285.5	139.7	941.1	165.2	11.8	2,212.8
1990-91	815.0	309.7	155.3	964.1	196.9	12.6	2,453.5

Chart 1.

Research and Development in Canadian Industry,
1983 to 1992

Graphique 1.

La recherche et le développement dans l'industrie canadienne,
1983 à 1992

in millions of $ en millions de $

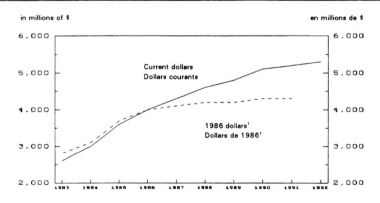

' The deflator for 1983-1991 is the implicit price index of the GDP: 1986 = 100.
' Le déflateur utilisé pour 1983-1991 est l'indice implicite des prix du PIB: 1986 = 100.

Appendix 3

Centres of Excellence

Federal Networks of Centres of Excellence

Centres of Excellence can be valuable information sources for leading edge technologies. The Federal and Ontario Networks are just some of the many organizations that bring together Canada's top researchers in industry, universities and government to focus on fundamental and long term applied research in selected areas.

NRC RESEARCH INSTITUTES

Biotechnology Research Institute National Research Council, 6100 Royalmount Avenue, Montreal, Quebec H4P 2R2 Tel:(514) 496–6100
Industrial Materials Institute National Research Council, 75 De Motagne Boulevard, Montréal, Québec J4B 6Y4 Tel: (514) 641–2280

Institute for Environmental Chemistry

The Institute will increase NRC's commitment to environmentally-related research and developmemt and maximize the impact of this work through extensive collaborations with industrial and government partners. It will focus on ways to help industry reduce the use of hazardous materials and improve methods for their control and destruction. It will also strengthen data bases and help spread information on clean technologies.
Dr Bryan Taylor, National Research Council

Institute for Information Technology

The Institute's program will include work in software engineering, knowledge-based systems, sensor-based automation, and systems integration. These fields account for many of the advances in manufacturing and processing techniques that define the productivity of major Canadian industries. The Institute will work with new research consortia and other partnership arrangements with clients in industry.
Mr. Alex Mayman, National Research Council

Institute for Marine Dynamics National Research Council, Kerwin Place and

Arctic Avenue, Memorial University St. John's, Newfoundland A1B 3T5 Tel: (709) 772–2479

Institute for Microstructural Sciences

The Institute will be a leader in the development of new materials and devices for applications in strategic information technologies. It will collaborate with Canadian industry and provide a critical mass in R&D through directed research projects, joint development of key advanced technologies, and consultation in product development. The program will encompass artificially structured materials, devices and processes, device integration, circuit architecture, advanced networks, and exploratory applications.
Dr Peter Dawson, National Research Council

Institute for National Measurement Standards

This Institute draws together all activities related to metrology to provide a basis for the National Measurement System of Canada.
 Dr Jacques Vanier, National Research Council

Institute for Research in Construction National Research Council Montreal Road Building M20 Ottawa, Ontario K1A 0R6 Tel:(613) 993–2607

Plant Biotechnology Institute National Research Council 110 Gymnasium Road Saskatoon, Saskatchewan S7N 0W9 Tel: (306) 975–5248

Steacie Institute for Molecular Sciences The Institute is devoted to basic research of very high quality in the area of molecular science. Research programs attract visiting researchers and graduate students from universities as well as Canadian and foreign research organizations and contributes scientific knowledge to support and reinforce other programs at NRC.
Dr Marek Laubitz, National Research Council

Alberta
CANADIAN NETWORK FOR SPACE RESEARCH

The network is concerned with increasing the understanding of how the hostile plasma of space affects satellites and other structures. It will use remote sensing to look at processes that take place in the middle atmosphere and in the polar environment. These findings may be critical to developing knowledge on global climate change and polar ozone depletion. The more applied aspects of the program will be directed at advances in spacecraft with the Canadian Space Agency, participation in international space science missions and close interaction with the industrial partners on long-term fundamental studies. Participants include the Atmospheric Environment Service, Toronto; Canadian Astronautics Ltd., Ottawa; Com Dev Ltd., Cambridge; the Institute for Space and Terrestrial Science, Toronto; Itres Research, Calgary; Myrias Corporation, Edmonton; the National Research Council; SED Systems Inc., Saskatoon, the Universities of Alberta, Calgary, Saskatchewan, Western Ontario and York.
 Dr Leroy Cogger, Canadian Network for Space Research Department of Physics, 2500 University Drive N.W., University of Calgary, Calgary, Alberta T2N 1N4 Tel: (403) 220–5386

British Columbia
BACTERIAL DISEASES: MOLECULAR STRATEGIES FOR THE STUDY AND CONTROL OF BACTERIAL PATHOGENS OF HUMANS, ANIMALS, FISH AND PLANTS

The network focuses on bacterial attack and host response in different biological systems. The causative agents of human diseases as well as those affecting the agricultural and aquaculture industries will be studied. Participants include Chembiomed, Edmonton; Connaught Labs, Toronto; International Broodstock Technologies, Vancouver; Microteck R&D Ltd., British Columbia; National Research Council; the Universities of Alberta, British Columbia, Calgary, Guelph, Victoria, Laval and VIDO Saskatoon.

Dr Robert Hancock, Bacterial Diseases: Molecular Strategies for the Study and Control of Bacterial Pathogens of Humans, Animals, Fish and Plants, Department of Microbiology, University of British Columbia, Vancouver, British Columbia, V6T 1W5 Tel: (604) 228–3489

CENTRE OF EXCELLENCE FOR MOLECULAR AND INTERFACIAL DYNAMICS

The network will focus on reaction dynamics, which is a key to understanding such important processes as ozone depletion and atmospheric pollution. It will also look at the properties of surfaces and promises to be of great industrial relevance for the fabrication of new materials. Participants include the National Research Council and the Universities of British Columbia, Guelph, Montreal, New Brunswick, Ottawa, Saskatchewan, Sherbrooke, Toronto, Waterloo, Western Ontario, as well as Carleton, Dalhousie, Laval, McMaster and Queen's.

Dr Terry Gough, Centre of Excellence for Molecular and Interfacial Dynamics, Department of Chemistry, P.O. Box 3055 University of Victoria, Victoria, British Columbia V8W 3P6 Tel: (604) 721–7150

GENETIC BASIS OF HUMAN DISEASE: INNOVATIONS FOR HEALTH CARE

The network will study the genes that directly cause or predispose humans to disease. The goal is to determine the biological function of each of the relevant genes and to discover how mutation in each causes disease. This research could lead to major commercial opportunities for Canada in the areas of DNA diagnostics and therapeutics. Participants include the Biomedical Research Centre, Vancouver; the Clinical Research Institute, Montreal; the Universities of British Columbia, Calgary, Manitoba, Montreal, Ottawa, Quebec, Toronto, and Laval, McGill and Queen's.

Dr Michael Hayden, Genetic Basis of Human Disease: Innovations for Health Care, Department of Medical Genetics, University of British Columbia, Vancouver, British Columbia V6T 2B5 Tel: (604) 228–7738

PROTEIN ENGINEERING: 3D STRUCTURE, FUNCTION AND DESIGN

Protein engineering uses a variety of techniques to understand the functioning of proteins, and then to improve them by making systematic changes to their building block structure. Improved proteins can be of enormous benefit in the treatment of infectious diseases, and are also used in the food industry, the development of disease resistant crops and other industrial products such as converting agricultural and forest wastes into high-grade chemicals. Close industrial ties give the network an exellent potential for technology transfer and spin-offs. Participants include the Biomedical Research Centre, Vancouver; the National Research Council; and the Universities of Alberta, British Columbia and Toronto.

Dr Michael Smith, Protein Engineering: 3D Structure, Function and Design, Department of Biochemistry, University of British Columbia, Vancouver, British Columbia V6T 1W5 Tel: (604) 228–4838

Ontario
BIOTECHNOLOGY FOR INSECT PEST MANAGEMENT

The network is concerned with developing environmentally acceptable methods of pest control. Participants include Agriculture Canada, Forestry Canada, the National Research Council, the Universities of British Columbia, Calgary, New Brunswick, Ottawa, Toronto, Waterloo, Western Ontario, as well as Laval, Queen's and York.

Dr Gerard Wyatt, Biotechnology for Insect Pest Management, Department of Biology, Queen's University, Kingston, Ontario K7L 3N6 Tel: (613) 545–6120

INSTITUTE FOR ROBOTICS AND INTELLIGENT SYSTEMS (IRIS)

The network will operate as a separate component of Precarn Associates Inc., a consortium of 32 companies whose mission is to carry out advanced R&D in robotics and artificial intelligence. Participants include École Polytechnique, INRS Telecommunications, Technical University of Nova Scotia, PRECARN Associates Inc., the Universities of Alberta, British Columbia, Guelph, Montreal, Saskatchewan, Toronto, Victoria, Waterloo, Western Ontario, as well as Concordia, Laval, McGill, McMaster, Queen's, Simon Fraser and York.

Mr. Gordon MacNabb, Institute for Robotics and Intelligent Systems (IRIS), Precarn Associates Inc., 30 Colonnade Road, Nepean, Ontario K2E 7J6 Tel: (613) 727–9576 Fax: (613) 727–5672

MICROELECTRONIC DEVICES, CIRCUITS AND SYSTEMS FOR ULTRA LARGE SCALE INTEGRATION (ULSI)

This technology is expected to become the mainstay of the next generation of telecommunications and computer systems and an area vital to the future of Canadian industry. The ULSI network ties together efforts in devices, circuits and systems, a truly coordinated and vertically integrated approach, with strong industrial participation. Participants include INRS-Energy, and the Universities of Calgary, Manitoba, Toronto, Victoria, Waterloo, Windsor, Carleton and McGill.

Dr André Salama, Microelectric Devices, Circuits and Systems for Ultra Large Scale Integration (ULSI), Department of Electrical Engineering, University of Toronto, Toronto, Ontario M5N 2L2 Tel: (416) 978–8658

Québec
HIGH PERFORMANCE CONCRETE

The network links materials experts, designers and practitioners from seven universities and two construction firms. Researchers will investigate the whole concrete-making process from colloidal phenomena in fluid concrete through to problems in the design of large structures. Participants include Bickley and Associates, Toronto; Hardy BBT, Vancouver; the Universities of Alberta, British Columbia, Ottawa, Sherbrook, Toronto, and Laval, and McGill. M. Pierre-Claude Aitcin, High Performance Concrete, Deparment of Civil Engineering, University of Sherbrooke, Sherbrooke, Québec J1K 2R1 Tel: (819) 821–7117

INSTITUTE FOR TELECOMMUNICATIONS RESEARCH

The network will focus on broad band and wireless communications, two rapidly growing areas that present the most important emerging markets for telecommunicatons over the next decade. Participants include Alberta Telecommunications Research Centre, INRS Telecommunications, the Universities of British Columbia, Montreal, Ottawa, Toronto, Victoria, Waterloo as well as Carleton, Concordia, Laval, McGill, McMaster and Queen's.

Dr Maier Blostein, Institute for Telecommunications Research, Department of Electrical Engineering, McGill University, Montréal, Québec H3A 2A7 Tel: (514) 398–7116

NEURAL REGENERATION AND FUNCTIONAL RECOVERY

The object of the research is to promote nervous system regeneration and recovery of functions lost as a result of trauma or disease. A major reason for the permanent disabilities caused by brain and spinal cord injuries or by neurological disorders such as Alzheimer's and Huntington's disease is that damaged cells are not replaced nor do they restore connections with their natural targets. Major advances have been made recently in promoting the regrowth of the nervous system after injury. The field is expected to undergo further advances with the application of new techniques in molecular biology and genetic engineering. Participants include the Ludwig Institute, Montreal; the National Research Council, the Universities of Alberta, British Columbia, Calgary, Lethbridge, Manitoba, Montreal, Ottawa, Saskatchewan, Toronto as well as Carleton, Concordia, Dalhousie, Laval, McGill and Queen's.

Dr Albert Aguayo, Neural Regeneration and Functional Recovery, Department of Neurology, McGill University, Montréal, Québec H3G 1A6 OR M. Yves Lamarre, Department of Physiology, University of Montréal, Montréal, Québec H3C 3J7 Tel: (514) 934–8060

OCEAN PRODUCTION ENHANCEMENT NETWORK (OPEN)

This is a network of university, industry and government scientists created to contribute to the economic health of Canada's fisheries. The initial focus of the program will be on two species of great commercial value - the sea scallop and the Atlantic cod. The scientists will investigate the processes which control the survival, growth, reproduction and distribution of fish and shellfish. The Network has the support of three of Canada's largest seafood companies, National Sea Products Ltd., Clearwater Fine Foods and Fishery Products International. Participants include Fisheries and Oceans Canda, the University of British Columbia, Dalhousie, Laval, McGill, and Memorial.

Dr William Leggett, Ocean Production Enhancement Network (OPEN), Department of Biology, McGill University, Montréal, Québec H3A 1B1 Tel: (514) 398–4211

RESPIRATORY HEALTH NETWORK OF CENTRES OF EXCELLENCE

The network is concerned with such things as improved patient care, reduced health care costs, new drugs to overcome airway blockage in cystic fibrosis and asthma, the development of new apparatus such as improved mechanical lung ventilators and lung testing kits, occupational safety and improvements to building ventilation and purification systems. Participants include Employment and Immigration Canada, Engineering Interface, North York; Graham and Bierman, Montreal; Merck Frosst, Pointe Claire; Trudeau Medical, London; the Universities of British Columbia, Calgary, Manitoba, Montreal, Saskatchewan, Laval, McGill, and McMaster.

Dr Peter Macklem, Respiratory Health Network of Centres of Excellence, Department of Medicine, McGill University, Montréal, Québec H2X 3J9 Tel: (514) 849–5201

SCIENCE AND ENGINEERING FOR HIGH-VALUE PAPERS FROM MECHANICAL WOOD-PULPS

The network's goal is to develop the mechanical pulping process so that it produces superior grade paper that will not yellow. The process relies on mechanical breakdown of wood fibres as opposed to chemical separation (Kraft process). It accepts a much wider range of common northern species, allows Canada to take advantage of its inexpensive hydro power, is less wasteful of trees and has lower start up costs. Participants include the National Research Council, the Pulp and Paper Research Institute of Canada, the Universities of British Columbia, Ottawa, Quebec, Toronto, Western Ontario, as well as Lakehead, McGill, McMaster, Mount Allison and Queen's.

Dr Henry Bolker, Science and Engineering for High-Value Papers From Mechanical Wood-Pulps, Pulp and Paper Institute of Canada, Montréal, Québec 570 St. John's Boulevard, Pointe-Claire, Québec H9R 3J9 Tel: (514) 630–4104

Appendix 4

Government Information and/or Funding Sources

Regional Financial Assistance Programs and Services Related to Technology Transfer

British Columbia
SCIENCE AND TECHNOLOGY DEVELOPMENT FUND AND ASSISTANCE GRANTS FOR APPLIED RESEARCH (STDF - AGAR)

Program is designed to stimulate the development of s&t applicable to economic development. Applied research and technology development projects are considered in various economic sectors. Eligible technologies include advanced materials and robotics.

Science and Technology Development Fund and Assistance Grants for Applied Research (STDF-AGAR) British Columbia Tel: (604) 438–2752 Fax: (604) 438–6564

TECHNOLOGY ASSISTANCE PROGRAM (TAP)

Technology Assistance Program is sponsored by the Ministry of Advanced Education, Training and Technology. It provides financial support for firms with up to 250 employees undertaking R&D projects with BC Research.

Technology Assistance Program (TAP), 3650 Wesbrook Mall, Vancouver, British Columbia V6S 2L2 Tel: (604) 224–4331 Fax: (604) 224–0540

Alberta
CALGARY RESEARCH AND DEVELOPMENT AUTHORITY

Provides facilities and programs which attract and facilitate the growth of research activities and research based advanced technology. Calgary Research and Development Authority, 100 3553–31st Street N.W. Calgary, Alberta T2L 2K7 Tel: (403) 282–0464, Fax: (403) 282–1238

GOVERNMENT OF ALBERTA – TECHNOLOGY, RESEARCH AND TELECOMMUNICATIONS

The Technology Transfer Division of Technology, Research and Telecommunications assists Alberta's advanced technology industry with three programs. The Inventor Assistance Program assists individual inventors in evaluationg, producing and marketing their inventions. Services available to inventors include concept development and testing, patent search and registration, product licensing and marketing. The Technology Commercialization Program provides financial suppport for projects which demonstrate and assist innovation in technology intensive fields. The First Purchase Program provides an approved means for the first purchase of advance technology products and processes by provincial government departments, crown corporations, agencies and boards.

Dr Ade Ajao, Director Technology Transfer Technology, Research and Telecommunications, Government of Alberta, 11th Floor, Pacific Plaza, 10909 Jasper Avenue, Edmonton, Alberta T5J 3L9 Tel: (403) 422–0567 Fax: (403) 422–2003 Tlx: 037–42687

THE CAPITAL EQUIPMENT PROGRAM

Saskatchewan government will provide up to 50% of cost specified high technology equipment required to perform specific research contracts initiated by industry.

The Capital Equipment Program, Industry Science and Technology Canada (ISTC), 105 - 21st Street East, 6th Floor, Saskatoon, Saskatchewan S7K 0B3 Tel: (306) 975–4400

INDUSTRIAL RESEARCH ASSISTANCE PROGRAM (IRAP)

Assistance may be provided to encourage research, development and new product activities by Saskatchewan firms. A complement to the NRC's IRAP program, the Industrial Research Program augments funding from other sources.

Industrial Research Program, Industry, Science and Technology Canada (ISTC), 105 - 21st Street East, 6th Floor, Saskatoon, Saskatchewan S7K 0B3 Tel: (306) 975–4400

Manitoba
MANUFACTURING ADAPTATION PROGRAM

Provides assistance, loans and technical advice for studies on the introduction of new process technology and advanced technology equipment respectively.

Manufacturing Adaptation Program Industry, Trade and Tourism, 5th Floor, 155 Carlton Street, Winnipeg, Manitoba R3C 3H8 Tel: (204) 945–2472 Fax: (204) 945–1193

West
WESTERN ECONOMIC DIVERSIFICATION CANADA

Program promotes the diversification of the Western Canadian economy by assisting the development of new products to Western Canada, new markets for Canadian goods and services, new technologies, import substitution projects and industry-wide productivity improvements. The government provides a path finding service to help entrepreneurs identify other sources of government assistance for their projects.

Western Economic Diversification Canada, Western Diversification Program, 1200–1055 Dunsmuir Street, Vancouver, British Columbia V7X 1L3

Regional Addresses: Box 777, 712–240 Graham Avenue, Winnipeg, Manitoba R3C 2L4

Suite 601, S.J. Cohen Building, 119–4th Avenue South, Saskatoon, Saskatchewan S7K 3S7

Suite 1500, Canada Place, 9700 Jasper Avenue, Edmonton, Alberta T5J 4H7 Tel: (906) 666–6256

Ontario
FEDNOR: CORE INDUSTRIAL PROGRAM

Programs which assist new and existing businesses in Northern Ontario. Has technology components.

FEDNOR: Core Industrial Program Industry, Science and Technology Canada (ISTC), 473–477 Queens Street East, Sault Ste. Marie, Ontario P6A 1Z5 Tel: 1–800–461–6021

Québec
AEROSPACE INDUSTRY PILOT PROJECT

A consultant has been hired to assist Quebec machine shops engaged in aerospace subcontracting, to enhance their competitiveness through diagnostic review and identification of appropriate technology.

Aerospace Industry Pilot Project Industry, Science and Technology Canada (ISTC), Québec Regional Office, Tour de la Bourse, 800 Place Victoria, Suite 3800, P.O. Box 247, Montréal, Québec H4Z 1E8 Tel: (514) 283–8185 Fax: (514) 283–3315

ENTERPRISE DEVELOPMENT PROGRAM

Encourages the establishment, expansion and modernization of enterprises in Quebec and has technology components.

Enterprise Development Program, Industry, Science and Technology Canada (ISTC), Québec Regional Office, Tour de la Bourse, 800 Place Victoria, Suite 3800, P.O. Box 247, Montréal, Québec H4Z 1E8 Tel: (514) 283–8185 Fax: (514) 283–3315

MANUFACTURING PRODUCTIVITY IMPROVEMENT PROGRAM

Provides assistance for consultants as well as for the purchase of production equipment involved with the acquisition of advanced technologies by Quebec manufacturing and processing firms located in urban areas.

Manufacturing Productivity Improvement Program Industry, Science and Technology Canada (ISTC), Québec Regional Office, Tour de la Bourse, 800 Place Victoria, Suite 3800, P.O. Box 247, Montréal, Québec H4Z 1E8 Tel: (514) 283–8185 Fax: (514) 283–3315

Maritimes
ATLANTIC OPPORTUNITIES AGENCY (ACOA)

There are several sub-programs administered by ACOA: Innovation Assistance, Capital Investment Assistance, Loan Insurance, Business Studies and Business Support.

Atlantic Canada Opportunities Agency (ACOA) Head Office Blue Cross Centre 644 Main Street Moncton, New Brunswick E1C 9J8

Nova Scotia Suite 3000, ''The Brewery'' 1489 Hollis Street Halifax Nova Scotia B3J 3M5

Enterprise Cape Breton 15 Dorchester Street Sydney, Nova Scotia B1P 6K7

Prince Edward Island 3rd Floor 75 Fitzroy Street Charlottetown, Prince Edward Island A1C 5M5

New Brunswick 590 Brunswick Street Fredericton New Brunswick E3B 5A6 Toll Free: 1–800-561–7862 Tel: 1–800-565–1228 Tel: 1–800-565–0228 Tel: 1–800-563–5767 Tel: 1–800-561–4030

Appendix 5

Selected list of Technology and Research Centres and Programmes in Canada: Public and Private

(We are grateful to Elizabeth Payne, James Kelly and Jan Caroleo of Industry, Science and Technology Canada for giving us permission to reproduce certain portions of the 1992 Technology Networking Guide Canada published by the Services to Business Branch. The following is by no means a comprehensive listing of federal, provincial, territorial and municipal infrastructure support for S&T in Canada. It serves merely to illustrate the broad range and scope of Canada's S&T system).

Specialized Technology Centres

These are Canadian non-private sector centres, which enter joint ventures or collaborative relationships with technology recipients. The technology which is transferred to the recipient need not be technology which has been created by the centre. This category also includes Canadian public sector organizations which offer significant programs or support in the form of seminars and workshops or training in skills relating to the business of technology transfer.

ALBERTA AND BRITISH COLUMBIA

ALBERTA MICROELECTRONIC CENTRE

The Alberta Microelectronic Centre help companies identify and adapt microelectronic applications to their products and operations by offering consulting and feasibility studies, microelectronic prototype development, technology assessment and ASIC design and fabrication.

Alberta Microelectronic Centre, # 200 - 1620 29th Street N.W., Calgary, Alberta T2N 4L7

OR Alberta Microelectronic Centre, # 118 - 11315 87th Avenue, Edmonton, Alberta T6G 2C2 Tel: (403) 289–2043 Fax: (403) 289–2047 Tel: (403) 432–3914 Fax: (403) 432–0626

ALBERTA RESEARCH COUNCIL

Major R&D establishment serving a wide range of firms in many areas of technology. Established in 1921, the Research Council's stated mission is to advance the economy of the Province of Alberta by promoting technology development, performing applied research and providing expert advice, technical information and scientific infrastructure that is responsive to the needs of the private sector and supports the activities of the public sector.

Alberta Research Council, 250 Karl Clark Road, P.O. Box 8330, Station ''F'' Edmonton, Alberta T6H 5X2 OR Alberta Research Council, 3rd Floor, Digital Building, 6815 Eighth Street N.E., Calgary, Alberta T2E 1H1 Tel: (403) 450–5200 Fax: (403) 461–2651 Tel: (403) 297–2600 Fax:(403) 275–3003

ALBERTA TELECOMMUNICATIONS RESEARCH CENTRE

The Alberta Telecommunications Research Centre links industry, university and government in a joint research venture for applied telecommunications. They use their own facilities and those of industry members and draw upon the research expertise of university faculty, graduate students and industry partners to develop market-driven telecommunications technology.

Alberta Telecommunications Research Centre, # 200 - 4245 97th Street, Edmonton, Alberta T6E 5Y4 Tel: (403) 461–3830 Fax: (403) 463–3010

CANADIAN ENERGY RESEARCH INSTITUTE (CERI)

An non-profit research organization, CERI is funded by Canadian and foreign governments and corporations involved in, or with, the energy industry to undertake studies in energy economics and policy and related subjects. The research program is determined by a board which represents the sponsors. All research reports are published. Energy models developed by the Institute are made available to the sponsors. CERI holds national and international conferences in energy topics.

Canadian Energy Research Institute, 3512 - 33 Street N.W., Calgary, Alberta T2L 2A6 Tel: (403) 282–1231 Fax: (403) 284–4181 Tlx: 03821545

CENTRE FOR FRONTIER ENGINEERING RESEARCH (C-FER)

C-FER conducts research related to the materials, designs and construction of energy resource development facilities operating in severe weather conditions, such as in the Arctic or off-shore. C-FER is funded by industry partners and government and offers industry members the facilities and staff to perform collaborative research in the sponsor's area of interest.

C-FER Building, University of Alberta, Edmonton, Alberta T6G 2H2 Tel: (403) 492–5108 Fax: (403) 433–5634

EDMONTON RESEARCH PARK

The Edmonton Research Park was established in 1980 to assist in the diversification of the Edmonton economy through advanced technology. The Park is managed by an independent Authority whose board of directors represents the major stakeholders in the community - the University of Alberta, the government of Edmonton (City) and Alberta (Province) and the private advanced technology sector. The Park is home to more than 900 scientists, engineers, technicians and support staff. While a variety of research-based technologies are found at the Park, the major areas of growth have been in medical biotechnology, telecommunications and electronics, petroleum research and frontier engineering. The Edmonton Research Park is a member of the TriMAC (Triad Market Access Consortium), an international network whose collective goal is to promote business transactions through mutual marketing and technology transfer.

Richard H. Hermes, Manager Advanced Technology Centre, Edmonton Research Park, Advanced Technology Centre, Suite 203 9650 - 20 Avenue, Edmonton, Alberta T6N 1G1 Tel: (403) 462–2121 Fax: (403) 428–5376

ELECTRONICS TEST CENTRE (ETC)

The ETC tests and evaluates electronics products including those used in telecommunications, medical electronics, data processing, office automation and process instrumentation. Product design, performance and certification testing is performed to national and international standards. Engineering assistance focuses on design analysis and test methodology. ETC is administered by the Alberta Research Council.

Electronics Test Centre (ETC), P.O. Box 8330, Postal Station "F", Edmonton, Alberta T6H 5X2 Tel: (403) 450–5370 Fax: (403) 461–2651

THE LASER INSTITUTE

A wholly owned subsidiary of the University of Alberta, the Laser Institute is a state-of-the-art facility providing a wide range of laser technology services. These include engineering and feasibility studies, process and procedure development, system specification and design, prototype and production runs, training and education, precision measurement using lasers, sensor development, medical applications, and contract research.

The Laser Institute, 9924–45 Avenue, Edmonton, Alberta T6E 5J1 OR #144, 6815 - 8th Street, N.E. Calgary, Alberta T2E 7H7

DISCOVERY INNOVATION CENTRE

Discovery Innovation Centre, Discovery Foundation, P.O. Box 24, 770 Pacific Blvd., South Vancouver, British Columbia V6B 5E7 Tel:(604) 684–0647

RESOURCE INDUSTRIES TECHNOLOGY CENTRE

Resource Industries Technology Centre, 1558 Quinn Street, Prince George, British Columbia V2N 1X3 Tel: (604) 564–2662

MANITOBA

ATOMIC ENERGY OF CANADA LIMITED – WHITESHELL LABORATORIES

Many facilities at Whiteshell Laboratories are made available to external users. The laboratories offer commercial R&D services based on AECL's unique multidisciplinary problem-solving capabilities. Access to AECL Research technologies is also available via cooperative development projects, the sale or licensing of technologies and products, and technology transfer agreements.

Whiteshell Laboratories, AECL Research, Atomic Energy of Canada Pinawa, Manitoba R0E 1L0 Tel: (204) 753–2311 Fax: (204) 753–8404

THE BIOMASS ENERGY INSTITUTE INC

The Biomass Energy Institute Inc., 1329 Niakwa Road East, Winnipeg, Manitoba R2J 3T4 Tel: (204) 257–3891

CANADIAN FOOD PRODUCTS DEVELOPMENT CENTRE

Canadian Food Products Development Centre, 810 Phillips Street, P.O. Box 1240, Portage la Prairie, Manitoba R1N 3J9 Tel: (204) 857–7861

MANITOBA RESEARCH COUNCIL (MRC)

MRC provides information and advisory, consulting, product development and testing services in food technology from its facilities in Portage la Prairie, Manitoba, and in environmental chemistry, mechanical engineering and electronics engineering from its Winnipeg facilities. MRC also has cooperative initiatives with the University of Manitoba in the fields of biotechnology and advanced industrial materials, and with the National Research Council for the delivery of its Industry Research Assistance Program.

Manitoba Research Council (MRC), 1329 Niakwa Road East, Winnipeg, Manitoba R2J 3T4 OR Manitoba Research Council (MRC), 810 Phillips Street, Portage la Prairie, Manitoba R1N 3J9

Tel: (204) 945–6000 Tel: (204) 857–7861

ONTARIO

FORINTEK CANADA CORPORATION

A wood products research and development institute, FCC is supported by government in Canada, the Canadian wood products industry and contracts. Alliances are used in transferring technology (both developed in-house and elsewhere) to industry. Networking is key to program delivery. Technical information is communicated to members through technical reports, newsletters and seminars. Head Office and Western Laboratory: Forintek Canada Corporation 6620 N.W. Marine Drive Vancouver, British Columbia V6T 1X2 and Eastern Laboratory: Forintek Canada Corporation 800 Montreal Road Ottawa, Ontario K1G 3Z5 Tel: (604) 224–3221 Fax: (604) 224–3229 Tlx: 04–508552 Tel: (613) 744–0963 Fax: (613) 744–0903 Tlx: 053–3606

MANUFACTURING TECHNOLOGY CENTRE NEW BRUNSWICK

An IRAP Technology Centre P.O. Box 4400 Fredericton, New Brunswick E3B 5A3 Tel: (506) 453–4513

INNOVATION CENTRE

The Centre helps bring the technologies of New Brunswick entrepreneurs, innovators and inventors to market. Its services include: inventorying and evaluating inventions and new concepts, facilitating potential entrepreneurs' access to business planning and to specialized marketing knowledge, identifying sources of financing and establishing contacts with these sources. The centre also publishes a newsletter entitled TEMPS INNOVATEURS three times a year.
N.B. Innovation Centre 124 Saint John Street Fredericton, New Brunswick E3B 4A7 Tel: (506) 453–7174 Fax: (506) 453–7956

NEW BRUNSWICK RESEARCH AND PRODUCTIVITY COUNCIL

A major R&D establishment serving a wide range of firms in many areas of technology. Numerous technical seminars are sponsored.

New Brunswick Research and Productivity Council, P.O. Box 6000, College Hill Road, Fredericton, New Brunswick E3B 5H1 Tel: (506) 452–8994 Tlx: 014–46115 Nfld.

N.S. ADVANCED MATERIALS ENGINEERING CENTRE (AMEC)

Provides consulting services, R&D and training with respect to engineering techniques involved in the production and fabrication of advanced materials. Advanced Materials Engineering Centre (AMEC), Business Relations, P.O. Box 1618, Station "M", Halifax, Nova Scotia B3J 2Y3 Tel: (902) 425–4500 Fax: (902) 422–7907

APPLIED MICROELECTRONICS INSTITUTE

An IRAP Technology Centre, 1127 Barrington Street, Halifax, Nova Scotia B3H 2P8 Tel: (902) 421–1250

ATLANTIC INSTITUTE OF BIOTECHNOLOGY

An organization that provides project development, project management, commercial intelligence and financial assistance to private sector firms and associations. Projects may be research and development, technology transfer or both. Atlantic Institute of Biotechnology, 6093 South Street, Halifax, Nova Scotia B3H 1T2 Tel: (902) 421–1072

CANADIAN INSTITUTE OF FISHERIES TECHNOLOGIES

An IRAP Technology Centre Technical University of Nova Scotia, 1360 Barrington Street, P.O. Box 1000, Halifax, Nova Scotia B3J 2X4 Tel: (902) 429–8300

NOVA SCOTIA RESEARCH FOUNDATION CORPORATION

Major R&D establishment serving a wide range of firms in many areas of technology. Numerous technical seminars are sponsored. Nova Scotia Research Foundation Corporation, 101 Research Drive, Woodside Industrial Park, Dartmouth, Nova Scotia B2Y 3Z7 Tel: (902) 424–8670 Fax: (902) 465–7384 Tlx: 019–22719

TECHNOLOGY TRANSFER OFFICE

The Technology Transfer Office facilitates interaction between the Nova Scotia industrial and scientific communities. It can provide small high tech firms from Nova Scotia with marketing and general business advice, and assistance with arranging financing, not only government programs but often thorough private venture capitalists. Technology Transfer Office World Trade and Convention Centre, 6th Floor, 1800 Argyle Street, P.O. Box 519, Halifax, Nova Scotia B3J 2R7 Tel: (902) 424–7381 Fax: (902) 424–5739 Ontario

CENTRE FOR COLD OCEAN RESOURCES ENGINEERING (C-CORE)

C-CORE, an engineering institute, undertakes research that contributes to the safe and economic development of Canada's ocean resources. Working collaboratively with industry and government, it has developed a number of technologies which are being used commercially, others which are in the process of commercialization.

Dr Jack I. Clark, Director Centre for Cold Ocean Resources Engineering (C-Core), Memorial University of Newfoundland, St. John's, Newfoundland A1B 3X5 Tel: (709) 737–8354 Fax: (709) 737–4706

NEWFOUNDLAND AND LABRADOR INSTITUTE OF FISHERIES AND MARINE TECHNOLOGY

Newfoundland and Labrador Institute of Fisheries and Marine Technology, P.O. Box 4920, St. John's, Newfoundland A1C 5R3 Tel: (709) 778–0200

NEWFOUNDLAND OCEANS RESEARCH AND DEVELOPMENT CORPORATION (NORDCO)

Newfoundland Oceans Research and Development Corporation, 34 Glencoe Drive, P.O. Box 8833, St. John's, Newfoundland A1B 3T2 Tel: (709) 364–1200

SEABRIGHT CORP. LTD.

Seabright Corp. Ltd., P.O. Box 4200, St. John's, Newfoundland A1C 5S7 Tel: (709) 737–4527

AGRICULTURE CANADA RESEARCH STATION

Conducts innovative research in crop breeding and new crop development, control of crop diseases and insect pests, cereals and oilseeds processing, crop management practices. The principal crops are oilseeds, cereals and forages. Integrated pest management systems rely on plant resistance and biological and chemical methods. The Experimental Farm at Scott conducts research on crop production systems, including weed control, and crop rotation practices.

Dr R. E. Howarth, Director, Agriculture Canada Research Station, 107 Science Place, Saskatoon, Saskatchewan S7N 0X2 Tel: (306) 975–7014 Fax: (306) 242–1839

ATOMIC ENERGY OF CANADA LIMITED – CHALK RIVER LABORATORIES (CRL)

CRL is involved in robotic technology to automate nuclear fuel processes. It is also involved in the development of computer-based systems for control of safety in non-nuclear applications with expertise in computer simulation and design. Many facilities at Chalk River Laboratories are made available to external users. The laboratories offers commercial R&D services based on AECL's unique multidisciplinary problem-solving capabilities. Access to AECL Research technologies is also available via cooperative development projects, the sale or licensing of technologies and products, and technology transfer agreements.

Chalk River Laboratories (CRL), Atomic Energy of Canada Ltd., Chalk River, Ontario K0J 1J0 Tel: (613) 584–3311

BIOSTAR INC.

BIOSTAR was established in 1983 to commercialize research developments made by the Veterinary Infectious Disease Organization. BIOSTAR is a market-driven technology company focused on the development and production of advanced biological products to regulate the immune system of animals and humans. Products include cattle vaccines for respiratory diseases.

Dr S. Acres, BIOSTAR INC., Box 1000, Sub P.O. #6, Saskatoon, Saskatchewan S7N 0W0 Tel: (306) 966–7473 Fax: (306) 996–7478

CANADA CENTRE FOR MINERAL AND ENERGY TECHNOLOGY (CANMET)

CANMET is the major technology research arm of the federal government Department of Energy, Mines and Resources Canada. Extensive library and computer databases are maintained on worldwide technological developments. Technology transfer seminars and demonstration projects are held. Laboratories are located at Ottawa, Edmonton, Calgary, Elliot Lake, Cape Breton, and at the Sudbury-Blackill Research Laboratory of Laurentian University School of Engineering. Areas of activity include coal, oil, gas, domestic heating, uranium, mining, mineral refining, explosives and metallurgical research.

Canada Centre for Mineral and Energy Technology (CANMET), Energy, Mines and Resources Canada, Technology Information Division, 555 Booth Street, Ottawa, Ontario K1A 0G1 Tel: (613) 995–4194

CANADIAN ADVANCED INDUSTRIAL MATERIALS FORUM (CAIMAF)

Provides information to firms on all aspects of international technical developments, processes and applications of advanced industrial materials.

Canadian Advanced Industrial Materials Forum (CAIMAF), One Yonge Street, 14th Floor, Toronto, Ontario M5E 1J9 Tel: (416) 363–7261 Fax: (416) 363–3779

CANADIAN GAS RESEARCH INSTITUTE

Performs research related to natural gas distribution and transmission. Includes research and development and general troubleshooting and problem solving. Canadian Gas Research Institute, 55 Scarsdale Road Don Mills, Ontario M3B 2R3 Tel: (416) 447–6465

CANADIAN INDUSTRIAL INNOVATION CENTRE (CIIC)

CIIC assists innovators and provides specialized support services to technology entrepreneurs and small and medium-sized businesses to promote the commercialization of innovations and the launching of new businesses. Services offered include: assessing technical strengths, conducting market research, evaluating commercial potential, managing development and testing, assisting in venture planning, and providing training.

Canadian Industrial Innovation Centre (CIIC), 156 Columbia Street West, Waterloo, Ontario N2L 3L3 Tel: (519) 885–5870 1–800–265–4559 Tlx: 069–55259

CANADIAN INSTITUTE OF GUIDED GROUND TRANSPORT

Focuses on strategic requirements related to productivity and technology advances, improving management efficiency of R&D and developing programs for the rail industry relating to guided ground transportation in the North American context.

Canadian Institute of Guided Ground Transport, Queen's University, Kingston, Ontario K7L 3N6 Tel: (613) 545–2810

CANADIAN MANUFACTURING ADVANCED TECHNOLOGY EXCHANGE (CAN-MATE)

CAN-MATE is a national not-for profit organization established by the Canadian Manufacturers' Association, the National Research Council and Industry, Science and Technology Canada. It is intended to provide advice and information to manufacturers relating to advanced technology for production and processing. With the growing emphasis on new technology as a competitive tool, access to relevant information is crucial. CAN-MATE's services provide you with powerful, customized information, while saving you time and money. In order to maintain their current levels of excellence, and to develop new resources to aid clients, CAN-MATE charges for some of its service on a fee-per-use basis. It also promotes cooperation between industry and appropriate CAD-CAM centres.

Canadian Manufacturing Advanced Technology Exchange (CAN-MATE),

One Yonge Street, 14th Floor, Toronto, Ontario M5E 1J9 Tel: (416) 363–7261 1–800-268–8177

CANADIAN PLASTICS INSTITUTE (CPI)

The Canadian Plastics Institute (CPI) explores the world for advanced plastics technologies and provides technical information on these opportunities to the plastics industry. Its focus is on small- and medium-sized plastics processors. The goal of the Institute is to help the plastics industry become more internationally competitive and responsive to environmental challenges. CPI is a program of the Society of the Plastics Industry of Canada in partnership with Industry, Science and Technology Canada. Technology is disseminated to industry in the following ways: computer data bases, a technical library, individual company visits, seminars and workshops, presentations to industry groups, participation in the NRC Field Advisory Service (IRAP), newsbriefs, technical information bulletins, monographs, studies, seminar proceedings and other publications. All the technologies endorsed by the CPI have the following three characteristics: they are ahead of current Canadian methods; they have commercial or near-commercial status; the technology source is willing to license the technologies, sell them, or enter into a joint venture.

Canadian Plastics Institute (CPI), 1262 Don Mills Road, Suite 48 Don Mills, Ontario M3B 2W7 Tel: (416) 441–3222 Fax: (416) 441–1208

CENTRE FOR INDUSTRIAL AND TECHNOLOGICAL COOPERATION (CITEC)

CITEC was formed in Toronto by the Japan External Trade Organization (JETRO) in 1985. It encourages investment of capital for technological and industrial development in Canada and Japan. It fosters industrial cooperation and the exchange of technology between Canada and Japan and encourages joint ventures by Canadian and Japanese companies in the areas of technology and industry. They sponsor seminars, arrange factory tours, provide consulting services funding.

Centre for Industrial and Technological Cooperation (CITEC), JETRO, 151 Bloor Street West, Suite 700 Toronto, Ontario M5S 1T7 Tel: (416) 962–5050

COMMUNICATIONS RESEARCH CENTRE (CRC)

A Communications Canada research program consisting of a number of laboratories in areas such as Communications Devices and Components including research in monolithic microwave integrated circuits, very high speed digital integrated circuits, phased-array antennas, EHF components, optoelectronics, optical communications, photonics, new compound semi-conductor devices and reliability studies. It also performs research in Communications Technologies including satellite, terrestial and wireless communications and radio propogation. Research is also carried out in Broadcast Technologies including advanced television and sound technologies and new broadcast media. This group has extensive testing facilities including an Advanced TV Lab with full HDTV testing capabilities. Also located at CRC is a full-service Technology Transfer office which provides the necessary linkages between CRC and the private sector. Communications Research Centre (CRC), 3701 Carling Avenue, P.O. Box 11490, Station "H", Ottawa, Ontario K2H 8S2 Tel: (613) 998–2321 Fax: (613) 998–3185

COMPUTER INTEGRATED MANUFACTURING

A CAD-CAM facility with a computerized machine shop. Designs and develops prototype parts. Provides training facilities for industry.

Computer Integrated Manufacturing, 1276 Sandhill Drive, P.O. Box 7317, Ancaster, Ontario L9G 3N6 Tel: (416) 648–5011

DATCO TECHNOLOGY LIMITED

The company provides technical consulting and research and development services in the field of high-temperature inorganic binders and coatings. It is also active in the development and application of advanced ceramics.

Datco Technology Limited, J.O. Bernt and Associates, 1220 Corporate Drive, Burlington, Ontario L7L 1J1 Tel: (416) 332–1807

DAVID FLORIDA LABORATORY

Assembly and environmental testing of spacecraft and related equipment. David Florida Laboratory, Canadian Space Agency, 3701 Carling Avenue, P.O. Box 11490, Station "H", Ottawa, Ontario K2H 8S2 Tel: (613) 998–2383 Fax: (613) 998–2433

DIVISION OF MECHANICAL ENGINEERING (DME) – NRC

Maintains nine laboratories dedicated to basic and applied research in mechanical engineering. Companies can use DME laboratories and testing facilities for prototype testing and other activities on a fee-for-service basis. DME also provides technical support to firms that commercialize laboratory innovations.

Division of Mechanical Engineering (DME), National Research Council Laboratories, Industrial Liaison Office, Building M-2, Montreal Road, Ottawa, Ontario K1A 0R6 Tel: (613) 993–2424

HARD MATERIALS RESEARCH INC.

Specializes in selection and development of abrasion and wear-resistant materials and products for the mining industry. Provides services related to wear problems, development of specialized hard materials, and testing of cemented carbide products.

Hard Materials Research Inc., Boart Canada Inc., 980 Pacific Gate, Unit 15, Mississauga, Ontario L5T 1Y1 Tel: (416) 564–0505

HYBRIDOMA CENTRE

Consults for industrial and other clients in this bio-tech area.

Hybridoma Centre, 401 Sunset Avenue, Windsor, Ontario N9B 3P4 Tel: (519) 253–4232

INNOVATION ONTARIO CORPORATION

Innovation Ontario Corporation, 56 Wellesley Street West, 7th Floor, Toronto, Ontario M7A 2E7 Tel: (416) 963–5717 Fax: (416) 963–2088

INSTITUTE FOR CHEMICAL SCIENCE AND TECHNOLOGY

Carries out collaborative pre-competitive research that will contribute to the international competitiveness and growth of the Canadian chemical, petrochemical and petroleum processing industries.

Institute for Chemical Science and Technology, 265 North Front Street, Suite 106H, Sarnia, Ontario N7T 7X1 Tel: (519) 336–8211

MANAGEMENT OF TECHNOLOGY AND INNOVATION INSTITUTE

Management of Technology and Innovation Institute, 1276 Sandhill Drive, Ancaster, Ontario L9G 4V5 Tel: (416) 648–7344 Fax: (416) 648–7311

MANAGEMENT OF TECHNOLOGY AND INNOVATION INSTITUTE (MTI)

MTI is an independent, not-for-profit centre dedicated to training in the management of technological innovation. Jointly funded by industry, the Federal Governemt and McMaster University, MTI's mandate is to respond to the crisis in international competitiveness and technological exploitation faced by Canadian industry. Assists firms to acquire new skills required for exploiting new technology. Management education, educational materials, research and consulting services, and access to international practices are offered.

Management of Technology and Innovation Institute (MTI), 1276 Sandhill Drive, Ancaster, Ontario L9G 4V5 Tel: (416) 648–7344 Fax: (416) 648–7311

NATIONAL RESEARCH COUNCIL CANADA – INTELLECTUAL PROPERTY SERVICES OFFICE

The Intellectual Property Services Office offers, for licence, technologies developed in the National Research Council of Canada laboratories as well as in the laboratories of some other selected government departments.

Dr Alec M. Bialski, Director, Intellectual Property Services Office, National Research Council Canada, Building M-58, Room EG-06 Ottawa, Ontario K1A 0R6 Tel: (613) 990–9547 (613) 993–9996 Fax: (613) 952–6082

OCM TECHNOLOGY INC.

An R&D Centre that provides technology transfer to industry, specializing in embedded microprocessor-based systems. Areas of work include: industrial control systems, data acquisition and control, and communications systems. OCM Technology Inc., 17 Grenfell Crescent, Nepean, Ontario K2G 0G3 Tel: (613) 723–7499 Fax: (613) 723–7761

ONTARIO CENTRE FOR RESOURCE MACHINERY TECHNOLOGY

Provides research assistance and venture capital for any industry involved in the development of innovative resource machinery.

Ontario Centre for Resource Machinery Technology, 127 Cedar Street, 4th Floor, Sudbury, Ontario P3E 1B1 Tel: (705) 673–6606

ONTARIO CENTRE FOR ADVANCED MANUFACTURING, ROBOTICS CENTRE

745 Monaghan Road, Peterborough, Ontario K9J 5K2 Tel: (705) 876–1611

ONTARIO CENTRE FOR ADVANCED MANUFACTURING

Consulting and training services related to advanced manufacturing technology to industry. The CAD-CAM Centre, Cambridge, as well as the former Robotics Centre, Peterborough, are now part of this Centre. Also manages the Advanced Manufacturing Centre in Windsor.

Ontario Centre for Advanced Manufacturing, 400 Collier MacMillan Drive, Cambridge, Ontario N1R 7H7 Tel: (519) 622–3100

ONTARIO CENTRE FOR AUTOMOTIVE PARTS TECHNOLOGY

Provides consulting services available to all industries. No longer limited to the auto industry.

Ontario Centre for Automotive Parts Technology, 46 Elderwood Drive, St. Catharines, Ontario L2S 3E7 Tel: (416) 688–2600

ORTECH INTERNATIONAL (formerly ONTARIO RESEARCH FOUNDATION)

Major R&D not-for-profit organization employing 380 staff. Founded in 1928, serves Canada and North America. Draws 70% of revenue from industry contracts. Serves a wide range of firms in many areas of technology. Numerous technical seminars are sponsored. Ortech International is a member of the TriMAC (Triad Market Access Consortium) network, whose collective goal is to promote business transactions through mutual marketing and technology transfer.

W. David Heaslip, Vice President, Corporate Marketing & Materials Technology ORTECH, International Sheridan Park, 2395 Speakman Drive, Mississauga, Ontario L5K 1B3 Tel: (416) 822–4111 Ext. 348 1–800-268–5390 Fax: (416) 823–1446

OTTAWA CARLETON RESEARCH INSTITUTE

Non-profit research co-operative involving two universities, a college and regional high technology industry. Major focus is on microelectronics, satellite communications, data communications, computer protocols, real-time microprocessing systems, artificial intelligence, CAD/CAM.

Ottawa Carleton Research Institute, 1150 Morrison Drive, Suite 302, Ottawa, Ontario K2H 8S9 Tel: (613) 726–8827

OTTAWA CIVIC HOSPITAL

Research investigators affiliated with the University of Ottawa conduct basic and clinical research in neuroscience, reproductive biology, and endocrinology and metabolism.

Robert Hanion, Director, Research Programs, Loeb Medical Research Institute, Ottawa Civic Hospital, 1053 Carling Avenue Ottawa, Ontario K1Y 4E9 Tel: (613) 761–5079 Fax: (613) 761–9151

PRECARN ASSOCIATES INC.

PRECARN Associates was incorporated as a non-profit corporation in 1987. Its mission is to develop a better awareness and competence within Canadian industry of the curent and future potential of "intelligent systems" of all kinds, from the simplest expert system to the development and application of autonomous robotic devices. Intelligent systems was selected as the focus of the effort because of the leadership in the field in a number of Canadian universities and because of its broad generic nature; there is hardly a sector of the economy that will not feel the impact of automated intelligent devices. The process chosen to increase our collective comptence was to bring a group of Canadian industries together to select and pursue a long-term, precompetitive program of research in cooperation with experts within our universities and governments. PRECARN members join together in the planning, funding, and in the actual performance of the research, therefore providing a built-in capacity for knowledge dissemination to our industrial structure.

Mrs Lise McCourt, Manager, Corporate and Public Relations, PRECARN Associates Inc., 30 Colonnade Road, Suite 300, Nepean, Ontario K2E 7J6 Tel: (613) 727–9576 Fax: (613) 727–5672

THE RIVER ROAD ENVIRONMENTAL TECHNOLOGY CENTRE (RRETC)

Primarily concerned with technologies to monitor and control air pollutants and spills of oil and hazardous materials. Develops and demonstrates equipment and processes and provides a highly accessible information base in support of sound environmental management decisions.

The River Road Environmental Technology Centre (RRETC), 3439 River Road, Ottawa, Ontario K1A 0H3 Tel: (613) 998–3671 Fax: (613) 998–0004

SAMUEL LUNENFELD RESEARCH INSTITUTE

Terry Donaghue, Mount Sinai Hospital, 600 University Avenue, Room 970, Toronto, Ontario M5G 1X5Tel: (416) 586–8225 Fax: (416) 586–8588

TECHNICAL INSTITUTE FOR MEDICAL DEVELOPMENT FOR CANADA (TIMEC)

100 Sparks Street, Suite 501, Ottawa, Ontario K1P 5B7 Tel: (613) 230–6370 Fax: (613) 235–5889

TRANSPORTATION TECHNOLOGY PROGRAM

This program encompasses all of NRC's transportation equipment and safety activities. The program relies on facilities and staff of NRC division and institutes. Research activities relate to all modes of transportation, to multimodal system analyses and to generic technologies including computational fluid dynamics and machinery condition monitoring.

Transportation Technology Program, National Research Council Laboratories, Building M-16, Montreal Road, Ottawa, Ontario K1A 0R6 Tel: (613) 993–4175 (613) 993–9900 Fax: (613) 954–1473

UNDERWRITERS' LABORATORIES OF CANADA

The laboratories test and certify devices, construction material and services to ensure that they meet firm, accident and construction standards.

Underwriters' Laboratories of Canada, Head Office, 7 Crouse Road, Scarborough, Ontario M1R 3A9; Regional Contacts, Suite, 204112–28th Street, S.E. Calgary, Alberta T2A 6J9; Suite 202, 5311 boul. de Maisonneuve Ouest, Montréal, Québec H4A 1Z5; Suite 350, 409 Granville Street, Vancouver, British Columbia V5C 1W5 Tel: (416) 757–3611 Fax: (416) 757–9540

WASTEWATER TECHNOLOGY CENTRE

Focuses on the identification, characterization and control of toxic substances in wastewater discharges. Approves designs for marine sanitation devices, and develops programs for treatment plant operators.

Wastewater Technology Centre, Environment Canada (CCIW), P.O. Box 5050, 867 Lakeshore Road, Burlington, Ontario L79 4A6 Tel: (416) 336–4599

WELDING INSTITUTE OF CANADA

An industry supported non-profit Institute based on corporate and individual membership with a structure of local chapters across Canada. Advises, consults, and provides information to firms on all aspects of welding technology. Undertakes sponsored research and development, laboratory investigations and structured training programs. Provides modular courses, publications and library services.

Welding Institute of Canada, Head Office, 391 Burnhamthorpe Road, East Oakville, Ontario L6J 6C9 OR Welding Institute of Canada, Québec Centre, 2399 rue de la Province, Longueuil, Québec J4G 1G3 Tel: (416) 845–9881 Fax: (416) 845–9886 Tel: (514) 651–5086 Fax: (514) 651–5089

WOMEN INVENTORS PROJECT

An organization to assist women inventors in successfully commercializing their inventions through such means as seminars, workshops and an information newsletter (Women Inventors Project - FOCUS). Stimulating more women to become inventors is also a major aim.

Women Inventors Project, 22 King Street South, Suite 500, Waterloo, Ontario N2J 1N8 Tel: (519) 746–3443 Québec

CANADIAN WORKPLACE AUTOMATION RESEARCH CENTRE (CWARC)

The Centre will act as a catalyst for technological developments in the important area of telecommunications. Its objectives are to provide leadership in applied research into computerized office systems and to utilize the related potential for enhanced productivity in the public and private sectors; to synthesize the needs of users and to contribute to problem-solving; to become the focal point of information exchange in the field of workplace automation and to foster cooperation between experts and different client groups. Research areas include the Organizational Research Directorate, the Integrated Systems Directorate, the Advanced Technology Directorate and the External Cooperation Directorate. The Centre also has a list of publications available.

Canadian Workplace Automation Research Centre (CWARC), 1575 Chomedey Blvd, Laval, Québec H7V 2X2Tel: (514) 682–3400

CENTRE DE DEVELOPMENT DES INDUSTRIES DE LA MODE

555 rue Chabanel Ouest, Bureau 800, Montréal, Québec H2N 2H8 Tel: (514) 384–9760

CENTRE DE PRODUCTION AUTOMATISEE

8475 avenue Christophe-Colomb, Montréal, Québec H2M 2N9 Tel: (514) 283–0348

CENTRE DE RECHERCHE INDUSTRIELLE DU MEUBLE ET DU BOIS OUVRE

765 rue Notre-Dame est C.P. 98 Victoriaville, Québec G6P 6S4 Tel: (819) 758–8129

CENTRE DE RECHERCHE INDUSTRIELLE DU QUÉBEC (CRIQ)

Major R&D establishment serving a wide range of firms in many areas of technology. Numerous technical seminars are sponsored.

Centre de recherche industrielle du Québec (CRIQ), 333, rue Franquet, C.P. 9038, Sainte-Foy, Québec G1V 4C7 OR Centre de recherche industrielle du Québec (CRIQ), 8475, rue Christophe Colomb, C.P. 2000, Montréal, Québec H2P 2X1 Tel: general info: (418) 659–1550 industry info: (418) 659–1558 Tel: general info: (514)383–1550 industry info: (514) 383–3240 Elsewhere in Québec: Tel: 1–800–463–3390

CENTRE DES TECHNOLOGIES TEXTILES

3000 rue Boulle, Saint-Hyacinthe, Québec J2S 1H9 Tel: (514) 778–1870

CENTRE FOR INDUSTRIAL INNOVATION/MONTREAL (CIIM)

A non-profit organization which provides assistance for evaluating and improving the commercial potential of innovative new products and processes, and for planning the marketing of such. Services include a low-cost invention evaluation service, assistance in setting up R&D programs, seminars on technological innovation, project management, financial counselling and assistance in locating capital, partners, etc., and technology brokerage.

Centre for Industrial Innovation/Montréal (CIIM), 6600, Ch. de la Côte-des-Neiges, Suite 500, Montréal, Québec H3S 2A9 Tel: (514) 737–9883 Fax: (514) 737–1535 Tlx: 05826852

CENTRE QUEBECOIS DE PRODUCTIVITE DU MEUBLE ET DU BOIS OUVRE

85 St-Paul ouest, 4e étage, suite B5, Montréal, Québec H2Y 3V4 Tel: (514) 844–2088

CENTRE SPECIALISE DE TECHNOLOGIE PHYSIQUE

140, 4e avenue, La Pocathier, Québec G0R 1Z0 Tel: (416) 856–1525 Ext. 389

CENTRE TECHNOLOGIQUE DES MATERIAUX COMPOSITES (CTMC)

Centre technologique des matériaux composites (CTMC), Cégep de Saint-Jérôme, 455, rue Fournier, Saint-Jérôme, Québec J7Z 4V2 Tel: (514) 436–3042 Fax: (514) 436–1756

FOOD RESEARCH AND DEVELOPMENT CENTRE

State of the art facilities and expertise are available to assist the food processing industry. Information on a range of programs and services, including cost-sharing and strategic research, 3600 Casavant Boulevard West, Saint-Hyacinthe, Québec J2S 8E3 Tel: (514) 773–1105 Fax: (514) 773–8461

GROUPEMENT QUÉBECOIS D'ENTREPRISES

99, rue Cormier, Drummondville, Québec J2C 2M5 Tel: (819) 477–7535

INDUSTRIAL MATERIALS RESEARCH INSTITUTE

(IMRI) Conducts basic and applied research in the development, behavior and durability of metals, polymers ceramics and composites. It has five research programs: Polymer and Composite Materials, Metals and Composites, Coatings and Ceramics, Instrumentation and Sensors, and Computer Integrated Material Processing. Technology transfer is emphasized and most projects are carried out with industrial partners.

Industrial Materials Research Institute (IMRI), National Research Council Laboratories, 75 de Montagne Boulevard, Boucherville, Québec J4B 6Y4 Tel: (514) 641–2280

INSTITUT DE SOUDAGE DU CANADA

2399, rue de la Province, Longueuil, Québec J4G 1G3 Tel: (514) 651–5086

INTERNATIONAL CENTRE GP

Provides firms with information and training on major project management methodology.

International Centre GP, 321, rue de la Commune ouest, Bureau 200, Montréal, Québec H2Y 2E1 Tel: (514) 848–6100 Fax: (514) 848–9992

TEXTILE TECHNOLOGY CENTRE (TTC)

The Textile Technology Centre was created to heighten the productivity of the Canadian textile industry. It provides a full range of high calibre services, and develops activities related to technology transfer.

C. Martel, Textile Technology Centre, 3000 Boulle Street, Saint-Hyacinthe, Québec J2S 1H9 Tel: (514) 778–1870 Fax: (514) 773–9971.

AGRICULTURE CANADA – REGINA RESEARCH STATION

Intergrated management of weeds in cereals, field crops and native pastures, using cultural, biological and chemical technologies. Maintains specialized facilities for quarantine, application technology, residue chemistry.

Raj Grover, Director, Regina Research Station, Agriculture Canada, Box 440, 5000 Wascana Parkway, Regina, Saskatchewan S4P 3AZ Tel: (306) 780–7453 Fax: (306) 780–7400

AG-WEST BIOTECH INC.

Ag-West Biotech is a non-profit company involved in Agricultural Biotechnology in Saskatchewan. Ag-West Biotech is involved in the commercializing of biotechnology by assisting with the transfer of technology to industry from research organizations in Saskatchewan. It also looks for new technologies from other areas of Canada and internationally, that could be commercialized in Western Canada. Its mission is "To facilitate the use of Agricultural Biotechnology as a means of advancing the economic competitiveness of Saskatchewan agriculture and agribusiness".

Dr Murray McLaughlin, President, Ag-West Biotech Inc., 105 - 15 Innovation Boulevard, Saskatoon, Saskatchewan S7N 2X8 Tel: (306) 975–1939 Fax: (306) 975–1966

NUVOTECH VENTURES INTERNATIONAL

Nuvotech Ventures International is the wholly-owned commercialization arm of the POS Pilot Plant Corporation, an internationally-recognized contract research and development facility serving the agri-food industry. Nuvotech forms strategic partnerships and/or arranges for licensing agreements to bring together the necessary managerial and marketing expertise, along with the supply, distribution, manufacturing and financial resources required to take technologies out of the lab and put them into the marketplace. Through POS and other research institutions, POS lends the technology, technical support, and small-scale production capabilities to the partnership. Product can be produced, test-marketed, and modified to meet the standards demanded by consumers before a major capital expenditure is required – substantially reducing the risk of commercializing a new technology.

Don Hrtyzak, Vice-President of Finance and Business Development; Ron Mantyka Business Development Officer; 118 Veterinary Road, Room #1000 Saskatoon, Saskatchewan S7N 2R4 Tel: (306) 975–3761 Fax: (306) 975–3766

PHILOM BIOS INC.

Philom Bios develops, manufactures and markets biological input products for the domestic and international plant-growing industries. Our main market is the commercial agricultural industry, with secondary emphasis on the vegetable and horticulture sectors. Products currently on the market include: preformulated rhizobium inoculants for peas and lentils, and for alfalfa, marketed under the trade name N-PROVE; and PROVIDE, a phosphate inoculant which improves the availability of soluble phosphate to growing plants.

John V. Cross, President, Philom Bios Inc., 104–110 Research Drive, Saskatoon, Saskatchewan S7N 3R3 Tel: (306) 975–7055 Fax: (306) 975–1215

SASKATCHEWAN RESEARCH COUNCIL

Major R&D organization, applying s&t for Saskatchewan development. Sponsors technical seminars, assists small business and serves a growing clientele in applied research.

Saskatchewan Research Council, 15 Innovation Blvd., Saskatoon, Saskatchewan S7N 2X8 Tel: (306) 933–5400 Fax: (306) 933–7446 Tlx: 074 2484.

Appendix 6

Other Sources of Technology

Specialized Technology Transfer Centres

This appendix describes technology transfer, licensing or commercialization offices.

THE TECHNOLOGY DEVELOPMENT GROUP

Specializes in finalizing the last step which transforms technology into a marketable product. They assist in prototype and process design, development and testing, calibration and testing facilities, market analysis, etc.

The Technology Development Group, Southern Alberta Institute of Technology, 1301–16 Avenue N.W., Calgary, Alberta T2M 1L4 Tel: (403) 284–8791

UNIVERSITY OF ALBERTA HOSPITALS

University of Alberta Hospitals Technology Commercialization Research and Technology Department, 8440–112th Street, Edmonton, Alberta T6G 2B7 Tel: (403) 492–6711 Fax: (403) 492–4351

INSTITUTE FOR TECHNOLOGICAL DEVELOPMENT

Institute for Technological Development, Faculty of Engineering, University of Manitoba, Winnipeg, Manitoba R3T 2N2 Tel: (204) 474–6689 Fax: (204) 261–3475

APPLIED ELECTROSTATICS RESEARCH CENTRE

The Applied Electrostatics Research Centre (AERC) provides research development and consulting services to assist industry in the development of better techniques for using electrical forces to propel fine particulate matter along predefined trajectories. AERC's research has resulted in precise methods for controlling the movement of particulates such as aerosols, ink toners, water droplets and other fine materials, by applying electrical charges to the particulates. Drawing upon its international experience, AERC offers industry assistance in the following areas agricultural spraying, coating & painting, hazard detection and prevention, electrophotography, medical science, and aerospace applications.

Prof. Ion Inculet, Director, Applied Electrostatics Research Centre, Faculty of Engineering Science, The University of Western Ontario, London, Ontario, Canada N6A 5B9 Tel: (519) 661–2002 Fax: (519) 661–3808 Tlx 064–7134

BOUNDARY LAYER WIND TUNNEL LABORATORY (BLWTL)

The Boundary Layer Wind Tunnel is one of the world's leading facilities for modelling and studying the effects of wind action on structures. The laboratory's two tunnels have helped to pioneer a new understanding of many fields including the dispersion of pollutants, snow drifting, wind energy, wind/wave loading and other phenomena affected by the turbulent "boundary layer" of the wind near the earth's surface. For over twenty years, the BLWTL team has applied innovative techniques and advanced skills to the areas of wind effect on buildings and special structures, environmental problems, wind/wave interaction, wind climate studies, and other dynamic problems.

Dr Alan G. Davenport, Director, The Boundary Layer Wind Tunnel Laboratory, The University of Western Ontario, London, Ontario, Canada N6A 5B9 Tel: (519) 661–3338 Fax: (519) 661–3339 Tlx 064–78585 LDN

CENTENNIAL COLLEGE CENTRE OF ENTREPRENEURSHIP

Intended to create an environment which is conducive to entrepreneurial activity the centre provides support facilities, seminars and conferences, and a small business consulting centre through which to serve the Scarborough-East York area.

Centennial College Centre of Entrepreneurship, Centennial College, 41 Progress Court, Room E1–20 P.O. Box 631, Station A, Scarborough, Ontario M1K 5E9 Tel: (416) 439–7180 Fax: (416) 439–0219

CENTRE FOR ADVANCED TECHNOLOGY EDUCATION (CATE)

The Centre offers seminar/workshops (1 to 5 days) in computer systems, light systems (lasers, fibre optics, etc.), most software packages, customized seminars, project management, and research projects.

Centre for Advanced Technology Education (CATE), Ryerson Polytechnical Institute, 350 Victoria Street, Toronto, Ontario M5B 2K3 Tel: (416) 979–5106 Fax: (416) 979–5047

CENTRE FOR MANUFACTURING STUDIES

Provides technical education and skills development ranging from entrepreneurship to highly specialized courses for working managers and industry executives focusing on integrated manufacturing. Has robotic equipment from former Robotic Centre.

Centre for Manufacturing Studies, Sir Sandford Fleming College, 743 Monaghan Road, Peterborough, Ontario K9J 5K2 Tel: (705) 876–1611 Fax: (705) 876–9423

CENTRE FOR RESOURCE STUDIES

Provides information and analysis on mineral resource policy issues, carries out research relative to alternatives in the development and use of mineral resources and promotes exchange of information on different segments of Canadian industry.

Centre for Resource Studies, Queen's University, Kingston, Ontario K7L 3N6 Tel: (613) 545–2553 Fax: (613) 545–6651

CENTRE IN MINING & MINERAL EXPLORATION RESEARCH (CIM-MER)

Provides research related to the minerals industries including new exploration, production and processing methods, worker health and safety, and the environment.

Centre in Mining and Mineral Exploration Research (CIMMER), Laurentian University, Ramsey Lake Road, Subdury, Ontario P3E 2C6 Tel: (705) 673–6572 Fax: (705) 675–1024

CHEMICAL REACTOR ENGINEERING CENTRE (CREC)

The Chemical Reactor Engineering Centre (CREC) provides the petrochemical, chemical and energy industries with innovative reactor technology for the transformation of petroleum feedstocks or conversion of intermediate petrochemical products. CREC uses advanced computer modelling and scaled-down prototype units to develop optimal chemical processes to increase the reactor's energy efficiency and decrease the volume of pollutants released by the various chemical reactions. Developing efficient, safe and clean chemical reactors is at the heart of Chemical Reactor Engineering Centre's research and development. CREC's personnel offer help in the following technologies fluidized bed reactors, catalytic cracking, hydrocarbon synthesis, catalyst preparation, particle classification, hydrocracking reactors, gasification, novel reactors, bubble columns and three phase fluidized beds.

Dr Hugo de Lasa, Director, Chemical Reactor Engineering Centre, Faculty of Engineering Science, The University of Western Ontario, London, Ontario Tel: (519) 661–2144 Fax: (519) 661–3808 Tlx 064–7134

COMPUTER AIDED PROCESS ENGINEERING LABORATORY (CAPE LAB)

This program brings the university, industry and government together to develop the next generation of design methods using computers. Industry can gain access to research through affiliate memberships and the visiting professionals program. CAPE Lab also undertakes confidential contract research for industry. Expertise of Cape Lab Faculty lie in these areas. process synthesis, process simulation, process control and expert systems.

Computer Aided Process Engineering Laboratory (CAPE Lab) c/o University of Waterloo, Department of Chemical Engineering, Waterloo, Ontario N2L 3G1 Tel: (519) 885–1211 Ext. 2913 Fax: (519) 746–4979 Tlx 069–552529

COMPUTER SYSTEMS RESEARCH INSTITUTE

Focuses on research and development in the design, implementation and operating characteristics of complex computing systems and promotes the use of such to business, governments and universities.

Computer Systems Research Institute, University of Toronto, Sanford Fleming Building, 10 King's College Road, Toronto, Ontario M5S 1A4 Tel: (416) 978–5034 Fax: (416) 978–4765

COMPUTER SYSTEMS GROUP

Computer Systems Group, University of Waterloo, 200 University Avenue West, Waterloo, Ontario N2L 3G1 Tel: (519) 888–4004 Fax: (519) 746–5422

DESIGN AUTOMATION AND MANUFACTURING RESEARCH LABORATORY (DA&M)

Design Automation and Manufacturing (DA&M) Research Laboratory is a research and consulting laboratory dealing with all aspects of design and manufacturing automation, computer aided design, computer aided manufacturing (CAD/CAM), applied artificial intelligence and expert systems. DA&M Research Laboratory conducts long-term basis research as well as applied research in the areas of design and manufacturing automation. Current activities of the group include design, robotics and manufacturing.

Prof. W.H. El Maraghy, Director, Design Automation & Manufacturing Research Laboratory, Faculty of Engineering Science, The University of Western Ontario, London, Ontario Tel: (519) 661–3121 Fax: (519) 661–3808 Tlx 064–7134 UWO TEL LDN

DURHAM COLLEGE PRODUCTIVITY IMPROVEMENT CENTRE

The Centre is designed to offer public seminars and in-house customized training. Areas of expertise include statistical process control, designed experiments, production/inventory control, Autocad, and small business consulting.

Durham College Productivity Improvement Centre, Durham College, P.O. Box 385, 2000 Simcoe Street North, Oshawa, Ontario L1H 7L7 Tel: (416) 576–0210 Fax: (416) 728–2530

EASTERN ONTARIO CENTRE OF ENTREPRENEURSHIP

Intended to help create an environment which is conducive to entrepreneurial activity in Eastern Ontario and affiliated with St. Lawrence College, Queen's University and Loyalist College, the Centre provides current information, expertise, support services and seminars for new entrepreneurs, college and university students and teachers of entrepreneurship in the region.

Eastern Ontario Centre of Entrepreneurship, 275 Ontario Street, Suite 100, Kingston, Ontario K7K 2X5 Tel: (613) 546–2777 Fax: (613) 546–2882

ENTERPRISE YORK

Intended to create an environment which is conducive to entrepreneurial activity. Programs and a private sector incubation centre are available to start-up firms and to growth firms beyond the start-up phase.

Enterprise York, York University, Administrative Studies Building, 4700 Keele Street, Room 230C, Downsview, Ontario M3J 1P3 Tel: (416) 736–5091 Fax: (416) 736–5772

FACSYM RESEARCH LIMITED

Faculty of Management Studies, University of Toronto, 246 Bloor Street West, Toronto, Ontario M5S 1V4 Tel: (416) 978–2826

GEORGE BROWN INNOVATION CENTRE

Focusing on the transfer and commercialization of cogeneration and solar energy. George Brown Innovation Centre, George Brown College, 146 Kendal Avenue, Toronto, Ontario M5R 1Z1 Tel: (416) 944–4520 Fax: (416) 944–4638

GEOTECHNICAL RESEARCH CENTRE (GRC)

The Geotechnical Research Centre is recognized internationally for its experience in solving soil- and rock-related problems ins civil, geotechnical and geological engineering. The group makes extensive use of computer modelling techniques, as well as laboratory and field study, to address problem in soil and rock mechanics as they affect structural design, environmental issues and geotechnical theory. The Geotechnical Research Centre offers highly specialized contract research, both theoretical and experimental, in the following areas of expertise underground structures and soft ground tunneling; soil and foundation dynamics; dynamic soil-structure interaction; mineralogical/geochemical/environmental, clay/leachate compatibility; contaminant transport through soil/wast disposal; foundation engineering; stability of concrete dams on rock foundations; geosythetics and reinforced soil.

Dr R.M. Quigley, Director, Geotechnical Research Centre, Faculty of Engineering Science, The University of Western Ontario, London, Ontario, Canada N6A 5B9 Tel: (519) 661–3344 Fax: (519) 661–3808

HUMBER COLLEGE BUSINESS AND INDUSTRY SERVICE CENTRE

Provides hands-on customized training and seminars in a wide range of technology including computer integrated manufacturing and computer engineering. Specializes in cross training and re- skilling employees as industrial clients integrate new technology into their manufacturing process. Fully operational FMS training facility on site.

Humber College Business and Industry Service Centre, 205 Humber College Boulevard, Etobicoke, Ontario M9W 5S7 Tel: (416) 674-BISC Fax: (416) 675–6681

HUMBER COLLEGE INNOVATION CENTRE

The activities of this centre focus primarily on printing, publishing, production and video products.

Humber College Innovation Centre, 205 Humber College Boulevard, Etobicoke, Ontario M9W 5L7 Tel: (416) 674–5052

INDUSTRIAL RESEARCH INSTITUTE

Contracts with companies, local industry and governments for testing and research related to air quality and environmental studies.

Industrial Research Institute, University of Windsor Windsor Hall Tower, Room 418 Windsor, Ontario N9B 3P4 Tel: (519) 973–7032 Fax: (519) 973–7050

INNOVATION NORTH (THUNDER BAY)

Provides bilingual evaluation and advisory services on marketing approaches and business plan development focusing on forestry and pulp and paper. (See also Northern Technology Transfer, sponsored by Laurentian University).

Innovation North (Thunder Bay), c/o Lakehead University, Lutheran House, Thunder Bay, Ontario P7B 5E1 Tel: (807) 343–8110

INNOVATION YORK

Focusing on the encouragement of links between York's research strengths and industrial development in the private sector. There is University lab space for rent to pursue R&D activities.

Innovation York, York University, Farquharson Building, Room 117, 4700 Keele Street, Downsview, Ontario M3J 1P3 Tel: (416) 736–5026

INSTITUTE FOR AEROSPACE STUDIES

Engages in consulting and contract work in most fields of aerospace.

Institute for Aerospace Studies, University of Toronto, 4925 Dufferin Street, Downsview, Ontario M3H 5T6 Tel: (416) 667–7712

INSTITUTE FOR COMPUTER RESEARCH

Promotes and facilitates collaboration between industry and computer researchers at Waterloo University and participates in University-industry research projects.

Institute for Computer Research, University of Waterloo, 200 University Avenue West, Waterloo, Ontario N2L 3G1 Tel: (519) 888–4530

INSTITUTE FOR ENVIRONMENTAL STUDIES

Focuses on the use of Inductively Coupled Plasma Emissions (ICP). Institute has developed a process which provides an accurate analytical capability for determination of up to 25 elements simultaneously.

Institute for Environmental Studies, University of Toronto, Toronto, Ontario M5S 1A4 Tel: (416) 978–6526

INTERFACE SCIENCE WESTERN (ISW)

Interface Science Western (ISW) is an interdisciplinary research unit specializing in the unique physical and chemical interactions that take place at the interface of differing materials. Applications to modern materials and processes include the study of corrosion and catalytic phenomena, adhesion and friction, along with the fabrication and operation of microelectronic materials. Clients of the group have included major industrial concerns in manufacturing processes, microelectroncis, thin-film technologies, communications and energy production. ISW operates a wide range of surface and near-surface characterization methods and is continually developing new ones. Through the connection between ISW and Ontario Centre for Materials Research, scientific collaborators and clients carry out long and short range research on the entire spectrum of modern materials microelectronic materials, biomaterials, polymers, and composites, catalysts, high performance metals and alloys.

Prof. P.R. Norton, Department of Chemistry, The University of Western Ontario, London, Ontario, Canada N6A 5B7 OR Prof I.V. Mitchell, Department of Physics, The University of Western Ontario, London, Ontario, Canada N6A 3K7 Tel: (519) 679–2111 Ext. 6349 Fax: (519) 661–3022 Tel: (519) 661–3393 Fax: (519) 661–2033

ISOTRACE LABORATORY

Analytical services relating to the micro-analysis of isotopes and rare atoms in environmental, geological, industrial and medical samples.

Isotrace Laboratory, University of Toronto, 60 St. George Street, Toronto, Ontario M5S 1A7 Tel: (416) 978–2258

LAKEHEAD UNIVERSITY-CONFEDERATION COLLEGE CENTRE OF ENTREPRENEURSHIP

Lakehead University-Confederation College Centre of Entrepreneurship Lakehead University, Confederation College, P.O. Box 398, Thunder Bay, Ontario P7C 4W1 Tel: (807) 475–6363

MANUFACTURING RESEARCH CORPORATION OF ONTARIO

A Centre of Excellence designed to stimulate world-class research and encourage transfer and diffusion of technology. Activities focus on pre-competitive manufacturing technology operations emphasizing computer integrated manufacturing.

Manufacturing Research Corporation of Ontario, University of Toronto, 5 King's College Road, Toronto, Ontario M5S 1A4 Tel: (416) 978–7198

McMASTER INSTITUTE FOR POLYMER PRODUCTION TECHNO-LOGY

Cooperates with private sector polymer producers to improve existing production processes. Also issues many technical reports available to member companies. Membership fee is $15,000 per year. Computer models are free of charge to member companies or can be purchased separately by non-member companies.

McMaster Institute for Polymer Production Technology, McMaster University, Building JHE, Room 136, 1280 Main Street West, Hamilton, Ontario L8S 4L7 Tel: (416) 523–1643

NATIONAL CENTRE FOR MANAGEMENT RESEARCH & DEVELOP-MENT

Involved in research into best practice approaches to management and to effectively disseminating these ideas to potential and practising managers. Activity areas include Entrepreneurship, Productivity and International Business.

National Centre for Management Research and Development, University of Western Ontario, London, Ontario N6A 3K7 Tel: (519) 661–3275

NORTH BAY CENTRE OF ENTREPRENEURSHIP

Intended to create an environment which is conducive to entrepreneurial activity. The Centre will collect and publish a listing of entrepreneurial curricula support materials, make entrepreneurship courses available, assist high schools with entrepreneurial programs, and provide business and technical information assistance.

North Bay Centre of Entrepreneurship, Canadore and Nipissing Colleges, P.O. Box 5001, North Bay, Ontario P1B 8K9 Tel: (705) 474–7600 Fax: (705) 474–2384

NORTHERN TECHNOLOGY TRANSFER (formerly INNOVATION NORTH (SUDBURY))

Provides research assistance, with emphasis on the minerals industries including feasibility analysis, market studies and matching funds proposals. Negotiates and manages research contracts with industry for the Centre in Mining and Mineral Exploration Research and the Faculty of Science and Engineering at Laurentian University.

Northern Technology Transfer, Laurentian University, Ramsey Lake Road, Sudbury, Ontario P3E 2C6 Tel: (705) 673–6572 Fax: (705) 675–1024

ONTARIO LASER AND LIGHTWAVE CENTRE

A Centre of Excellence designed to stimulate advanced research and encourage the transfer and diffusion of technology in the laser and lightwave field. A unique feature is the Central Facility which is a state-of-the-art facility providing a wide range of laser and lightwave technology services. These include engineering, feasibility studies, process and quality control procedure development, precision measurements using lasers, training and education, consulting and contract research.

Ontario Laser and Lightwave Centre, University of Toronto, Department of Physics, 60 St. George Street, Toronto, Ontario M5S 1A7 Tel: (416) 978–2948 Fax: (416) 978–3939

OTTAWA CARLETON RESEARCH INSTITUTE

Non-profit research co-operative involving two universities, a college and regional high technology industry. Major focus is on microelectronics, satellite communications, data communications, computer protocols, real-time microprocessing systems, artificial intelligence, CAD/CAM.

Ottawa Carleton Research Institute, 1150 Morrison Drive, Suite 302, Ottawa, Ontario K2H 8S9 Tel: (613) 726–8827

PARTEQ INNOVATIONS

Manages technology transfer between the academic and research community and the industrial and business sector. Patents, licensing and research contacts all included.

PARTEQ Innovations, Queen's University, Kingston, Ontario K7L 3N6 Tel: (613) 545–2342 Fax: (613) 545–6300

PARTEQ RESEARCH & DEVELOPMENT INNOVATIONS

PARTEQ is the Technology Transfer arm of Queen's University and is charged with identifying, protecting and commercializing technology developed at the University and its associated centres of excellence. Parteq may also offer similar services to entrepreneurs in Eastern Ontario.

John P. Molloy, Director of Commercial Development; Richard J. Hicks, Director of Patents & Licensing, PARTEQ Research & Development Innovations, Queen's University, Kingston, Ontario K7L 3N6 Tel: (613) 545–2342 Fax: (613) 545–6853

PIEZOELECTRICITY RESEARCH LABORATORY

Piezoelectricity Research Laboratory, York University, Ross Building, Room 531 North, 4700 Keele Street, Downsview, Ontario M3J 1P3

POWER ENGINEERING ANALYSIS AND RESEARCH LABORATORY (PEARL)

The Power Engineering Analysis and Research Laboratory (PEARL) undertakes research in the application of power electronics and new control techniques to improve the operation of electrical power generation and transmission systems. Utilizing the latest in Flexible AC Transmission Systems (FACTS) technology, PEARL develops innovative high powered electronic devices which allow its clients to operate electrical power transmission systems with greater efficiency, reliability, stability and security. FACTS developed by PEARL help prevent black-outs and brown-outs within public utility and privately owned transmission systems. PEARL offers expertise in the following areas static reactive power compensation, HVDC transmission, thyristor controlled phase shifters, AC systems, and simulations.

Prof. R.M. Mathur, Director, Power Engineering Analysis and Research Laboratory, Faculty of Engineering Science, The University of Western Ontario London, Ontario, Canada N6A 5B9 Tel: (519) 661–2128 Fax: (519) 661–3808 Tlx 064–7134 E-Mail

THE RESEARCH CENTRE FOR MANAGEMENT OF NEW TECHNO-LOGY (REMAT)

Assists small to medium sized manufacturing companies in effectively adopting and implementing new technologies. Committed to assisting industry in coping with change through a focus on strategic planning, and technological justification and training.

The Research Centre for Management of New Technologies (REMAT), Wilfrid Laurier University, Waterloo, Ontario N2L 3C5 Tel: (514) 884–1970 Ext. 2662 Fax: (514) 884–8853

RYERSON INNOVATION CENTRE

Offers to the public, prototype development, contracts, research, patent searches and business plan preparation through a small business consulting service. It is an affiliate organization of the Canadian Innovation Centre in Waterloo.

Ryerson Innovation Centre, Ryerson Polytechnical Institute, 350 Victoria Street, Toronto, Ontario M5B 2K3 Tel: (416) 979–5071

SENECA COLLEGE OF APPLIED ARTS AND TECHNOLOGY

Assists technology transfer through educational services in technology, health sciences, business and applied arts. Strong international program activity, training Canadians and foreign nationals in trade, management and technical matters. Information services available include technology review, business opportunity assessment, start-up planning and cultural and language advice. The College will arrange focused tours and study sessions in Canada and abroad. The College is affiliated with the Ontario Construction Innovation Centre.

Seneca College of Applied Arts and Technology, Senior Dean, International Programs, 1750 Finch Avenue East, Toronto, Ontario M2J 2X5 Tel: (416) 491–5050 Fax: (416) 491–3081

SURFACE SCIENCE WESTERN (SSW)

Surface Science Western (SSW) is a research and consulting laboratory dealing with all aspects of material surface properties. Processes such as corrosion, wear, adhesion, catalysis, electrical switching, mineral dressing and fixation all depends on surface chemical and physical properties. The team of research engineers, scientists and technologists at Surface Science Western carry out contract research for major Canadian and foreign industries in the microelectronic, metallurgical and mining sectors. In addition, the laboratory frequently undertakes shorter term consulting work for many local companies. Typical applications for the analyses performed by the lab include analyzing defective painted surfaces; evaluating substrate cleaning procedures prior to painting, welding or thin film deposition; detecting contaminants on electrical contacts and semiconductor devices; measuring thin film composition and thickness on electronic devices; conducting corrosion and failure analysis of steam generator components; performing assays of problem mining ores for precious element distribution; and detecting process contaminants in plastics.

Mr. R. Davidson, Manager of Scientific Operations, Surface Science Western, The University of Western Ontario, London, Ontario, Canada N6A 5B7 Tel: (519) 661–2173 Fax: (519) 661–3486

UNIVERSITY OF GUELPH INNOVATION CENTRE

A Centre for enhancing the innovation process within the community. Provides advice, counselling, technical and research support services. Also markets university technology and innovation developments. A listing of available technologies entitled Innovations can be obtained.

University of Guelph Innovation Centre, University of Guelph, Guelph, Ontario N1G 2W1 Tel: (519) 824–4120 Ext. 2776 Fax: (519) 824–5236 Tlx UOFG INAT GLPH 069–56645

UNIVERSITY OF OTTAWA INNOVATION CENTRE

A Centre for enhancing the innovation process within the community. The Centre provides counselling, especially to inventors outside the university and helps to commercialize their inventions.

University of Ottawa Innovation Centre, University of Ottawa, Vanier Hall, 136 Jean-Jacques Lussier, Ottawa, Ontario K1N 6N5 Tel: (613) 564–5463

UNIVERSITY OF OTTAWA HEART INSTITUTE

The Institute was established to provide patient care and undertake clinical and basic research in the field of heart disease. Areas of expertise include cardiac drugs, surgical techniques, cardiac devices including artificial hearts, biomaterials and biotechnology of cardiac systems.

University of Ottawa Heart Institute, Director of Industrial Relations, Ottawa Civic Hospital, 1053 Carling Avenue, Ottawa, Ontario K1Y 4E9 Tel: (613) 761–4721 Fax: (613) 729–3937

UNIVERSITY OF TORONTO INNOVATION FOUNDATION

Focuses on patenting and licensing inventions from the University, and licensing software authored in the University.

University of Toronto Innovation Foundation, University of Toronto, 203 College Street, Suite 205, Toronto, Ontario M5T 1P9 Tel: (416) 978–5117 Fax: (416) 978–6052

UNIVERSITY OF WATERLOO CENTRE FOR INTEGRATED MANUFACTURING (WatCIM)

Mission is to advance state-of-the-art technologies through research and contractual endeavours with industry. Areas of expertise include automated assembly and robotics, computer control of production processes, computer vision, discrete and continuous systems, machine intelligence and knowledge engineering, management systems etc.

University of Waterloo Centre for Integrated Manufacturing (WatCIM), University of Waterloo, 2724 Davis Centre, Waterloo, Ontario N2L 3G1 Tel: (519) 888–4599 Fax: (519) 888–6197

Index